Direct Methods for Limit and Shakedown Analysis of Structures

Paolo Fuschi · Aurora Angela Pisano
Dieter Weichert
Editors

Direct Methods for Limit and Shakedown Analysis of Structures

Advanced Computational Algorithms and Material Modelling

Springer

Editors
Paolo Fuschi
PAU—Patrimonio, Architettura e Urbanistica
Università Mediterranea di Reggio Calabria
Reggio Calabria
Italy

Aurora Angela Pisano
PAU—Patrimonio, Architettura e Urbanistica
Università Mediterranea di Reggio Calabria
Reggio Calabria
Italy

Dieter Weichert
Institute of General Mechanics
RWTH-Aachen University
Aachen
Germany

ISBN 978-3-319-12927-3 ISBN 978-3-319-12928-0 (eBook)
DOI 10.1007/978-3-319-12928-0

Library of Congress Control Number: 2014955792

Springer Cham Heidelberg New York Dordrecht London

Printed on acid-free paper

Springer International Publishing AG Switzerland is part of Springer Science+Business Media (www.springer.com)

Foreword

The methods of plastic analysis result in designs based on strength and have proven to be very useful in solving many engineering problems. Limit states design is in widespread use in the design of civil structures, where it has been instrumental in improving the efficient use of materials. One of the early proponents of plastic analysis, Lord Baker, came to consider elastic solutions to be an improper design tool for structural frames after stresses measured in frames of real buildings bore little relation to those calculated using elastic analysis. The analysis of shakedown is more limited to specialized applications, typically involving components that operate at high temperature with periodic cooling cycles, such as pressurized pipes and vessels in elevated temperature service. Components of high temperature boilers or reactors, and coke drums are examples where operational limits may be imposed by the excessive deformation (ratcheting) that occurs when shakedown is not achieved.

One of the first tentative applications of elements of plastic design dates back almost four centuries to Galileo's calculation of the collapse load of a cantilever beam. Throughout the next centuries, there were further isolated applications of various fundamental ideas of plastic analysis. A true analysis of plastic failure comprising the concept of plastic slip and a yield condition is found in Coulomb's study of earth-retaining walls for military fortifications at the end of the eighteenth century. The nineteenth and the first half of the twentieth centuries saw the development of the criteria of plastic failure and the associated flow rules, along with applications to design such as the theory of slip lines.

The theory of plasticity and associated static and kinematic theorems were developed in the twentieth century starting in the 1930s when Gvozdev proved the static and kinematic theorems of limit analysis. At about the same time, Melan derived the static shakedown theorem, which includes the static theorem of limit analysis as a special case. Curiously, this fact was not realized until quite a while later. The fundamental theorems of plasticity, such as the consistency condition of plastic flow and the principle of maximum dissipation, became available in the 1940–1950s through the work of Horne, Greenberg, Hodge and Prager. The theory is developed the furthest for elastic perfectly plastic material behaviour.

One of the earliest and most seminal comprehensive contributions is due to Koiter. He provided a clear formulation of the lower bound (static) and upper bound (kinematic) shakedown theorems, which gave a necessary and sufficient condition for the energy dissipation under cyclic loading to remain bounded over time. That situation is referred to as (elastic) shakedown and is equivalent to stating that the structure approaches purely elastic cycling. A later extension of the concept of shakedown includes the regime of alternating plasticity where the deformation is bounded over time, but the plastic dissipation is not. The formal extension of the shakedown theorems to this case is still the subject of active research. Of great practical value is the path-independent property of the plastic limit state and of the shakedown limit state, i.e. these limit states are independent of the order and history of load application, although the plastic deformation on the way to a limit state is generally path dependent. Direct methods use these concepts to arrive at a fully plastic state. Their use for simple problems is well established, but their generalization to arbitrary loading and general plastic constitutive behaviour still represent an area of very fertile research.

The papers collected in this volume are concerned with the plastic limit and shakedown solutions obtained through direct methods. Three paths may be followed when solving limit states or shakedown problems to obtain the desired load level at limit collapse or at the shakedown boundary. Firstly, a solution may be achieved by solving the full set of continuum mechanics equations along with the constitutive equations of nonlinear elastic-plastic material behaviour. Direct methods, on the other hand may obtain statically admissible stress fields or kinematically admissible deformation fields and then improve the solution by iteration. Another type of direct methods formulates and solves a (typically nonlinear) optimization problem based on the upper or lower bound theorem.

In the recent past, plastic solutions for practical design applications have been obtained mostly by finite element software, which solves the full set of continuum mechanics equations. Such solutions are computationally relatively expensive when the objective is to obtain a plastic collapse or shakedown limit state. Direct methods, on the other hand, solve only the part of the problem that is required to approach the desired limit state. Particularly for shakedown problems, which require cyclic analysis when the full load history is applied, direct solutions can be much more efficient than the solutions of the full continuum mechanics problem. Tantalizingly, it has been difficult to derive direct methods that can consistently solve all types of problems more efficiently. The promise of direct methods has thus not yet been fully realized. There are efficient methods for special problems and there are general optimization algorithms that could solve any well-posed problem, but which have often been computationally inefficient. For this reason, direct methods continue to be an area of active research.

The present collection showcases important results from the latest research on direct methods of limit and shakedown analysis that were presented at the Fourth International Workshop on Direct Methods (DM2013), held at the Mediterranea University of Reggio Calabria, Italy, in October 2013. The papers in this volume will be valuable to any readers working in the field of direct methods who want to get an appreciation of recent advances, as well as any others interested in getting an overview of active research topics in this field.

Mississauga, Canada Wolf Reinhardt

Preface

Over the last decades powerful numerical methods have been developed to carry out one of the oldest and most important tasks of design engineers, which is to determine the load carrying capacity of structures and structural elements. Particularly attractive among these methods are the so-called "Direct Methods", embracing Limit—and Shakedown Analysis because they allow rapid and direct access to the requested information in mathematically constructive manners without cumbersome step-by-step computation.

This collection of papers is devoted to this subject. It is the outcome of a workshop hosted by the University of Reggio Calabria in October 2013, in line with previous workshops at RWTH-Aachen University, University of Technology and Sciences of Lille, and National Technical University of Athens and give an excellent insight into the state of the art in this broad and growing field of research.

The individual contributions stem namely from the areas of new numerical developments rendering the methods more attractive for industrial design, extensions of the general methodology to new horizons of application, probabilistic approaches and specific technological applications. The papers are arranged in the order as presented in the workshop.

It might be worth noting that the success of the workshops and the growing interest in Direct Methods in the scientific community were motivations to create the association IADiMe (http://www.iadime.unirc.it/) as a platform for exchange of ideas, advocating scientific achievements and not least, promotion of young scientists working in this field. It is open for all interested researchers and engineers.

The editors warmly thank all the scientists who have contributed by their outstanding papers to the quality of this edition.

—We hope you enjoy reading it!

Reggio Calabria, August 2014 — Paolo Fuschi
Reggio Calabria — Aurora Angela Pisano
Aachen — Dieter Weichert

Contents

A Stress-Based Variational Model for Ductile Porous Materials and Its Extension Accounting for Lode Angle Effects

Long Cheng, Vincent Monchiet, Géry de Saxcé and Djimédo Kondo

Abstract The aim of this work is to derive by homogenization techniques a macroscopic plastic model for porous materials with von Mises matrix. In contrast to the Gurson's well known kinematical approach [19] applied to a hollow sphere, the proposed study proceeds by means of a statical limit analysis procedure, for which a suitable trial stress field is proposed. In the first part, the formulation of the stress variational model is developed, by considering the Hill's variational principle, and introducing a Lagrange's multiplier to solve the resulting saddle-point minimization problem. This methodology being opposite to the Gurson's kinematical approach, complements the limit analysis methods for porous materials. The second part is devoted to an application of the proposed approach to the porous materials with von Mises matrix. To this end, an axisymmetric model is first studied by adopting a suitable trial stress field, which is composed by a heterogeneous part corresponding to the exact solution of hydrostatic loading and a homogeneous part for capturing the shear effects. We derive closed form formula which depends not only on the first and second invariant of the macroscopic stress tensor but also on the sign of the third invariant of the stress deviator. Moreover, an extension of the above axisymmetric model to the general case of non-axisymmetric loadings by introducing a more general trial stress field is studied. The established new yield locus explicitly depends on the effect of the third invariant (equivalently the Lode angle). The obtained results are fully

L. Cheng (✉) · G. de Saxcé
Laboratoire de Mécanique de Lille (UMR CNRS 8107), Université de Lille 1,
Cité Scientifique, 59655 Villeneuve d'Ascq, France
e-mail: long.cheng@ed.univ-lille1.fr

G. de Saxcé
e-mail: gery.desaxce@univ-lille1.fr

V. Monchiet
Laboratoire Modlisation et Simulation Multi Echelles (UMR CNRS 8208),
Université Paris-Est, 5, boulevard Descartes, 77454 Marne la Vallée, France
e-mail: vincent.monchiet@univ-mlv.fr

D. Kondo
Institut Jean Le Rond d'Alembert (UMR CNRS 7190), Université Pierre et Marie Curie,
4 place Jussieu, 75005 Paris, France
e-mail: djimedo.kondo@upmc.fr

P. Fuschi et al. (eds.), *Direct Methods for Limit and Shakedown Analysis of Structures*,
DOI 10.1007/978-3-319-12928-0_1

discussed and compared to existing models, available numerical data and to Finite Elements results obtained from cell calculation carried out during the present study.

1 Introduction

More than 30 years ago, Gurson [19] proposed a kinematically-based limit analysis approach of a hollow sphere and hollow cylinder having a von Mises rigid plastic matrix. This approach delivered an upper bound of the macroscopic criterion which depends on the pressure and on the von Mises equivalent stress. Several extensions of Gurson's model have been further proposed in the literature, the probably most important developments being those accounting for void shape effects [15, 16, 25]. Plastic anisotropy was treated by Benzerga and Besson [2], Monchiet et al. [26]. Other extensions take into account the plastic compressibility of the matrix through associated Drucker-Prager model for applications to polymer and cohesive geomaterials [1, 18, 21, 22, 24].

In Gurson's footsteps, all the above limit analysis-based models of ductile porous media are obtained by using kinematical approach which requires the choice of a suitable trial velocity field. On the other hand, few works have been made to develop a theoretical dual stress based model. One may mention the pioneering study of [17], even it has been phenomenologically inspired. A statical limit analysis attempt has been first done for ductile porous media by Sun and Wang [31] who developed a semi-analytical approach which aimed to deliver a lower bound criterion. Despite the interest of the above approaches by Sun and Wang [31], the resulting criteria are in fact obtained by some fitting procedure based on numerical computations.

From a more general point of view, it must be noted that, although the direct and accurate knowledge of the stress field is of great interest in plasticity due to the fact that the yield criterion is often expressed in terms of stress components, the main reason which probably explains the preference in past studies of the kinematical approach which leads to upper bounds is technical: the dissipation function is non smooth but only for null plastic strains. As it is generally the case for limit analysis of microporous ductile materials, the reference cell (the hollow sphere in the present study) is completely plastified at limit state and the dissipation functional is smooth, differentiable with respect to the trial velocity field parameters. And as it is well known in duality theory, the more a functional is smooth, the more its dual one is non smooth. It is exactly what occurs in plasticity where the stress functional is much more difficult to manage due to its non smoothness concentrated in the satisfaction of the yield criterion.

The principal aim of this study is to face this difficulty and to open a new way—alternative and complementary to Gurson like models—to build macroscopic yield criteria for ductile porous materials thanks to a stress model leading to a macroscopic criterion. The developed approach also enters in the framework of limit analysis, which is a general method to determine the plastic collapse of structures under proportional loading [30]. The variational formulation of the lower bound theorem is based on Hill's functional [20] of which we present a specialized version adapted to the homogenization techniques by applying it to the hollow sphere model. The lower bound is conserved only if the trial stress field is statically and plastically admissible. This condition is very difficult to strictly satisfy in a hollow sphere. In order to obtain a full analytical model, the key idea is to satisfy only the equilibrium equations, relaxing the plastic criterion with Lagrange's multipliers. Moreover, the stress condition at the void boundary is also difficult to satisfy by simple trial stress fields capturing the shear effects that break the central symmetry; it will be also relaxed. A priori, the final picture could seem too rough but, although the trial stress field is rather simple with a strict number of field parameters able to fit the hydrostatic and deviatoric macro-stress components, the present approach provides a rather accurate model, as it will be shown. Indeed, the lower bound will be lost but, but by comparison to accurate numerical data, the interest and the validity of the new results will be demonstrated.

2 Formulation of the Statical Limit Analysis for Porous Materials

In many theories of Mechanics, one of underlying mathematical structure consists in a constitutive law, that is a graph $M \subset X \times Y$ from a linear space X into its dual one Y. The dual pairing between these spaces will be denoted

$$X \times Y : (\boldsymbol{x}, \boldsymbol{y}) \mapsto \langle \boldsymbol{x}, \boldsymbol{y} \rangle \tag{1}$$

where $\langle \boldsymbol{x}, \boldsymbol{y} \rangle$ is the inner product of the dual variable fields $\boldsymbol{x}$ and $\boldsymbol{y}$, which can be, for instance, the strain rate field $\boldsymbol{d}$ and stress field $\boldsymbol{\sigma}$.

Although general, the mathematical structure of graph is poverty-stricken for applications to the science of materials and the continuum mechanics. A fruitful idea is representing a constitutive law by a numerical function. The advantage is double:

- the constitutive laws can be classified in a convenient manner for theoretical and numerical purposes,
- but—maybe above all—powerful variational methods can be developed for the solving of boundary value problems by building functional from these functions.

In this work, we focus our attention on the constitutive laws of associated plasticity, which can be derived from the framework of classic limit analysis.[1]

[1] It should be noted here that an extended limit analysis approach aiming at solving the problem of non associated plasticity has already been proposed in literature [10] (see also [4, 7–9, 11]).

2.1 Associated Plasticity and Superpotentials

The theory of associated plasticity is concerned by a class of materials (typically metals and alloys) for which:

- the elastic domain is a smooth, convex and closed.
- the plastic strain rate is an exterior normal to the yield surface.

In mechanics, this class of materials can be represented with a generalized model based on the existence of two convex superpotentials conjugating one to the other $\pi(\boldsymbol{d})$ and $\phi(\boldsymbol{\sigma})$ satisfying Fenchels inequality [13]:

$$\forall(\boldsymbol{d}', \boldsymbol{\sigma}'), \quad \pi(\boldsymbol{d}') + \phi(\boldsymbol{\sigma}') \geq \langle \boldsymbol{d}', \boldsymbol{\sigma}' \rangle \tag{2}$$

A couple $(\boldsymbol{d}, \boldsymbol{\sigma})$ is called extremal when variables are related by the constitutive law. So equality is achieved in (2):

$$\pi(\boldsymbol{d}) + \phi(\boldsymbol{\sigma}) = \langle \boldsymbol{d}, \boldsymbol{\sigma} \rangle \tag{3}$$

The materials generated by superpotentials are called generalized standards materials. For instance, the associated plasticity is obtained by taking ϕ as the indicator function of the elastic domain K (equal to zero when the stress field is statically admissible and $+\infty$ otherwise) and by considering the normal flow rule

$$\boldsymbol{d} \in \partial\, \phi\, (\boldsymbol{\sigma}) \tag{4}$$

Its Fenchel's conjugate

$$\pi(\boldsymbol{d}) = \sup_{\boldsymbol{\sigma} \in K} (\boldsymbol{d} : \boldsymbol{\sigma}) \tag{5}$$

is called the support function of K. It is positively homogeneous of order 1. The converse law reads

$$\boldsymbol{\sigma} \in \partial\, \pi\, (\boldsymbol{d})$$

and the elastic domain K is nothing else $\partial\, \pi\, (\boldsymbol{0})$. Taking into account (3), the normal yielding rule (4) is satisfied if and only if

$$\boldsymbol{\sigma} \in K \quad \text{and } \pi(\boldsymbol{d}) = \boldsymbol{d} : \boldsymbol{\sigma} \tag{6}$$

It is worthy to recall that the kinematical limit analysis theory is derived from Eq. (6) simultaneously by considering the homogenization technique. While for the statical limit analysis in this study, the start point must be the relation (4), which is nothing else Hill's inequality

$$\forall \boldsymbol{\sigma}' \in K, \quad (\boldsymbol{\sigma}' - \boldsymbol{\sigma}) : \boldsymbol{d} \leq 0 \tag{7}$$

Let us take a step back and recall a basic concept of convex analysis, the subdifferential of the superpotential ϕ at a point $\boldsymbol{\sigma}$ which is the (possibly empty) set:

$$\partial\phi(\boldsymbol{\sigma}) = \left\{ \boldsymbol{d} \mid \forall \boldsymbol{\sigma}', \quad \phi(\boldsymbol{\sigma}') - \phi(\boldsymbol{\sigma}) \geq (\boldsymbol{\sigma}' - \boldsymbol{\sigma}) : \boldsymbol{d} \right\}. \tag{8}$$

For more details on convex analysis, the reader is referred for instance to [12, 27, 29]. Since $\boldsymbol{\sigma} \in K$ entails $\phi(\boldsymbol{\sigma}) = 0$, the inequality in (8) degenerates into (7) when $\boldsymbol{\sigma}' \in K$ and is true otherwise, the left hand member being infinite while the left hand one is finite. Hence, the indicator function of the elastic domain can be perfectly expressed as introducing the semicontinuous and convex indicator function:

$$\phi(\boldsymbol{\sigma}) = \left\{ \begin{matrix} 0 & \text{if } \boldsymbol{\sigma} \in K \\ +\infty & \text{otherwise} \end{matrix} \right\}, \tag{9}$$

which will be considered for the statical limit analysis formulation in the following sections.

2.2 Stress-Based Variational Approach: Application of the Statical Limit Analysis to Porous Materials

In the perspective of limit analysis application to ductile porous materials, let us consider a reference unit volume or macro-element Ω composed of a void ω and matrix $\Omega_M = \Omega - \omega$. The macro-element Ω is bounded by surface $\partial\Omega$ and the void ω by $\partial\omega$. The matrix is made of a rigid plastic material with a yield criterion:

$$F(\boldsymbol{\sigma}) \leq 0, \tag{10}$$

where F is a lower semicontinuous and convex function of the cauchy stress tensor $\boldsymbol{\sigma}$. As classically, the normality law is assumed:

$$\mathbf{d} = \dot{\varepsilon}^p \frac{\partial F}{\partial \boldsymbol{\sigma}}, \tag{11}$$

where $\mathbf{d}$ is the strain rate tensor, while $\dot{\varepsilon}^p$ is the equivalent plastic strain rate.

It is worth to note again that, equivalently, the strain rate and stress tensor satisfy Hill's inequality (7), with the set of plastically admissible stress fields:

$$\mathscr{S}_p = \{\boldsymbol{\sigma} \quad s.t. \quad \boldsymbol{\sigma} \in K \quad \text{or} \quad F(\boldsymbol{\sigma}) \leq 0\} \tag{12}$$

and the stress fields must be statically admissible being elements of the following set:

$$\mathscr{S}_a = \{\boldsymbol{\sigma} \quad s.t. \quad div\, \boldsymbol{\sigma} = 0 \quad \text{in } \Omega, \quad \boldsymbol{\sigma} \cdot \boldsymbol{n} = 0 \quad \text{on } \partial\omega, \quad \boldsymbol{\sigma} = 0 \quad \text{in } \omega\}.$$

where $\boldsymbol{n}$ is the unit outward normal vector.

The set of kinematical admissible velocity fields classically reads:

$$\mathscr{K}_a = \{\mathbf{v} \quad s.t. \quad \mathbf{v}(\mathbf{x}) = \mathbf{D} \cdot \mathbf{x} \quad \text{on } \partial\Omega\}.$$

And, the strain rate field, symmetric part of the velocity gradient, is $\mathbf{d}(\mathbf{v}) = grad_s \mathbf{v}$.

From the classical Hill lemma, the macroscopic stress $\boldsymbol{\Sigma}$ and macroscopic strain rate $\mathbf{D}$ are obtained as volume averages of their microscopic counterparts $\boldsymbol{\sigma}$ and $\mathbf{d}$:

$$\boldsymbol{\Sigma} = \frac{1}{\mid \Omega \mid} \int_{\Omega} \boldsymbol{\sigma} \, dV, \qquad \mathbf{D} = \frac{1}{\mid \Omega \mid} \int_{\Omega} \mathbf{d} \, dV. \tag{13}$$

Let us state Hill's variational principal [20]:

Among the statically admissible stress fields, the true one makes the functional

$$\int_{\Omega_M} \phi(\boldsymbol{\sigma}) \, dV - \int_{S_v} (\boldsymbol{\sigma} \cdot \boldsymbol{n}) \cdot \boldsymbol{v} dS, \tag{14}$$

an absolute minimum.

In (14), $\boldsymbol{v}$ is the imposed velocity on the part S_v of the boundary of Ω_M. Adapted to the context of the present homogenization problem, e.g. the problem of a hollow sphere subjected to uniform strain rate boundary conditions $\boldsymbol{v}(\boldsymbol{x}) = \boldsymbol{D} \cdot \boldsymbol{x}$ on its boundary $\partial \Omega$, this principle requires to introduce the the following average functional for the hollow sphere:

$$\Phi = \min_{\boldsymbol{\sigma} \in \mathcal{S}_a} \left(\frac{1}{\mid \Omega \mid} \int_{\Omega_M} \phi(\boldsymbol{\sigma}) \, dV - \mathbf{D} : \boldsymbol{\Sigma} \right), \tag{15}$$

where $\boldsymbol{\Sigma}$ depends on the stress field $\boldsymbol{\sigma}$ through:

$$\boldsymbol{\Sigma} = \frac{1}{\mid \Omega \mid} \int_{\partial \Omega} (\boldsymbol{\sigma} \, \mathbf{n}) \otimes \boldsymbol{x} \, dS,$$

or equivalently (13), provided that the stress field $\boldsymbol{\sigma}$ is statically admissible.

Indeed, let $\boldsymbol{v}$ and $\boldsymbol{\sigma}$ be the velocity and stress fields at limit state. Taking into account Hill's lemma and $\boldsymbol{d}(\boldsymbol{v}) \in \partial \phi(\boldsymbol{\sigma})$, it holds for any statically admissible fields $\boldsymbol{\sigma}'$:

$$\frac{1}{\mid \Omega \mid} \int_{\Omega_M} (\phi(\boldsymbol{\sigma}') - \phi(\boldsymbol{\sigma}) - (\boldsymbol{\sigma}' - \boldsymbol{\sigma}) : \boldsymbol{d}((\boldsymbol{v})) \, dV \geq 0,$$

which proves that the limit stress field $\boldsymbol{\sigma}$ realizes the minimum of the functional among all the statically admissible stress fields $\boldsymbol{\sigma}'$. Defining the set of licit stress fields:

$$\mathcal{S}_l = \{\boldsymbol{\sigma} \in \mathcal{S}_a \quad s.t. \quad \boldsymbol{\sigma} \in \mathcal{S}_p \quad a.e. \quad \text{in} \quad \Omega_M\}, \tag{16}$$

let us notice that if $\boldsymbol{\sigma}$ is licit, the value of the functional in (15) is finite, infinite otherwise. The minimum being finite, it is realized only for licit fields. Because ϕ

vanishes almost everywhere for the licit fields, the above variational principle (15) is equivalent to the following one:

$$\min_{\boldsymbol{\sigma} \in \mathcal{S}_l} \left(-\mathbf{D} : \boldsymbol{\Sigma} \right), \tag{17}$$

The limit analysis approach consists in finding non trivial solutions qualified as failure mechanisms. It is expected that they exist only under an equality condition on $\boldsymbol{\Sigma}$ that can be interpreted as the equation of the macroscopic yield surface.

In the same spirit, as in the work of [18], a first approximation consists in relaxing the yield criterion (10). Introducing Lagrange's multiplier field $\mathbf{x} \mapsto \dot{\Lambda}(\mathbf{x})$, this constrained minimization problem is transformed into an equivalent saddle-point problem:

$$\max_{\dot{\Lambda} \geq 0} \min_{\boldsymbol{\sigma} \in \mathcal{S}_a} \left(\mathcal{L}(\boldsymbol{\sigma}, \dot{\Lambda}) = \frac{1}{\mid \Omega \mid} \int_{\Omega_M} \dot{\Lambda} F(\boldsymbol{\sigma}) \, dV - \mathbf{D} : \boldsymbol{\Sigma} \right),$$

We perform a new approximation by imposing Lagrange's multiplier field to be uniform in Ω_M:

$$\max_{\dot{\Lambda} \geq 0} \min_{\boldsymbol{\sigma} \in \mathcal{S}_a} \left(\mathcal{L}(\boldsymbol{\sigma}, \dot{\Lambda}) = \dot{\Lambda} \frac{1}{\mid \Omega \mid} \int_{\Omega_M} F(\boldsymbol{\sigma}) \, dV - \mathbf{D} : \boldsymbol{\Sigma} \right). \tag{18}$$

that is equivalent to minimize the functional Φ under the following condition:

$$\frac{1}{\mid \Omega \mid} \int_{\Omega_M} F(\boldsymbol{\sigma}) \, dV = 0. \tag{19}$$

Satisfying the condition (10) only in an average sense (Eq. 19) but not locally anywhere in Ω_M is a strong approximation but required here in order to make the calculation possible. The minimum principle allows then to obtain the "best" solution within the framework imposed by the adopted approximations. Hence, a stress variational macroscopic model (which will be called SVM in the following) can be obtained from Eq. (19). Additionally, it should be emphasized that this model could be seen as a quasi-lower bound due to the adoption of the relaxed licit stress fields which appears as an uncontrollable approximation in the Hill's variational principle (15). Note that Gurson's kinematically-based model is a true upper bound, all the approximation that it had required being controllable.

The final and crucial step, detailed in the next section, is the choice of a trial stress field depending on some parameters. After expressing it with respect to the invariants of the macro-stress, the macroscopic loading function is:

$$\mathcal{F}(\Sigma) = \frac{1}{\mid \Omega \mid} \int_{\Omega_M} F(\boldsymbol{\sigma}(\Sigma)) \, dV = 0. \tag{20}$$

Thus, the saddle-point problem (18) reads:

$$\max_{\dot{\Lambda}\geq 0} \min_{\Sigma} \left(\mathscr{L}(\boldsymbol{\Sigma}, \dot{\Lambda}) = \dot{\Lambda}\mathscr{F}(\Sigma) - \mathbf{D} : \boldsymbol{\Sigma} \right),$$

Performing the variation with respect to $\dot{\Lambda}$ provides the macroscopic yield condition:

$$\mathscr{F}(\Sigma) \leq 0,$$

and with respect to Σ gives the macroscopic plastic flow rule:

$$\mathbf{D} = \dot{\Lambda} \frac{\partial \mathscr{F}}{\partial \boldsymbol{\Sigma}}(\boldsymbol{\Sigma}). \tag{21}$$

where $\dot{\Lambda}$ turns out then to be the plastic multiplier and must satisfy Kuhn-Tucker conditions:

- $\dot{\Lambda} = 0$ if $\mathscr{F} < 0$, or if $\mathscr{F} = 0$ and $\dot{\mathscr{F}} < 0$
- $\dot{\Lambda} > 0$ if $\mathscr{F} = 0$ and $\dot{\mathscr{F}} > 0$

3 Stress-Based Variational Approach of Ductile Porous Materials in Axisymmetric Loading Case [6]

Let us consider a hollow sphere (Fig. 1), made up of a spherical void embedded in a homothetic matrix of a rigid-plastic isotropic and homogeneous material with von Mises model:

$$F(\boldsymbol{\sigma}) = \sigma_e(\boldsymbol{\sigma}) - \sigma_0 \leq 0,$$

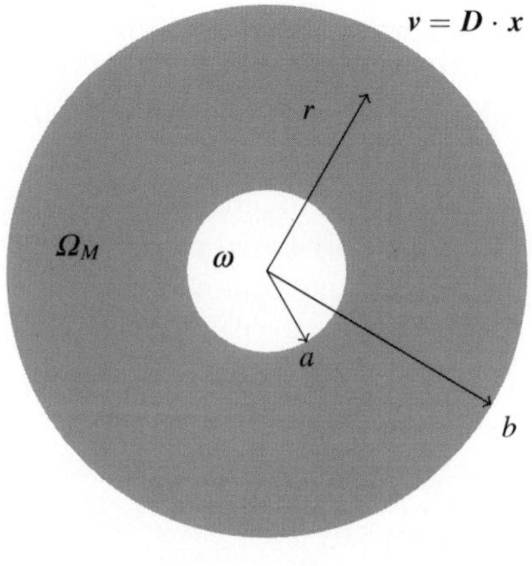

Fig. 1 *Hollow sphere* model

where $\sigma_e = \sqrt{\frac{3}{2}\mathbf{s} : \mathbf{s}}$ is the von Mises equivalent stress defined from the deviatoric part $\mathbf{s}$ of the stress tensor $\boldsymbol{\sigma}$. As usually, the quantity $\sigma_0 > 0$ represents the yield stress of the matrix material. The inner and outer radii of the hollow sphere are respectively denoted by a and b, giving the void volume fraction $f = (a/b)^3 < 1$.

3.1 Proposed Axisymmetric Trial Stress Field

In this subsection we first consider the hollow sphere model subjected to an axisymmetric loading. Let us propose a trial stress field as the sum of the following fields:

- A heterogeneous part corresponding to the exact field under isotropic loadings, which in spherical coordinates with orthonormal frame $\{\mathbf{e}_r, \mathbf{e}_\phi, \mathbf{e}_\theta\}$ reads:

$$\boldsymbol{\sigma}^{(1)} = -A_0 \left(\ln\left(\frac{a}{r}\right) \mathbf{1} - \frac{1}{2}(\mathbf{e}_\theta \otimes \mathbf{e}_\theta + \mathbf{e}_\phi \otimes \mathbf{e}_\phi) \right), \tag{22}$$

  where A_0 is a constant parameter.
- A homogeneous deviatoric part which is taken in the following form, in the cylindrical coordinates with orthonormal frame $\{\mathbf{e}_\rho, \mathbf{e}_\phi, \mathbf{e}_z\}$:

$$\boldsymbol{\sigma}^{(2)} = A_1(\mathbf{e}_\rho \otimes \mathbf{e}_\rho + \mathbf{e}_\phi \otimes \mathbf{e}_\phi - 2\mathbf{e}_z \otimes \mathbf{e}_z), \tag{23}$$

where $\mathbf{1}$ is the second order unit tensor, while A_1 is also constant parameter.

Consequently, in the matrix Ω_M, the resultant two parameters-based trial stress field in the matrix can be written as:

$$\boldsymbol{\sigma} = \boldsymbol{\sigma}^{(1)} + \boldsymbol{\sigma}^{(2)}, \tag{24}$$

Note here that a vanishing stress field is considered in the void ω. Moreover, it should be point out that the choice of the above stress field, defined by (24), together with (22) and (23), implies that

$$\Sigma_m^{void} = \frac{1}{\mid \Omega \mid} \int\limits_{\omega} \sigma_m \, dV = \frac{1}{3 \mid \Omega \mid} \int\limits_{\partial\omega} \mathbf{x} \cdot (\boldsymbol{\sigma}\, \mathbf{n}) \, dS = 0. \tag{25}$$

where $\sigma_m = tr(\boldsymbol{\sigma})/3$ is the microscopic mean stress. Consequently, Eq. (25) appears as a relaxed form of the void boundary condition, for which it is difficult to be satisfied by such a simple stress field, and then the statically variational principle (17) must then be considered with the following relaxed set of licit stress fields:

$$\mathcal{S}_r = \{\boldsymbol{\sigma} \;\; s.t. \;\; div\, \boldsymbol{\sigma} = 0, \;\; \boldsymbol{\sigma} \in \mathcal{S}_p \;\text{ in}\, \Omega_M, \;\; \boldsymbol{\sigma} = 0 \;\text{ in}\, \omega \;\text{ and}\; \Sigma_m^{void} = 0\}. \tag{26}$$

at the place of $\mathcal{S}_l$ defined by (16).

In order to determine the stress-based variational model, it is convenient to transform both two parts of the trial stress field (24) in the same coordinates. For instance, in cylindrical coordinates, (24) can be derived as:

$$\begin{aligned}\boldsymbol{\sigma} = &\left[\left(-\ln\left(\frac{a}{r}\right)+\frac{1}{2}\cos^2\theta\right)A_0 + A_1\right](\mathbf{e}_\rho\otimes\mathbf{e}_\rho)\\ &+\left[\left(-\ln\left(\frac{a}{r}\right)+\frac{1}{2}\right)A_0 + A_1\right](\mathbf{e}_\phi\otimes\mathbf{e}_\phi)\\ &+\left[\left(-\ln\left(\frac{a}{r}\right)+\frac{1}{2}\sin^2\theta\right)A_0 - 2A_1\right](\mathbf{e}_z\otimes\mathbf{e}_z)\\ &-\frac{A_0}{2}\sin\theta\cos\theta\left(\mathbf{e}_\rho\otimes\mathbf{e}_z+\mathbf{e}_z\otimes\mathbf{e}_\rho\right)\end{aligned}$$

3.2 Macroscopic Yield Criterion Under Axisymmetric Loading

It is then readily to calculate the microscopic equivalent stress:

$$\sigma_e = \sqrt{\left(\frac{A_0}{2}\right)^2 + \frac{3}{2}A_0A_1\left(3\cos^2\theta - 1\right) + (3A_1)^2} \tag{27}$$

and it follows that the macroscopic stress tensor takes the form:

$$\boldsymbol{\Sigma} = -\frac{A_0\ln f}{3}\mathbf{1} + (1-f)A_1(\mathbf{e}_\rho\otimes\mathbf{e}_\rho + \mathbf{e}_\phi\otimes\mathbf{e}_\phi - 2\mathbf{e}_z\otimes\mathbf{e}_z), \tag{28}$$

It can be calculated from Eq. (28):

$$\Sigma_m = -\frac{A_0\ln f}{3}, \qquad \Sigma_e = 3(1-f)\mid A_1\mid, \qquad J_3 = -2\,(1-f)^3A_1^3. \tag{29}$$

where Σ_m, Σ_e and J_3 are the macroscopic mean stress, the macroscopic equivalent stress and the macroscopic stress deviator, respectively. For simplification, let us introduce the following stress quantities:

$$\tilde{\Sigma}_e = \frac{\Sigma_e}{1-f}, \qquad \tilde{\Sigma}_m = -\frac{3\Sigma_m}{2\ln f}, \qquad \tilde{J}_3 = \frac{J_3}{(1-f)^3}, \tag{30}$$

The microscopic equivalent stress (27) can be recast into:

$$\sigma_e = \sqrt{\tilde{\Sigma}_e^2 + \tilde{\Sigma}_m^2 - sign(J_3)\tilde{\Sigma}_e\tilde{\Sigma}_m(3\cos^2\theta - 1)},$$

in which $sign(J_3)$ is the sign of the macroscopic third invariant of stress tensor obtained from:

$$sign(J_3) = \frac{27}{2}\frac{\tilde{J}_3}{\tilde{\Sigma}_e^3} = -sign(A_1).$$

The macroscopic yield condition (20) can be then calculated as:

$$\frac{1}{2}\int_{0}^{\pi}\sqrt{\tilde{\Sigma}_e^2+\tilde{\Sigma}_m^2-sign(J_3)\tilde{\Sigma}_e\tilde{\Sigma}_m(3\cos^2\theta-1)}\sin\theta\,d\theta=\sigma_0 \tag{31}$$

Consequently, the macroscopic yield criterion takes the final form:

$$\mathscr{F}(\boldsymbol{\Sigma},f)=\sqrt{\tilde{\Sigma}_e^2+\tilde{\Sigma}_m^2}\,\mathscr{J}(\zeta)-\sigma_0\ \leq 0, \tag{32}$$

where:

$$\zeta=-sign(J_3)\frac{2\tilde{\Sigma}_e\tilde{\Sigma}_m}{\tilde{\Sigma}_e^2+\tilde{\Sigma}_m^2}=-sign(J_3)\frac{2\tilde{T}}{1+\tilde{T}^2} \tag{33}$$

with

$$\tilde{T}=\frac{\tilde{\Sigma}_m}{\tilde{\Sigma}_e}=-\frac{3(1-f)}{2lnf}T, \tag{34}$$

T being the stress triaxiality classically defined as $T=\dfrac{\Sigma_m}{\Sigma_e}$.

It should be noted that ζ depends not only on the sign of the third invariant of the stress deviator but also on that of the stress triaxiality, and on the porosity.

The function $\mathscr{J}(\zeta)$ in Eq. (32) is then defined as:

$$\mathscr{J}(\zeta)=\frac{1}{2}\int_{0}^{\pi}\sqrt{1+\frac{1}{2}(3\cos^2\theta-1)\zeta}\ \sin\theta\,d\theta. \tag{35}$$

and has the following closed-form expression:

- for $-1\leq\zeta\leq 0$:

$$\mathscr{J}(\zeta)=\frac{1}{2}\left(\sqrt{1+\zeta}+\frac{2-\zeta}{\sqrt{6\mid\zeta\mid}}\arcsin\sqrt{\frac{3\mid\zeta\mid}{2-\zeta}}\right), \tag{36}$$

- for $0\leq\zeta\leq 1$:

$$\mathscr{J}(\zeta)=\frac{1}{2}\left(\sqrt{1+\zeta}+\frac{2-\zeta}{\sqrt{6\zeta}}\ln\frac{\sqrt{3\zeta}+\sqrt{2(1+\zeta)}}{2-\zeta}+\frac{\sqrt{6}}{12}\frac{(2-\zeta)\ln(2-\zeta)}{\sqrt{\zeta}}\right). \tag{37}$$

Eq. (32) together with (36) and (37) constitute a stress-based variational yield function (noted as SVM) for ductile porous material of which the matrix obeys the von Mises criterion. It is clearly shown that this macroscopic criterion depends not only on Σ_m and Σ_e, but also on the sign of the third invariant $sign(J_3)$. This kind of dependence has been clearly noted in [3] in the context of the Gurson-like kinematical approach.

3.3 Flow Rule and Porosity Evolution Under Axisymmetric Loading

According to the SVM criterion (32), it is interesting to derive the macroscopic flow rule giving the plastic deformation from normality rule:

$$D_e = \dot{\Lambda}\frac{\partial \mathscr{F}}{\partial \Sigma_e} = \dot{\Lambda}\left[\frac{\mathscr{J}(\zeta)\Sigma_e}{(1-f)^2\sqrt{\tilde{\Sigma}_e^2 + \tilde{\Sigma}_m^2}} - \kappa sign(J_3)\frac{d\mathscr{J}(\zeta)}{d\zeta}\frac{\Sigma_m\sqrt{\tilde{\Sigma}_e^2 + \tilde{\Sigma}_m^2}}{4\,(\ln f)^2\,\Sigma_e^2 + 9\,(1-f)^2\,\Sigma_m^2}\right]$$

$$D_m = \frac{1}{3}\dot{\Lambda}\frac{\partial \mathscr{F}}{\partial \Sigma_m} = \dot{\Lambda}\left[\frac{3\mathscr{J}(\zeta)\Sigma_m}{4(\ln f)^2\sqrt{\tilde{\Sigma}_e^2 + \tilde{\Sigma}_m^2}} + \frac{\kappa}{3}sign(J_3)\frac{d\mathscr{J}(\zeta)}{d\zeta}\frac{\Sigma_e\sqrt{\tilde{\Sigma}_e^2 + \tilde{\Sigma}_m^2}}{4\,(\ln f)^2\,\Sigma_e^2 + 9\,(1-f)^2\,\Sigma_m^2}\right]$$

where $\dot{\Lambda}$ is the plastic multiplier and

$$\kappa = \frac{12\,(1-f)\ln f\,\left[4\,(\ln f)^2 - 9\,(1-f)^2\,T^2\right]}{4\,(\ln f)^2 + 9\,(1-f)^2\,T^2} \tag{38}$$

while the expression of $d\mathscr{J}(\zeta)/d\zeta$ can be developed for $-1 \le \zeta \le 0$ as:

$$\frac{d\mathscr{J}(\zeta)}{d\zeta} = \frac{|\zeta|-1}{4\,|\zeta|\,\sqrt{1+\zeta}} - \frac{|\zeta|-2}{4\,|\zeta|\,\sqrt{6\,|\zeta|}}\arcsin\left(\sqrt{\frac{3\,|\zeta|}{2-\zeta}}\right) \tag{39}$$

and for $0 \le \zeta \le 1$ as:

$$\frac{d\mathscr{J}(\zeta)}{d\zeta} = \frac{1}{4\sqrt{1+\zeta}} - \frac{\sqrt{6}}{24\zeta^{3/2}}(2+\zeta)\ln\frac{\sqrt{3\zeta}+\sqrt{2\,(1+\zeta)}}{2-\zeta} - \frac{\sqrt{6}}{48\sqrt{\zeta}}\left[\frac{2+\zeta}{\zeta}\ln(2-\zeta)+2\right] + \frac{1}{12}\frac{3\sqrt{1+\zeta}\,(2+\zeta)+\sqrt{6\zeta}\,(4+\zeta)}{\zeta\sqrt{6\zeta\,(1+\zeta)}+2\zeta\,(1+\zeta)}$$

In fact, the macroscopic flow rule can be recast into the following form:

$$D_e = \frac{\dot{\Lambda}}{1-f}\left[\frac{\mathscr{J}(\zeta)}{\sqrt{1+\tilde{T}^2}} - 2sign(J_3)\frac{d\,\mathscr{J}(\zeta)}{d\zeta}\frac{\tilde{T}(\tilde{T}^2-1)}{(\tilde{T}^2+1)^{\frac{3}{2}}}\right] \tag{40}$$

$$D_m = -\frac{\dot{\Lambda}}{2\ln f}\left[\frac{\mathscr{J}(\zeta)\tilde{T}}{\sqrt{1+\tilde{T}^2}} + 2sign(J_3)\frac{d\,\mathscr{J}(\zeta)}{d\zeta}\frac{\tilde{T}^2-1}{(\tilde{T}^2+1)^{\frac{3}{2}}}\right] \tag{41}$$

It can be obviously found that the macroscopic yield rule (40) and (41) obtained from the stress-based variational approach under axisymmetric loading case depends only on the sign of third invariant of the stress deviator and on the the stress triaxiality.

Finally, the void growth equation can be also derived as classically from the mass balance equation, $\dot{f} = 3\,(1-f)\,D_m$, which can be suitably rewritten in the form:

$$\frac{\dot{f}}{D_e} = 3\,(1-f)\,\frac{D_m}{D_e} \tag{42}$$

Consequently, the void growth is sensitive not only to stress triaxiality but also to the sign of J_3.

For completeness, we provide in Appendix (Sect. 6) the illustration and validation of the stress variational model from the comparison with the numerical computations and with the available numerical bounds in literature.

4 Extension to a Load Depended Stress Variational Model

In this section, we aim at deriving a new stress variational model for ductile porous media with a non-axisymmetric trial stress field. A macroscopic criterion depending not only on the macroscopic mean and equivalent stresses (Σ_m and Σ_e), but also on the third invariant of the stress deviator J_3 (or Lode angle θ_L) will be expressed, while in the Sect. 3.1 only the sign of J_3 was taken into account under the axisymmetric loading case.

4.1 Proposed Non-axisymmetric Trial Stress Field

Owing to the central symmetry of the hollow sphere model, we propose a trial non-axisymmetric trial stress field, which contains two part as follows,

- The same heterogeneous part corresponding to the exact solution under pure hydrostatic loadings as expressed by Eq. (22) in Sect. 3.1.
- A homogeneous part which is non axisymmetric and taken for capturing the shear effect:

$$\boldsymbol{\sigma}^{(2)} = \boldsymbol{B}, \qquad tr\boldsymbol{B} = 0 \tag{43}$$

Hence, the final trial stress field in the matrix can be written as

$$\boldsymbol{\sigma} = \boldsymbol{\sigma}^{(1)} + \boldsymbol{\sigma}^{(2)} \tag{44}$$

which turns to be null in the void $\boldsymbol{\omega}$.

It follows that the non-axisymmetric macroscopic stress field read:

$$\boldsymbol{\Sigma} = -\frac{1}{3}A\ln f \cdot \mathbf{1} + (1-f)\boldsymbol{B} \tag{45}$$

Next, let us compute that in mechanics, there are three invariants for defining the plastic limit state. From (45) and (43), they can be respectively calculated:

- Macroscopic mean stress,

$$\Sigma_m = -\frac{1}{3}A\ln f \tag{46}$$

- Macroscopic equivalent stress,

$$\Sigma_e = (1-f)\,B_{eq} \tag{47}$$

where B_{eq} is the equivalent quantity associated to the deviator $\boldsymbol{B}$ (or microscopic stress deviator of $\boldsymbol{\sigma}^{(2)}$):

$$B_{eq} = \sqrt{\frac{3}{2}\boldsymbol{B}:\boldsymbol{B}} \tag{48}$$

- Third invariant of the macroscopic stress deviator,

$$J_3 = (1-f)^3\det(\boldsymbol{B}) \tag{49}$$

For convenience, let us introduce the stress based quantities:

$$\tilde{\Sigma}_m = -\frac{3\Sigma_m}{2\ln f} = \frac{A}{2}, \quad \tilde{\Sigma}_{eq} = B_{eq} = \frac{\Sigma_e}{1-f}, \quad \tilde{J}_3 = \frac{J_3}{(1-f)^3} \tag{50}$$

from which, the macroscopic Lode angle θ_L can be defined as:

$$\cos(3\theta_L) = \frac{27\tilde{J}_3}{2\tilde{\Sigma}_{eq}^3} = \frac{27J_3}{2\Sigma_e^3}, \qquad 0 \leq \theta_L \leq 60° \tag{51}$$

4.2 Macroscopic Yield Criterion Under Non-axisymmetric Loading

From Eqs. (22), (43) and (44) the deviator $\boldsymbol{s}$ of the local stress field can be written as:

$$\boldsymbol{s} = \boldsymbol{s}^{(1)} + \boldsymbol{s}^{(2)} = \boldsymbol{s}^{(1)} + \boldsymbol{B} \tag{52}$$

where $s^{(1)}$ is the deviator calculated from (22). Hence, the equivalent stress can be obtained from:

$$\sigma_e = \sqrt{\frac{3}{2}\left[s^{(1)} : s^{(1)} + 2s^{(1)} : s^{(2)} + s^{(2)} : s^{(2)}\right]} \tag{53}$$

It can be calculated from (52) and (53) that

$$s^{(1)} : s^{(1)} = \frac{A^2}{6}, \quad s^{(1)} : s^{(2)} = -\frac{A}{2}\tilde{B}, \quad s^{(2)} : s^{(2)} = \frac{2}{3}B_{eq}^2 \tag{54}$$

for which expression of $\tilde{B}$ is determined in the following.

Indeed, in order to compute the stress quantity $\tilde{B}$ in spherical coordinates, let us first express the principal stress tensor of $\boldsymbol{B}$ in cartesian coordinates with orthonormal frame $\{\boldsymbol{e}_x, \boldsymbol{e}_y, \boldsymbol{e}_z\}$:

$$\boldsymbol{B} = B_1(\boldsymbol{e}_x \otimes \boldsymbol{e}_x - \boldsymbol{e}_z \otimes \boldsymbol{e}_z) + B_2(\boldsymbol{e}_y \otimes \boldsymbol{e}_y - \boldsymbol{e}_z \otimes \boldsymbol{e}_z), \tag{55}$$

for which the components can be expressed (without loss of generality) in the form:

$$B_1 = -\frac{\Sigma_e}{3}\cos(\theta_L) + \frac{\Sigma_e}{\sqrt{3}}\sin(\theta_L), \quad B_2 = -\frac{\Sigma_e}{3}\cos(\theta_L) - \frac{\Sigma_e}{\sqrt{3}}\sin(\theta_L) \tag{56}$$

Consequently, one can then reexpress $\boldsymbol{B}$ in spherical coordinates; it follows immediately that

$$\tilde{B} = \frac{1}{1-f}\left[\frac{\Sigma_e}{3}\cos(\theta_L)(3\cos^2(\theta) - 1) + \frac{\Sigma_e}{\sqrt{3}}\sin(\theta_L)\sin^2(\theta)\cos(2\phi)\right] \tag{57}$$

where θ and ϕ are the polar angle and azimuthal one in spherical coordinates system.

As a result, the microscopic equivalent stress (53) can be written as

$$\sigma_e = \sqrt{\frac{A^2}{4} - \frac{3A}{2}\tilde{B} + B_{eq}^2} \tag{58}$$

Taking into account (46), (47) and (54), (58) can be recast in the form:

$$\sigma_e = \sqrt{\tilde{\Sigma}_m^2 - 3\tilde{\Sigma}_m\tilde{B} + \tilde{\Sigma}_e^2} = \sqrt{\frac{9\Sigma_m^2}{4\ln^2 f} + \frac{9\Sigma_m\Sigma_L}{2(1-f)\ln f} + \frac{\Sigma_e^2}{(1-f)^2}} \tag{59}$$

where Σ_L is the macroscopic counterpart of $\tilde{B}$; it reads:

$$\Sigma_L = (1-f)\tilde{B} = \frac{\Sigma_e}{3}\cos(\theta_L)(3\cos^2(\theta) - 1) + \frac{\Sigma_e}{\sqrt{3}}\sin(\theta_L)\sin^2(\theta)\cos(2\phi) \tag{60}$$

Hence, the local von Mises yield criterion can be expressed as

$$F(\boldsymbol{\sigma}(\boldsymbol{\Sigma})) = \sigma_e - \sigma_0 = \sqrt{\tilde{\Sigma}_m^2 + \tilde{\Sigma}_e^2 - \frac{3\tilde{\Sigma}_m \Sigma_L}{1-f}} - \sigma_0 \leq 0 \tag{61}$$

Let us recall that for obtaining the macroscopic criterion from (20), one need to integrate (61) over the matrix. However, due to the presence of the azimuth angle ϕ in the expression of Σ_L (60), there is no closed form solution. In order to overcome this difficulty, a simple idea consists in performing a Taylor series expansion (around 0) till the third order, this leads to the following approximation:

$$\sigma_e = \sqrt{\tilde{\Sigma}_m^2 + \tilde{\Sigma}_e^2} \cdot \sqrt{1 - \frac{3\tilde{\Sigma}_m \Sigma_L}{(1-f)(\tilde{\Sigma}_m^2 + \tilde{\Sigma}_e^2)}}$$

$$\sigma_e \simeq \sqrt{\tilde{\Sigma}_m^2 + \tilde{\Sigma}_e^2} \cdot \left[1 - \frac{3\tilde{\Sigma}_m \Sigma_L}{2(1-f)(\tilde{\Sigma}_m^2 + \tilde{\Sigma}_e^2)} - \frac{9\tilde{\Sigma}_m^2 \Sigma_L^2}{8(1-f)^2(\tilde{\Sigma}_m^2 + \tilde{\Sigma}_e^2)^2} - \frac{27\tilde{\Sigma}_m^3 \Sigma_L^3}{16(1-f)^3(\tilde{\Sigma}_m^2 + \tilde{\Sigma}_e^2)^3}\right]$$

Next, the final integration includes the computation of the following integrals:

$$\frac{1}{4\pi} \int_{S(r)} \Sigma_L dS = 0$$

$$\frac{1}{4\pi} \int_{S(r)} \Sigma_L^2 dS = \frac{4\Sigma_e^2}{45}$$

$$\frac{1}{4\pi} \int_{S(r)} \Sigma_L^3 dS = \frac{16\Sigma_e^3}{945} \cos(\theta_L)(4\cos^2(\theta_L) - 3)$$

Finally, from (20), the macroscopic criterion is obtained as:

$$\mathscr{F} \simeq D\left(1 - \frac{C^2 \Sigma_e^2}{90 D^4} + \frac{C^3 \Sigma_e^3}{945 D^6} \cos(\theta_L)(4\cos^2(\theta_L) - 3)\right) - \sigma_0 \leq 0 \tag{62}$$

where we have denoted C and D the following functions of $\boldsymbol{\Sigma}$

$$D(\boldsymbol{\Sigma}) = \sqrt{\tilde{\Sigma}_m^2 + \tilde{\Sigma}_e^2}$$

$$= \sqrt{\frac{9\Sigma_m^2}{4\ln^2 f} + \frac{\Sigma_e^2}{(1-f)^2}}, \quad C(\boldsymbol{\Sigma}) = -\frac{3\tilde{\Sigma}_m}{1-f} = \frac{9\Sigma_m}{2(1-f)\ln f} \tag{63}$$

It should be underlined that the established criterion (62) depends not only on the the macroscopic mean stress and equivalent stress, but also explicitly on the Lode angle (or the third invariant of the stress deviator).

4.3 Macroscopic Flow Rule Under Non-axisymmetric Loading

We aim now at deriving the plastic strain rate from the normality rule. Unlike the conventional modeling, the three invariants of the macroscopic criterion (62) are taken into account. Not only the mean strain rate D_m and the equivalent one D_e have to be computed, but also the contribution D_{III} related to the third invariant of deviator J_3 will be provided. It is worthy to interpret that D_{III} can indicated the influence of the Lode angle upon the π-plane of principal stress space to the macroscopic plastic flow rule. Let us first define the macroscopic stress,

$$\Sigma_{III} = \sqrt[3]{J_3} \tag{64}$$

Hence, the dissipation power Π can be written as

$$\Pi = \boldsymbol{D} : \boldsymbol{\Sigma} = 3\Sigma_m D_m + \Sigma_e D_e + \Sigma_{III} D_{III} \tag{65}$$

Moreover, considering the macroscopic criterion (62), the macroscopic strain rate can be obtained from the associated flow rule

$$D_e = \dot{\Lambda}\frac{\partial \mathscr{F}_{3D}}{\partial \Sigma_e}, \quad D_m = \frac{1}{3}\dot{\Lambda}\frac{\partial \mathscr{F}_{3D}}{\partial \Sigma_m}, \quad D_{III} = \dot{\Lambda}\frac{\partial \mathscr{F}_{3D}}{\partial \Sigma_{III}}. \tag{66}$$

It follows that D_e, D_m and D_{III} can be expressed in explicit forms:

$$\begin{aligned} D_e = \dot{\Lambda}\Bigg[& \frac{\partial D}{\partial \Sigma_e} - \frac{C^2}{90}\frac{2D\Sigma_e - 3\frac{\partial D}{\partial \Sigma_e}\Sigma_e^2}{D^4} \\ & - \frac{C^3 J_3}{70}\left(\frac{5}{D^6}\frac{\partial D}{\partial \Sigma_e}(\frac{729 J_3^2}{\Sigma_e^6} - 3) - \frac{6}{D^5}\frac{729 J_3^2}{\Sigma_e^7}\right)\Bigg] \end{aligned} \tag{67}$$

$$\begin{aligned} D_m = \frac{1}{3}\dot{\Lambda}\Bigg[& \frac{\partial D}{\partial \Sigma_m} - \frac{\Sigma_e^2}{90}\cdot\frac{2CD\frac{\partial C}{\partial \Sigma_m} - 3C^2\frac{\partial D}{\partial \Sigma_m}}{D^4} \\ & + \frac{J_3}{70}\left(\frac{729 J_3^2}{\Sigma_e^6} - 3\right)\left(\frac{3C^2 D^2\frac{\partial C}{\partial \Sigma_m} - 5D^4\frac{\partial D}{\partial \Sigma_m}}{D^7}\right)\Bigg] \end{aligned} \tag{68}$$

$$D_{III} = \dot{\Lambda}\left[\frac{C^3}{70D^5}\left(\frac{729 J_3^2}{\Sigma_e^6} - 1\right)J_3^{-\frac{2}{3}}\right] \tag{69}$$

where

$$\frac{\partial C}{\partial \Sigma_m} = \frac{9}{2(1-f)\ln f}, \quad \frac{\partial D}{\partial \Sigma_m} = \frac{9\Sigma_m}{2D\ln^2 f}, \quad \frac{\partial D}{\partial \Sigma_e} = \frac{2\Sigma_e}{D(1-f)^2}$$

Finally, the plastic void growth rate can be obtained from the mass balance equation. It expression can be obtained from (42) by taking into account the plastic flow rule (67) and (68) and eliminating the plastic multiplier $\dot{\Lambda}$. It obviously shows that the void growth rate depends on the third invariant of the stress deviator J_3 (or the Lode angle θ_L).

4.4 Illustration of the Macroscopic Yield Criterion and Void Growth Under Non-axisymmetric Loading

We provide in the subsection the illustration of the established criterion (62) and its comparison with Gurson model and the stress-based variational model (SVM) under axisymmetric loading provided in Sect. 3.1. It is worthy to note again that the later one has been derived from a closed-form formulation. A value of porosity $f = 0.01$ are adopted for the later illustration and comparisons.

First, five yield loci obtained from (62) are illustrated on Fig. 2 with different values of Lode angle: $\theta_L = 0°, 15°, 30°, 45°$ and $60°$, while the first and the last ones are corresponding to (but not equivalent to) the macroscopic model (32) obtained from the axisymmetric trial stress field (see Fig. 3). It can be observed that the yield surfaces obtained from other values of the Lode angle are absolutely between the

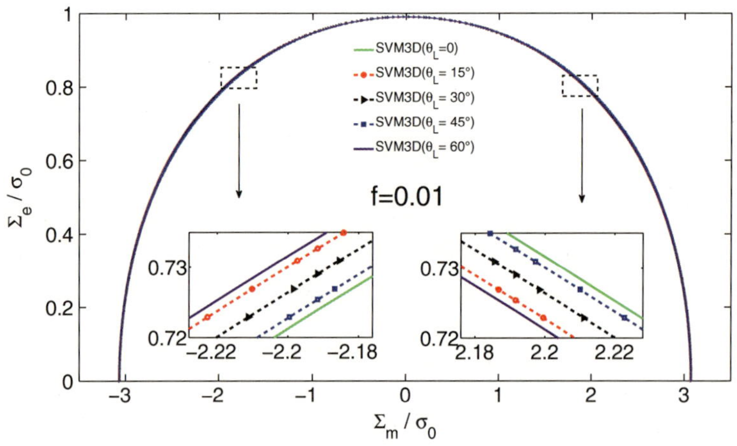

Fig. 2 Illustrations of the yield surfaces obtained from the new established criterion SVM3D (62) with *five* values of Lode angle: $\theta_L = 0°$, $15°$, $30°$, $45°$ and $60°$. Porosity: $f = 0.01$

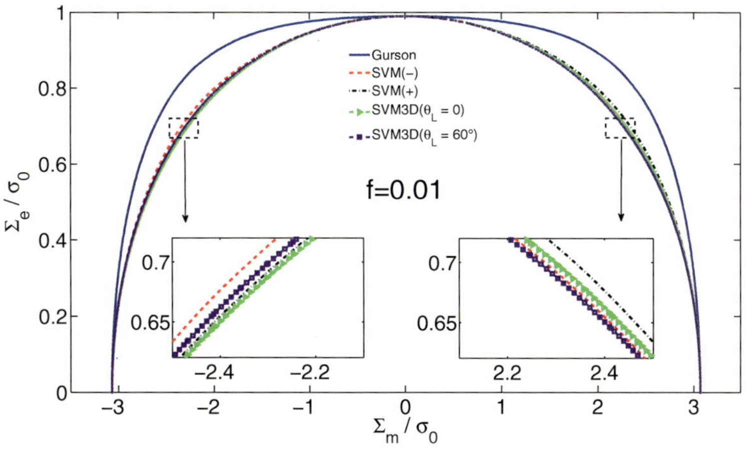

Fig. 3 Comparison between the yield surfaces obtained from the established criterion SVM3D (62) with the *closed* form criterion of stress-based variational model (SVM(+) and SVM(−)) (32) and the Gurson criterion [19]. Porosity: $f = 0.01$

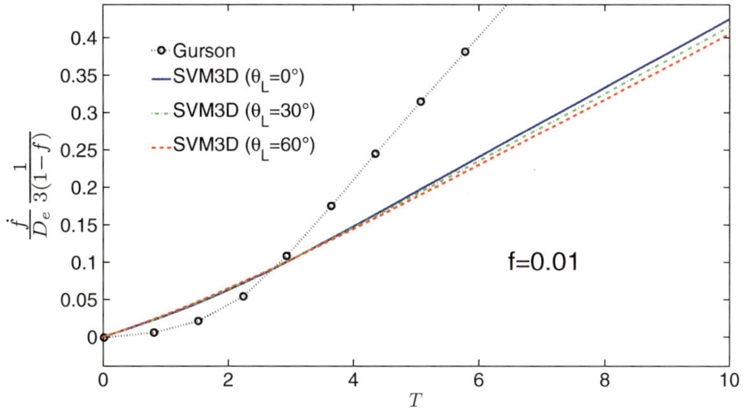

Fig. 4 Evolution of porosity as function of the stress triaxiality for initial porosity $f = 0.01$ with *three* values of Lode angle: $\theta_L = 0°$, 30° and 60°

above two ones. Consequently, it can be concluded that the yield surface displays a slight asymmetry due to the value of Lode angle (or due to the value of third invariant).

Figure 4 illustrates the evolution of porosity given as function of stress triaxiality $T = \frac{\Sigma_m}{\Sigma_e}$ for the case $f = 0.01$. It can be observed that, the ones with axisymmtric loadings ($\theta_L = 0$ and 60°) give two extremal values of the void evolution for a fixed value of triaxiality, while for another case with non axisymmetric state ($\theta_L = 30°$) is exactly between the two extremal ones. Slight differences due to the Lode angle (or the third invariant of stress deviator) can be observed.

5 Conclusion

In this study, we have proposed a stress-based variational approach of ductile porous materials in the framework of Limit Analysis. This has been done by applying homogenization theory combined with the statical limit analysis approach. The stress variational model under axisymmetric loading and the one under non-axisymmetric loading, fully described in the work, takes advantage of Hill's variational principle for which relaxed licit stress fields have been adopted. The established results clearly provide expressions of the statically-based macroscopic criteria. Due to the relaxed internal boundary condition resulting from the chosen trial stress fields, the criterion could be seen only as a quasi-lower bound. An interesting feature of the established criterion is its dependence not only on the two stress invariants Σ_m and Σ_e, but also on the third invariant of the stress deviator (or on the Lode angle); this leads to specific asymmetries of the macroscopic criterion.

The results derived from the obtained criterion are fully assessed by means of comparison with existing analytical criteria, with available numerical bounds and finally with our Finite Elements results. This has allowed to demonstrate the interest of the new theoretical results. For completeness, we also provide voids growth equations which clearly show the effects of the third invariant in addition to that of the stress triaxiality. This topic of the effect of stress states on ductile fracture is a growing and is deserving attention in several recent studies: see for instance [14, 23, 28].

6 Appendix: Illustration and Validation for the Macroscopic Model Under Axisymmetric Loading

Due to assumptions on the stress fields which have been introduced for the analytical derivation, the new criterion SVM (32) could be seen just as a quasi-lower bound. However, it still preserves the exact solution of the hollow sphere subjected to a hydrostatic loading, $\Sigma_m/\sigma_0 = -2/3ln(f)$, and leads to the same expression of the limit pure shear load as that given by the Gurson criterion, $\Sigma_e/\sigma_0 = 1-f$. Moreover, as mentioned before, due to the presence of the third invariant J_3 in the SVM criterion, the yield surface exhibits an asymmetry about the axis $\Sigma_m = 0$. For instance, the yield surfaces with two values of porosity $f = 0.01$ and $f = 0.064$ are deliberately plotted on Fig. 5 for negative and positive Σ_m. It is noted that the SVM criterion presents some relative small differences with the Gurson one, the surfaces predicted by SVM being strictly "below" the Gurson's ones, simultaneously coincident with them for hydrostatic loading (exact result) and pure deviatoric one. Finally, the slight asymmetry of the SVM yield surfaces can also be observed on Fig. 5, with the notations of SVM(+) and SVM(−) for the yield surfaces corresponding to $J_3 > 0$ and $J_3 < 0$, respectively (Fig. 6).

Noticeable difference with Gurson criterion is observed for small porosities. This can be explained by that in the context of the SVM, the observed difference may found

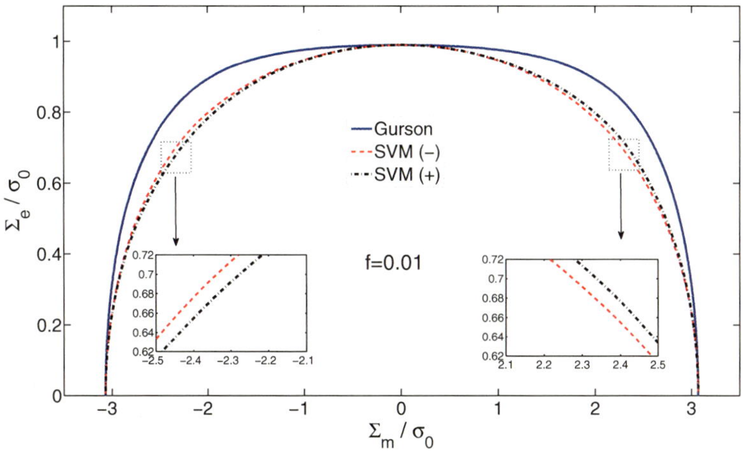

Fig. 5 Comparison between the yield surfaces obtained from the SVM (32) and the Gurson criterion [19]. Porosity $f = 0.01$

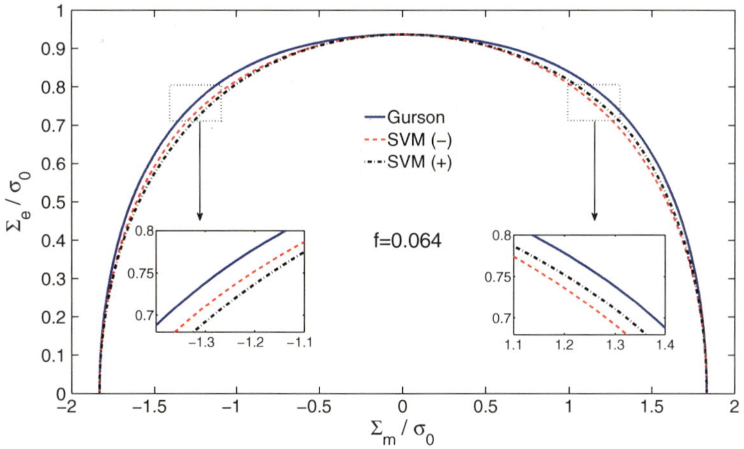

Fig. 6 Comparison between the yield surfaces obtained from the SVM (32) and the Gurson criterion [19]. Porosity $f = 0.064$

its origin in the inaccuracy resulting from the relaxation of the yield condition in the matrix, this condition being enforced only in the mean. For small porosities, large plastic strain heterogeneities may occur in the vicinity of the cavities and the above procedure consisting to relax the yield condition should be inaccurate. It is clear that the proposed model can be improved by considering a more refined admissible stress field able to avoid such procedure.

Next, the yield surfaces obtained by means of the stress variational approach will be compared with Finite Element Method (FEM) solutions obtained during the present study. The computations are carried out by means of ABAQUS/Standard software and a user subroutine MPC (Multi-Points Constraints). The main reason for which we need to enforce MPC conditions in the code is that we have to impose the velocity field $\boldsymbol{v}$ from $\boldsymbol{v} = \boldsymbol{D} \cdot \boldsymbol{x}$ (on the external boundary of the hollow sphere) such that the constraint of constant macroscopic stress triaxiality ($T = \Sigma_m/\Sigma_e$) be fulfilled. In practice, as in [18], this is done by applying a constant macroscopic stress ratio Σ_ρ/Σ_z corresponding to the desired Σ_m/Σ_e. Note that the implementation of this procedure is the one that is already described in [5] for the study of voids interaction and coalescence.

The comparisons between the surfaces obtained from the SVM, the Gurson's model and the FEM solutions are illustrated on Fig. 7 only for the first quadrant (the slight asymmetry of the SVM surface is disregarded) and for the relative smaller porosity $f = 0.01$. We can observe that in this cases, the FEM solutions are almost between the upper bound (Gurson's model) and the proposed yield criterion (SVM). This fact shows that the SVM could be seemed as a quasi lower bound for the porous materials which have relatively small values of porosity, especially for porous metal materials.

Moreover, Fig. 8 displays the comparisons between the predictions of the SVM criterion and the numerical bounds for $f = 0.01$. It should be noted that, even the FEM solution is between the numerical bounds, the yield surface of SVM is generally below the numerical LB, except for the loadings with high values of stress triaxiality $T = \Sigma_m/\Sigma_e$ for which it interestingly lies between the two bounds and coincides with the exact value of the hydrostatic loading.

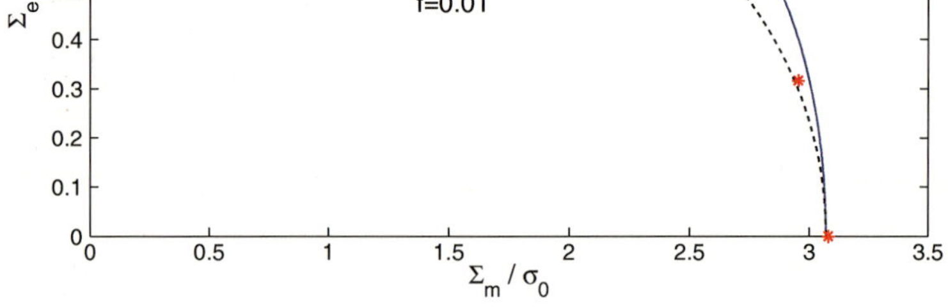

Fig. 7 Comparison between the yield surfaces obtained from the SVM (32), the Gurson criterion [19] and the FEM solution. Porosity: 0.01

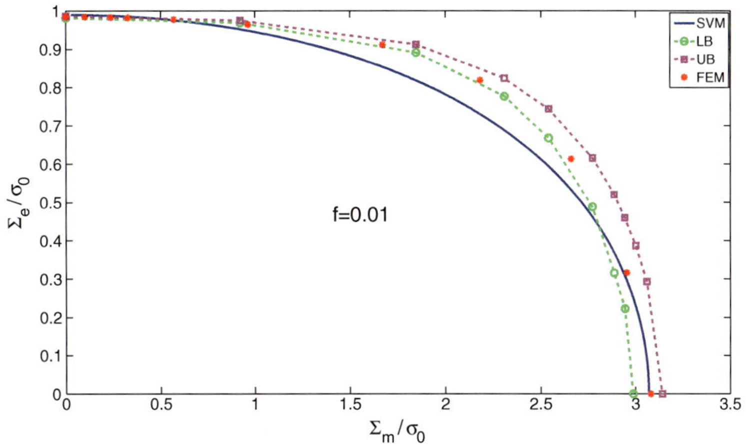

Fig. 8 Comparison between the yield surfaces obtained from the SVM (32), the numerical bounds [32] and the FEM solution. Porosity: 0.01

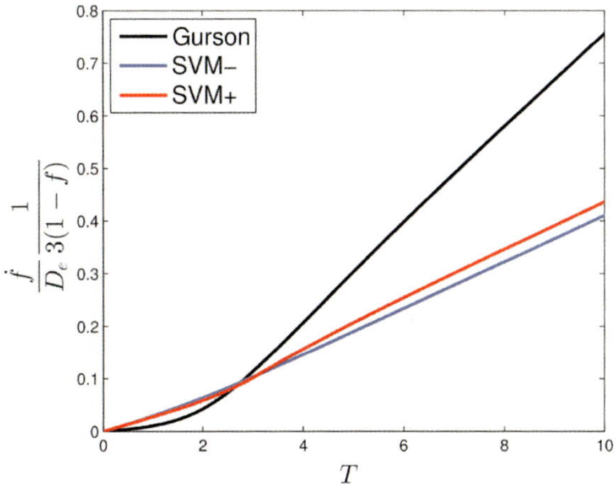

Fig. 9 Evolution of porosity as function of the stress triaxiality for initial porosity $f = 0.01$. Comparison between SVM predictions and that of Gurson model

Finally, Fig. 9 illustrates the evolution of porosity $f = 0.01$ given as function of stress triaxiality for three values of initial porosity. It is noted that despite the few influence of the sign of third invariant on the macroscopic criterion, a noticeable effect is noted for the porosity variation. The results are also compared with that predicted by the Gurson model. Clear differences are observed, particularly for high stress triaxialities for which the Gurson model is known to overestimate the variation of the porosity.

References

1. Barthélémy JF, Dormieux L (2003) Détermination du critère de rupture macroscopique d'un milieux poreux par homogénéisation non linéaire. Comptes Rendus Mécanique 331:271–276
2. Benzerga AA, Besson J (2001) Plastic potentials for anisotropic porous solids. Eur J Mech A/Solids 20:397–434
3. Cazacu O, Revil-Baudard B, Lebensohn R, Garajeu M (2014) On the combined effect of pressure and third invariant on yielding of porous solids with von Mises matrix. J Appl Mech 80(6):064501
4. Cheng L (2013) Homogenization of porous media with plastic matrix and non-associated flow rule by variational methods, Ph.D. thesis, University Lille 1
5. Cheng L, Guo TF (2007) Void interaction and coalescence in polymeric materials. Int J Solids Struct 44:1787–1808
6. Cheng L, de Saxcé G, Kondo D (2014) A stress variational model for ductile porous materials. Int J Plast 55:133–151
7. Cheng L, Jia y, Oueslati A, de Saxcé G, Kondo D (2014) A bipotential-based limit analysis and homogenization of ductile porous materials with non-associated Drucker-Prager matrix, submitted to J Mech Phys Solids
8. de Saxcé G (1992) Une généralisation de l'inégalité de Fenchel et ses applications aux lois constitutives. C R Acad Sci Paris Sér II 314:125–129
9. de Saxcé G, Feng ZQ (1991) New inequality and functional for contact friction: the implicit standard material approach. Mech Struct Mach 19:301–325
10. de Saxcé G, Bousshine L (1993) On the extension of limit analysis theorems to the non-associated flow rules in soils and to the contact with Coulomb's friction. In: XI Polish conference on computer methods in mechanics. Kielce, pp 815–822
11. de Saxcé G, Bousshine L (1998) Limit analysis theorems for the implicit standard materials: application to the unilateral contact with dry friction and the non associated flow rules in soils and rocks. Int J Mech Sci 40(4):387–398
12. Ekeland I, Temam R (1975) Convex analysis and variational problems. North Holland Publisher, Amsterdam
13. Fenchel W (1949) On conjugate convex functions. Can J Math 1:73–77
14. Gao X, Zhang T, Hayden M, Roe C (2009) Effects of the stress state on plasticity and ductile failure of an aluminum 5083 alloy. Int J Plast 25:2366–2382
15. Garajeu M, Suquet P (1997) Effective properties of porous ideally plastic or viscoplastic materials containing rigid particles. J Mech Phys Solids 45:873–902
16. Gologanu M, Leblond JB, Perrin G, Devaux J (1997) Recent extensions of Gurson's model for porous ductile metals. In: Suquet P (ed) Continuum micromechanics. Springer, New York
17. Green RJ (1972) A plasticity theory for porous solids. Int J Mech Phys Solids 14:215–224
18. Guo TF, Faleskog J, Shih CF (2008) Continuum modeling of a porous solid with pressure-sensitive dilatant matrix. J Mech Phys Solids 56:2188–2212
19. Gurson AL (1977) Continuum theory of ductile rupture by void nucleation and growth—part I: yield criteria and flow rules for porous ductile media. J Eng Mater Technol 99:2–15
20. Hill R (1950) Mathematical theory of plasticity. Oxford University Press, London
21. Jeong HY (2002) A new yield function and a hydrostatic stress-controlled model for porous solids with pressure-sensitive matrices. Int J Mech Phys Solids 32:3669–3691
22. Jeong HY, Pan J (1995) A macroscopic constitutive law for porous solids with pressure-sensitive matrices and its applications to plastic flow localization. Int J Mech Phys Solids 39:1385–1403
23. Li Z, Fu MW, Lua J, Yang H (2011) Ductile fracture: experiments and computations. Int J Plast 27:147–180
24. Maghous S, Dormieux L, Barthélémy JF (2009) Micromechanical approach to the strength properties of frictional geomaterials. Eur J Mech A/Solids 28:179–188
25. Monchiet V, Charkaluk E, Kondo D (2007) An improvement of Gurson-type models of porous materials by Eshelby-like trial velocity fields. Comptes Rendus Mécanique 335:32–41

26. Monchiet V, Cazacu O, Kondo D (2008) Macroscopic yield criteria for plastic anisotropic materials containing spheroidal voids. Int J Plast 24:1158–1189
27. Moreau JJ (2003) Fonctionnelles convexes. Istituto Poligrafico e Zecca dello Stato, Rome
28. Nahshon K, Hutchinson JW (2008) Modification of the Gurson model for shear failure. Eur J Mech A/Solids 27:1–27
29. Rockafellar RT (1970) Convex analysis. Princeton University Press, Princeton
30. Save MA, Massonnet CE, de Saxcé G (1997) Plastic limit analysis of plates, shells and disks. Elsevier, New York
31. Sun Y, Wang D (1989) A lower bound approach to the yield loci of porous materials. Acta Mechanica Sinica 5:237–243
32. Trillat M, Pastor J (2005) Limit analysis and Gurson's model. Eur J Mech A/Solids 24:800–819

Limit Analysis and Macroscopic Strength of Porous Materials with Coulomb Matrix

Franck Pastor, Djimedo Kondo and Joseph Pastor

Abstract The paper is devoted to the numerical Limit Analysis of a hollow spheroidal model with a Coulomb solid matrix. In a first part the hollow spheroid model is presented, together with its axisymmetric FEM discretization and its mechanical position. Then, after an adaptation of a previous static code, an original mixed (but fully kinematic) approach dedicated to the axisymmetric problem was elaborated with a specific quadratic velocity field associated to the triangular finite element. Despite the less good conditioning inherent to the axisymmetric modelization, the final conic mixed code appears very efficient, allowing to take into account numerical meshes highly refined. After a first validation in the case of spherical cavities and isotropic loadings, for which the exact solution is known, numerical bounds of the macroscopic strength are provided for both cases of spherical and spheroidal voids. Effects of the friction angle as well as that of the void aspect ratio are fully illustrated.

1 Introduction

Focused to the axisymmetric problem, the present paper aims at responding to two purposes. First it generalizes the linear programming formulations of the limit analysis (LA) in the axisymmetric case for Coulomb materials of [22, 31]: this is partly motivated by recent papers about this subject that only give, in fact, estimates of the resulting bounds, as recently pointed out in [26]. The second purpose deals with the determination of the macroscopic plasticity criterion of the porous Coulomb material with spheroidal voids, never investigated up to now, on the basis of the

F. Pastor
Athénée Royal Victor Horta, Bruxelles, Belgium
e-mail: franck.pastor@skynet.be

D. Kondo
Institut D'Alembert, Université P. M Curie, Paris, France
e-mail: djimedo.kondo@upmc.fr

J. Pastor (✉)
Laboratoire LOCIE, Université de Savoie, Le Bourget du Lac, France
e-mail: joseph.pastor@univ-savoie.fr

P. Fuschi et al. (eds.), *Direct Methods for Limit and Shakedown Analysis of Structures*,
DOI 10.1007/978-3-319-12928-0_2

Gurson-like hollow spheroid problem. To overcome the conditioning difficulties of the classic upper bound (or kinematic) approach for solving this problem, a new axisymetric mixed formulation is presented which preserves the rigorous kinematic character of the resulting solutions. This study was also motivated by the recent results about porous materials with Drucker-Prager matrices in [18]; in this paper the corresponding kinematic mixed approach has allowed—with very much lower computation times—significative improvements of the upper bounds obtained by previous fully 3D numerical methods.

The famous Gurson plasticity criterion [9] is based on the consideration of a hollow von Mises sphere or cylinder in the framework of the Limit Analysis (LA) kinematic approach. Recent studies were devoted to porous materials with a Drucker-Prager matrix and spherical voids ([8, 11, 28], etc.). Even in this relatively simple case of spherical voids, there is no theoretical attempt to micromechanically derive a macroscopic criterion for the Coulomb porous material. This is undoubtedly due to difficulties inherent to the Coulomb criterion which involves all the three stress invariants.

Several extensions of the Gurson model taking into account void shape effects have been also proposed in order to solve various practical cases (see for example [5–7, 15, 16]). Up to our knowledge, similar theoretical studies with pressure-sensitive matrices and non-spherical voids do not exist in the literature: clearly, in this case of spheroidal voids also, there is a strong need of an appropriate research effort to obtain efficient estimates.

On the other hand, using finite element discretizations, both static and kinematic methods of LA have been elaborated for Gurson's problems with cylindrical cavities ([4, 23, 27]). In [30] the Gurson criterion (then with a von Mises matrix) is shown to be relevant for materials with spherical voids by using the same tools and an new three-dimensional numerical model. Using also the hollow sphere model, a recent paper [19] was devoted to spherically porous materials with pressure-sensitive matrices obeying the Drucker-Prager, Mises-Schleicher and Green criteria. On the other hand, in the case of a von Mises matrix, these numerical studies have been extended to take into account the void shape effects by considering a central spheroid void in a matrix with a confocal boundary [21]. They concluded on the relevance of the criteria proposed by the above mentioned studies of Gologanu and Leblond, at least for the investigated porosity cases. Very recently, as mentioned before, these approaches have been extended to the hollow spheroid problem with a Drucker-Prager matrix in [18]. They provided very original bounds which are expected to be used as reference results for validation of future theoretical investigations.

Therefore, the second purpose of the present paper is concerned with providing lower/upper bound results to be used as reference values for forthcoming attempts to determine approximate criteria for porous Coulomb materials in the framework of limit analysis (LA) applied to the model of the hollow sphere or spheroid. Indeed, these numerical approaches provide rigorous bounds (*a posteriori* controllable) to the macroscopic criterion. Moreover, due to the selected projection approach, the direction of the kinematically admissible (macroscopic) strain rates can be easily obtained as the normal to the kinematic approach in the projection plane, since the homogenized material complies with the associated flow law.

First, we briefly present the hollow spheroid problem and its formulation in terms of LA. Then, we recall the basis of the LA static and mixed methods, and the corresponding expressions for the Coulomb material necessary for obtaining the bounds, or to assess them by post-analysis of the optimal fields. The next step presents the *ad hoc* extended static approach, and details the new formulation of the mixed (but rigorously kinematic) method, in the axisymmetric case for the Coulomb problem. Let us note that it is the first time that the LA mixed approach is applied to an axisymmetric finite element problem with this frictional material. After a validation on the hollow sphere problem under isotropic loading whose the exact solution is known, we finally provide the results of the axisymmetric tests for two values of the friction angle of the Coulomb criterion and three aspect ratios of the cavity. A comparison with Drucker-Prager results of [18] is also given, in order to point out the differences of the two porous material criteria, all porous Coulomb test exhibiting corners, not the Drucker-Prager ones.

2 The Hollow Spheroid Model

The hollow spheroid model is made up of a spheroidal cavity embedded in a confocal spheroidal cell. The solid matrix is a Coulomb material, homogeneous and isotropic. Figure 1 presents the geometric model, where the given aspect ratio a_1/b_1 and the porosity f allow to determine the parameters (a_2 and b_2) of the confocal spheroidal boundary. Let us consider first the three-dimensional point of view, and note Σ and D the macroscopic stress and strain rate tensors. These quantities are related to

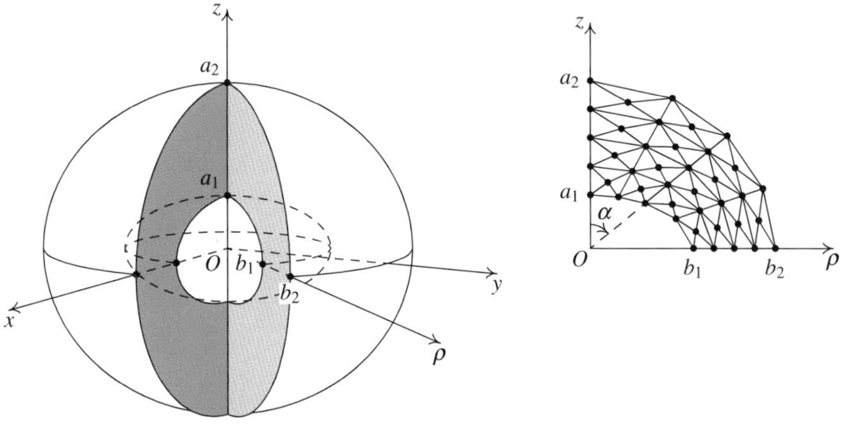

Fig. 1 The hollow spheroid model and its FEM axisymmetric discretization ($a_1/b_1 = 0.5, f = 0.1$)

the microscopic fields by the averages over the model of volume V:

$$\Sigma_{ij} = \frac{1}{V}\int_V \sigma_{ij}\, dV, \quad D_{ij} = \frac{1}{2V}\int_{\partial V} (u_i n_j + u_j n_i)\, dS, \tag{1}$$

where u denotes the velocity vector and n the normal vector to the boundary ∂V of the model.

Under the uniform strain rate boundary conditions, i.e. $u_i = D_{ij}x_j$ (where x represents the position vector), on the external boundary, the virtual dissipated power $P_{\text{tot}} = V\Sigma_{ij}D_{ij}$ can be written as follows:

$$P_{\text{tot}} = V\, Q \cdot q, \tag{2}$$

where the loading vector Q and the generalized velocity q here are defined as:

$$Q_1 = \Sigma_m = \frac{1}{3}\left(\Sigma_x + \Sigma_y + \Sigma_z\right), \quad Q_2 = \frac{\Sigma_x + \Sigma_y}{2} - \Sigma_z,$$

$$Q_3 = \frac{\sqrt{3}}{2}\left(\Sigma_x - \Sigma_y\right), \quad Q_4 = \Sigma_{yz}, \quad Q_5 = \Sigma_{zx}, \quad Q_6 = \Sigma_{xy};$$

$$q_1 = D_x + D_y + D_z, \quad q_2 = \frac{2}{3}\left(\frac{D_x + D_y}{2} - D_z\right), \quad q_3 = \frac{1}{\sqrt{3}}\left(D_x - D_y\right),$$

$$q_4 = 2D_{yz}, \quad q_5 = 2D_{zx}, \quad q_6 = 2D_{xy}.$$

From the matrix isotropy and the spheroidal geometry of the model, the resulting material is transversally isotropic around the axis z. Hereafter is investigated the macroscopic criterion $g(\Sigma)$ in the $(Oxyz)$ anisotropy frame of Fig. 1. We search the projection of $g(\Sigma)$ in the (Q_1, Q_2) plane by optimizing Q_2 for fixed $Q_1 = \Sigma_m$, other stress components defined in (2) being free. Consequently, $\frac{\partial g}{\partial \Sigma_{ij}} = 0 = 2D_{ij}$ for $i \neq j$, and $\frac{\partial g}{\partial Q_3} = 0 = q_3$ since the macroscopic material complies with the normality rule. Loadings can then be restricted to principal strain rates D as well as Σ ($(Oxyz)$ is a transverse isotropy frame), with $D_x = D_y$ and $\Sigma_x = \Sigma_y$. Finally, the problem involves only two non-zero loading parameters, Q_1 and Q_2.

Then, from the symmetries of the whole model, the quarter of the meridian plane of the hollow spheroid is meshed into triangular elements as shown in Fig. 1, right. Note that the macroscopic equivalent stress Σ_{eq} is, in the present case, linked to Q_2 by:

$$\Sigma_{eq} = \sqrt{\frac{3}{2}\,\Sigma' : \Sigma'} = |Q_2| = |\Sigma_x - \Sigma_z| = |\Sigma_\rho - \Sigma_z|, \tag{3}$$

where Σ' is the deviatoric part of Σ. In this case we can substitute in the above relations Σ_ρ to Σ_x and to Σ_y (Σ_z unchanged), and D_ρ to D_x and D_y. The loading conditions $u_i = D_{ij}x_j$ here become $u_\rho = D_\rho \rho$, $u_z = D_z z$.

To obtain the static approach, we have first extended the axisymmetric finite element code and the process defined in [24]. For the kinematic approach, we have elaborated an original axisymmetric version of the mixed kinematic approach based on specific quadratic velocities in the triangular elements. These *ad hoc* formulations avoids the possible singularities of the axisymmetric equations in order to preserve the rigorous bounding character of the solutions. Hereafter, we present the static process and we detail the new axisymmetric mixed method, and their application to the investigated problem. The final resulting graphs are presented for the usual values of the friction angle $\phi = 10°$ and $20°$, and for the aspect factor $a_1/b_1 = 1$, 0.5 and 0.2.

3 LA Methods and Coulomb Criterion

The main goal of limit analysis is the determination of the locus of the limit loads which corresponds to the macroscopic plasticity criterion in the present micro-macro problem. Classically, the limit loads can be determined by using the static (or lower bound) and the kinematic (or upper bound) methods.

3.1 The Static Method

The first one is the static method which is in terms of stresses and leads to a lower bound to the limit loads. A stress field is said admissible if it is statically admissible (SA), and plastically admissible (PA), i.e. verifying the (convex) plasticity criterion $f(\sigma)$; a loading vector $Q(\sigma)$ is admissible if the corresponding σ is admissible. Let us denote K the set of the admisible loading vectors. The final problem reads:

$$Q_{lim} = (Q_1^d, \ldots, \lambda_0 Q_i^d, \ldots, Q_n^d); \tag{4.i}$$

$$\lambda_0 = \max\left\{\lambda\,,\ Q(\sigma) = (Q_1^d, \ldots, \lambda Q_i^d, \ldots, Q_n^d)\right\} \tag{4.ii}$$

where the stress tensors σ are admissible, and Q_d is a fixed admissible loading vector. In fact, Relation (4.i and 4.ii) holds when all admissible fields σ can be taken into account, which is not the case in general. Then, by varying the direction of Q_d, a set of admissible Q, located near or on ∂K, is obtained: the smallest convex envelope of the corresponding points in K is an inner approach of the boundary ∂K, i.e. a lower bound to the exact macroscopic criterion investigated here.

3.2 The Mixed Kinematic Method

On the other hand, a so-called mixed kinematic formulation was pioneered by Anderheggen and Knopfel [1] for finite element continuous velocity and a linearized von Mises criterion, resulting in linear programming (LP) problems. An extension to the discontinuous velocity case, based on the assumption that the LP duality properties remain valid in non-linear programming, was proposed in [10]. A general extension to discontinuous velocity fields and convex optimization was successfully experienced in [17, 20] for homogeneous von Mises and Gurson materials in plane strain. This mixed formulation is here modified as:

$$\max_{Q,\sigma,T'} F = Vq_d \cdot Q \tag{5.i}$$

$$\text{s.t.} \int_V d : \sigma \, dV + \int_{S_d} [u] \cdot T' \, dS = Vq(u) \cdot Q \quad \forall \, \text{KA} \, u, \tag{5.ii}$$

$$f(\sigma) \leq 0, \quad f_{nt}(T') \leq 0, \tag{5.iii}$$

where d is the strain rate tensor, σ the stress tensor, S_d the union of the velocity discontinuity surfaces, T' the stress vector on these surfaces and $f_{nt}(T')$ the projection of $f(\sigma')$ on the Mohr plane associated to the discontinuity surface of normal n. In (5.i–5.iii) the velocity field u must be kinematically admissible (KA), i.e. piecewise continuous with bounded discontinuities $[u]$ and verifying the boundary conditions and the loading condition $q(u) = q_d$. It can be seen in the above mentioned papers that the optimal velocity field will also be PA (plastically admissible), i.e. there exist everywhere a stress tensor σ or a stress vector T' associated to the strain rate tensor or to the velocity jump by the normality law corresponding to $f(\sigma) = 0$ or $f_{nt}(T') = 0$, respectively.

The previous formulation gives the exact solution if all velocity and stress fields could be taken into account. In general this is not the case when we consider a discretization of the mechanical system in finite elements. However, the following axisymmetric formulation is formulated for preserving the fully kinematic character of the final result by using convexity properties concerning the set of the PA strain rates (and of the PA velocity jumps) and the unit dissipated powers.

3.3 The Coulomb Material

The original Coulomb criterion reads:

$$f(\sigma) = \left|\sigma_i - \sigma_j\right| - 2\,c\,\cos\phi + (\sigma_i + \sigma_j)\sin\phi \leq 0 \tag{6}$$

where σ_i and σ_j refer to the principal stresses without any *a priori* order ($i, j = 1, 2, 3$, $i \neq j$), c is the cohesion of the material and ϕ the internal friction angle. In [22], this criterion is written in the axisymmetric case in the form of three conic inequalities that are linearized to obtain a linear programming problem. From [24] the Coulomb criterion finally results in one conic inequation and three linear inequations:

$$\begin{cases} \sqrt{(\sigma_\rho - \sigma_z)^2 + 4\tau_{\rho z}^2} \leq U, \\ U \leq -\alpha \sin\phi + 2c \cos\phi, \\ U \leq -\left(\alpha - 2\sigma_\theta \frac{1-\sin\phi}{1+\sin\phi}\right) + 4c \frac{\cos\phi}{1+\sin\phi}, \\ U \leq \left(\alpha - 2\sigma_\theta \frac{1+\sin\phi}{1-\sin\phi}\right) + 4c \frac{\cos\phi}{1-\sin\phi}, \end{cases} \tag{7}$$

with $\alpha = \sigma_\rho + \sigma_z$.

In the Mohr plane related to a velocity discontinuity facet of normal n, the Coulomb criterion classically reads:

$$f_{nt}(T) = |\sigma_{nt}| + \sigma_n \tan\phi - c \leq 0, \tag{8}$$

Let us recall the PA (plastic admissibility) condition for the strain rates and the volumic dissipated power [25]:

$$(|d_1| + |d_2| + |d_3|) \sin\phi \leq tr(d)\,; \; \pi_{vol}(d) = \frac{c}{\tan\phi} \mathrm{tr} d \tag{9}$$

where d_1, d_2 and d_3 are the principal strain rates. From (9), the domain of the PA strain rates is a convex cone whose apex corresponds to the null tensor.

The corresponding relations about the velocity discontinuities read:

$$[u_n] \geq |\,[u_t]\,| \tan\phi\,; \; \pi_{disc}([u]) = c\frac{[u_n]}{\tan\phi} = c|\,[u_t]\,| \tag{10}$$

The expressions (9) and (10) here are only used in the post-analysis of the optimal solution of the mixed method.As mentioned before, two numerical implementation methods were carried out for this study. We first begin with the static method, more precisely the static iterative process which will deliver a lower bound for the solution of the Limit Analysis problem.

4 Numerical Implementation of the Axisymmetric Static Method

4.1 The Stress Field

Since the affine formulation is too "poor" to give satisfactory results, the FEM discontinuous stress field is expressed with *quadratic* functions, as follows:

$$\begin{cases} \sigma_\rho &= A \ + B\rho \ + Cz \ + H\rho z \ + I\rho^2 \ + Jz^2, \\ \sigma_\theta &= A \ + B_\theta\rho + Cz \ + H_\theta\rho z + I_\theta\rho^2 \ + Jz^2, \\ \sigma_z &= A_z + B_z\rho + C_z z + H_z\rho z + I_z\rho^2 \ + J_z z^2, \\ \tau_{\rho z} &= \qquad B_\tau\rho \qquad\quad + H_\tau\rho z + I_\tau\rho^2. \end{cases} \tag{11}$$

It is worth noting that the formulation has been adjusted so as to eliminate ρ from the fraction denominators containing ρ, in the equilibrium equations.

4.2 The SA Conditions

From (11), the SA conditions give rise to linear relations satisfying the definition of the macroscopic stresses, the equilibrium equations inside the triangles, the stress vector continuity across the element sides, the boundary and symmetry stress conditions in the horizontal plane of the (hemispherical) mesh.

4.3 The PA Conditions

The conic inequation of (7) was first used as such, and enhanced with the conic code MOSEK. Unfortunately, this conic formulation did not appear robust enough with significative meshing sizes. In a second attempt the cone was "PieceWise Linearized" (as in [24]), using the classic "PWL" method generating a polyhedric cone and a system of m linear inequations; the resulting problem was solved with the (interior point) linear programming code XA [3]. Due to the necessity of a m value sufficient to obtain precise results, the conditioning problems persisted for the spheroidal case.

Finally, only the so-called "BTN" linearization (based on a projection of cones) of Ben Tal and Nemirovski [2] (see [23]) induced a sufficiently robust formulation for the oblate cavity tests. This method gives an equilibrated number of additional rows $(3m + 1)$ and auxiliary columns $(2m)$ so that using $m = 6$ is equivalent to the PWL approach with a polyhedron of $2^m = 64$ sides, i.e. a value inducing an unreachable size of the final problem when using the PWL approach.

Moreover, because of the non-affine character of the stress field, the criterion cannot be verified everywhere by imposing it at the three apexes of the triangle only. Hence, the criterion is imposed at fifteen points (or more) regularly located in the element. Fortunately, the quadratic variation and the discontinuities of the stress field made that only moderately sizes of the mesh were necessary in the tests, due also to efficiency of the iterative process below.

4.4 The Post-analysis Process

After the optimization, a rigorous post-analysis is carried out:

- verification of the stress vector continuity across every boundary between adjacent elements; in fact, the stress vector jump is always smaller than 10^{-5};
- subdivision of each element into a large number of "subtriangles" (more than 200); accurate computation of Q_1 and Q_2 values, by integral calculation on each subtriangle, followed by a summation on the whole domain;
- verification of the Coulomb criterion inside each subtriangle of each element; this verification is performed with the three original inequations (6), since the stress field is known at this stage.
  To set an example, let us verify the following inequation in a subtriangle:

  $$|\sigma_1 - \sigma_2| - 2\,c\,\cos\phi + (\sigma_1 + \sigma_2)\sin\phi \leq 0;$$

  One has to calculate the k ratio: $k = \frac{|\sigma_1 - \sigma_2| + (\sigma_1 + \sigma_2)\sin\phi}{2\,c\,\cos\phi}$ and verify the inequality: $k \leq 1$;
  if $k > 1$, the current "faulty" element's number is stored and, if k is greater than the value previously stored during the current post-analysis step, k replaces this previous smaller value;
- if at least one element is non-PA, the optimization is reiterated, after modification of the PA conditions in each "faulty" element: the original cohesion, c, or its previously modified value, c_c is replaced with a new smaller fictitious c_c value: $c_c = c_c/k$, in order to make the PA conditions in this element more severe during the next optimization step;
- the whole process is reiterated, as many times as necessary, as long as the solution is not found admissible *everywhere* in the domain, with respect to the original Coulomb criterion.

5 Numerical Implementation of the Axisymmetric Mixed Method

The axisymmetric mesh in the frame (ρ, z) is given on the right of Fig. 1 with 4 sectors ($n_s = 4$) and 4 layers ($n_\rho = 4$). For each aspect ratio a_1/b_1 of the cavity and for a given porosity, the matrix boundaries of the mesh are adapted to obtain their confocal forms. Since these boundaries are not homothetic, the porosity of the resulting mesh is not exactly the same as the input one. Therefore, in a first step for each case of porosity and aspect ratio, the distribution of the angle α is optimized to obtain the desired porosity by progressively concentrating this distribution towards the more curved zone.

5.1 The Virtual Velocity Field

In each triangle, the FEM discontinuous displacement velocity field is expressed as follows:

$$\begin{cases} u_\rho = \rho\,(A + B\,\rho + C\,z) \\ u_\theta = 0 \\ u_Z = D + E\,\rho + F\,z + H\,\rho z + I\,\rho^2 + J\,z^2 \end{cases} \tag{12}$$

$$d_\rho = \frac{\partial u_\rho}{\partial \rho}\,;\; d_\theta = \frac{u_\rho}{\rho}\,;\; d_z = \frac{\partial u_z}{\partial z}\,;\; d_{\rho z} = \frac{1}{2}\Big(\frac{\partial u_\rho}{\partial z} + \frac{\partial u_z}{\partial \rho}\Big) \Rightarrow \{d\} = [B]\{X\} \tag{13}$$

Thus, each triangle generates nine constants ($X_i = A, B, \ldots, J$) which are the final virtual variables of the discretized model. The resulting strain rate field inside the triangle can be easily cast into the form $\{d(\rho, z)\} = [B(\rho, z)]\{X\}$ where the components of X are the (virtual) variables of the triangle. From this choice, the resulting strain rate tensor d cannot become singular and it varies linearly in the element: this linearity will be used later to (strictly) upper bound the (convex) dissipated power on the triangle.

Along a inter-element side the velocity jump $[u]$ is quadratic. To upper bound as above the corresponding dissipated power, the jump is linearized by enforcing the jump at the side middle to be equal to half the sum of its values at the ends of the discontinuity. Note that another interesting methodology for maintaining the PA character of a quadratic $[u]$ along the discontinuity is given in [14], but it is not easily applicable when using the present mixed method.

Note also that, in the recent LA numerical literature for axisymmetric problems, the drawback of the singularity at the origin is avoided by enforcing the PA conditions only at the centroid of the element, giving by this way only an estimate of the real solution.

5.2 Formulation of the Mixed Kinematic Method

5.2.1 Contribution of the Element Velocity Fields

The first part of the integral in (5.ii) becomes here, for a triangle of volume V:

$$P_V = \int_V d : \sigma\, dV = 2\pi \int_V \{d\}^t\{\sigma\}\,\rho\, d\rho\, dz \tag{14}$$

As detailed in [17, 19], from the Karush-Kuhn-Tucker optimality conditions, the product $\{d\}^t\{\sigma\}$ becomes the convex unit dissipated power $\pi_V(d)$ in the optimal solution. Then, we substitute to this product its linear interpolation $\mathscr{L}(\{d\}^t\{\sigma\})$ between its values at each vertex of the triangular element. Since the product $\rho\mathscr{L}$

is quadratic, the integral $\int \rho \mathcal{L}\, d\rho\, dz$ is classically calculated from its values at the medium of the sides of the triangle. Since each term of the product is linear, the final result depends only on the product values at the element vertices. Consequently, a stress tensor $\{\sigma\} = (\sigma_\rho, \sigma_z, \sigma_\theta, \sigma_{\rho z})^T$ is assigned at each vertex of the triangle, *without any variation assumption over the triangle.*

We finally obtain for the triangle with vertices $i = 1$ to 3 and area A:

$$P_V \leq \frac{\pi A}{6} \sum_{i=1,3} C_i \{d\}_i^t \{\sigma\}_i, \tag{15}$$

where $C_i = 2\rho_i + \rho_{i+1} + \rho_{i+2}$, with $\rho_4 = \rho_1$ and $\rho_5 = \rho_2$.

It is important to note that the strain rate d of the optimal solution will be PA at the vertices of the triangle; from the convexity of the set of PA strain rates associated to the present (convex) criterion, and from the linear variation of d in (13), it can be deduced (as first noticed in [13]) that the strain rate will be PA all over the element, a *sine qua non* condition for preserving the upper bound character of the result.

5.2.2 Contribution of the Velocity Discontinuities

The second part of the integral in (5.ii) is the sum of the power contribution of each discontinuity surface L_{1-2} (of ends noted 1 and 2):

$$P_d = \int_{L_{1-2}} [u] \cdot T'\, dS = 2\pi \int_{L_{1-2}} \{[u]\}^t \{T'\}\; \rho d\rho dz \tag{16}$$

According to [25], the product $[u] \cdot T'$ becomes the dissipated power $\pi_d([u])$ when the stress vector T' and the velocity jump vector $[u]$ are associated (by the normality law) relatively to the $f_{nt}\left(T'\right)$ criterion. Here also, we can use the convexity of $\pi_d([u])$ since $[u]$ is constrained to vary linearly along the discontinuity side: to the product $\{[u]\}^t\{T'\}$ we substitute its linear interpolation $\mathcal{L}\left(\left\{[u]\}^t\{T'\right\}\right)$ between its values at each end of the discontinuity side of normal n. Then, we can upper bound P_d by calculating the quadratic expression with Simpson formula, and by using the linearity of each term of the product $\rho\mathcal{L}$. By allocating an auxiliary stress vector $T' = (\sigma_{nn}, \sigma_{nt})$ (expressed in the orthonormal (n, t) frame of the side) at each of the two apices of the discontinuity side L_{1-2}, we finally obtain:

$$P_d \leq \frac{2\pi l}{6} \sum_{i=1,2} C_i \left(\left\{[u]_i\right\}^t \{T_i'\}\right), \tag{17}$$

where l is the length of L_{1-2} , $C_i = 2\rho_i + \rho_{i+1}$ with $\rho_3 = \rho_1$. It can be noted that the optimal jump $[u]$ will be PA all along L_{1-2} from the convexity of the PA jump set associated to the criterion f_{nt}.

5.2.3 Expression of the External Power

To the global vector of the virtual u-velocities are added the generalized velocities q_1, q_2, q_3 in order to form the final virtual vector $\{u\}$. Thus, from (2), the external power can be written as:

$$P_{ext} = V(q \cdot Q) = V\{q\}^T\{Q\} = \{u\}^T V[\beta]\{Q\} \tag{18}$$

where $\{q\} = [\beta]^T\{u\}$.

5.2.4 The PA Stress Conditions

In the global (x, y, z) reference frame, the Coulomb yield condition is written as in (7), where the cone can be easily cast into the Lorentz form required by the conic optimizer MOSEK [12]:

$$\sqrt{x_1^2 + x_2^2} \leq U \tag{19}$$

together with the three inequalities for each vertex of the triangular elements.

The criterion (8) for the stress vector T' gives rise to the following inequalities:

$$\sigma'_{nt} + \sigma'_n \tan\phi \leq c, \quad -\sigma'_{nt} + \sigma'_n \tan\phi \leq c, \tag{20}$$

which results in two linear constraints in terms of the real variables σ'_{nn} and σ'_{nt} for each ends of the discontinuity sides.

5.2.5 The Final Mixed Problem and the KA Conditions

Finally, the numerical form of the variational mechanical problem (5.i–5.iii) is as following:

$$\begin{aligned} &\text{Max } V\{q^d\}^T\{Q\} \\ &\text{s. t. } -[\alpha]\{\sigma\} - [\alpha']\{T'\} + V[\beta]\{Q\} = 0, \\ &f(\sigma) \leq 0 \quad \forall\sigma\,;\, f_{nt}(T') \leq 0 \quad \forall T', \\ &+\text{KA velocity conditions.} \end{aligned} \tag{21}$$

Indeed, a systematic change of d and u in terms of the final $\{X\}$ virtual variables is performed through specific subroutines of the Fortran code which generates the final problem in the MPS format required by MOSEK.

As shown in the detailed analysis of [17, 19], we can identify the dual variables of the solution of this optimization problem with the $\{X\}$ components. This analysis also details how the resulting velocity field is plastically admissible and how, by adding auxiliary columns, the kinematically admissible character of the optimal velocity field can be ensured, as in the following.

- We previously defined two supplementary rows (constraints) whose associated virtual variables are q_1, q_2 and two new columns for the corresponding macroscopic stresses Q_1, Q_2. At each apex and at the middle of the boundary triangle sides, we impose the loading conditions $u_\rho = D_{\rho}\rho$, $u_z = D_{zz}z$. This is done by adding one additional column (i.e., an additional variable) for each of these conditions. Let us note the chosen kinematic parameters as $\{q\} = [A_D]\{D\}$ with $\{D\}^T = (D_\rho, D_z)$; then, for example, the terms of a condition $u_\rho - \rho D_\rho = 0 \Longrightarrow u_\rho - \rho[A_D]^{-1}\{q\} = 0$ are dispatched on the (X, q) components (corresponding to u_ρ and q) of the additional column.
- A similar technique is used to impose the null symmetry value to the u_z components on the ρ axis, and to make that the jump $[u]$ of the middle of the discontinuity side is equal to half the sum of its end values.

6 Application to Sphere and Spheroid Models

In the following tests, the porosity f is taken equal to 0.1, and the friction angles to 10° and 20°, for the sake of homogeneity.

6.1 Comparison with Exact Results for Spherical Voids

Table 1 gives the values obtained with the static and mixed codes for isotropic loadings (i.e. $Q_2 = 0$) in tension and compression, together with the exact values given in [29]. It can be seen that the exact solution was always located between the numerical bounds that are very close; this also points out that the linearization of the matrix boundaries has no real influence with the selected n_s values. All the calculations were made on an Apple Mac Book Pro with a 2.7 GHz Core i7 and 16 gigabytes of RAM.

In the case of the static approach, the final numerical LP problem involved 1,024 triangles giving rise to 384,000 constraints and 249,000 free variables. For one point of the searched macroscopic criterion, 5–6 iterations, of approximately one minute each, were needed to obtain the convergence of the process described in Sect. 4.4.

Table 1 Comparison to Σ_m exact values [29]—Coulomb matrix—spherical void—porosity $f = 0.1$—cohesion $c = 1$—$(n_s, n_\rho) : (16, 16)$/static, (48, 48)/kinematic

	Friction angle ϕ	10°	20°
Compression	Present-static	−5.0970	−10.705
	Exact value	−5.1397	−10.805
	Present-kinematic	−5.1531	−10.844
Tension	Present-static	2.0543	1.4851
	Exact value	2.0704	1.4911
	Present-kinematic	2.0717	1.4916

The final mixed mesh of 9,200 triangles gives a conic problem with a constraint matrix involving 276,100 rows, 276,350 columns, and 27,650 Lorentz cones; the problem is solved in about 20 s by the version 5 of Mosek, using the four cores of the Intel Core i7 processor. This remarkable performance is mainly due to the quadratic variation of the velocity field and to the relatively small size of the final problem even for a very refined meshing of the problem. It can be also noted that the linearization of the velocity discontinuities, in order to preserve the kinematic character of the method, has not a significative influence on the performance in terms of upper bound solutions.

6.2 *Numerical Results for Axisymmetric Loadings in the Spherical Void Case*

In the reference [24], the hollow sphere problem with a Coulomb matrix was investigated, unfortunately only for non-negative Q_2 only. The corresponding results are recalled (JCAM dashed lines) in Figs. 2 and 3, for the friction angles $\phi = 10°$ and $20°$, respectively.

The JCAM static approach (in blue color) was obtained with a PWL version of the static code: this explains the better results (continuous lines) by using the present (conic projection) algorithm of Ben Tal and Nemirovski of Sect. 4.3. It can be seen

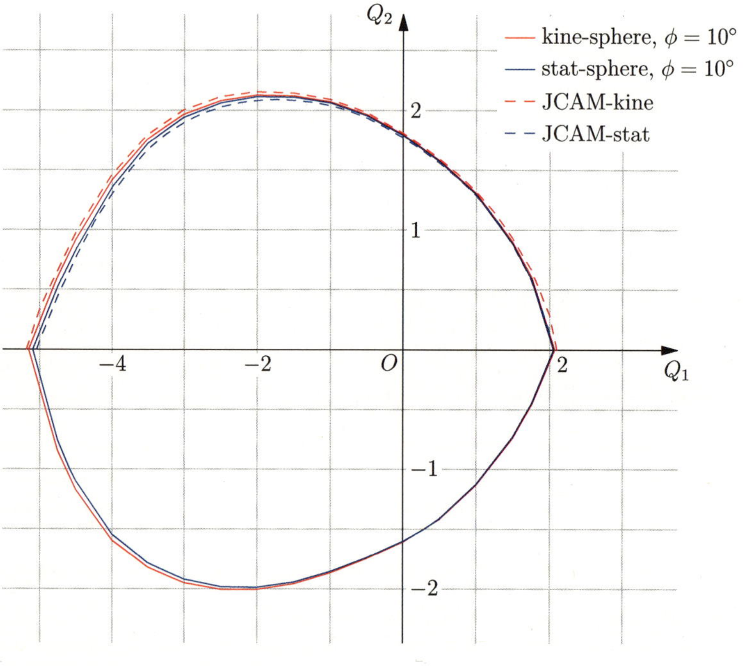

Fig. 2 Present results—Coulomb matrix—$f = 0.1$—spherical cavity—$\phi = 10°$; $c = 1$

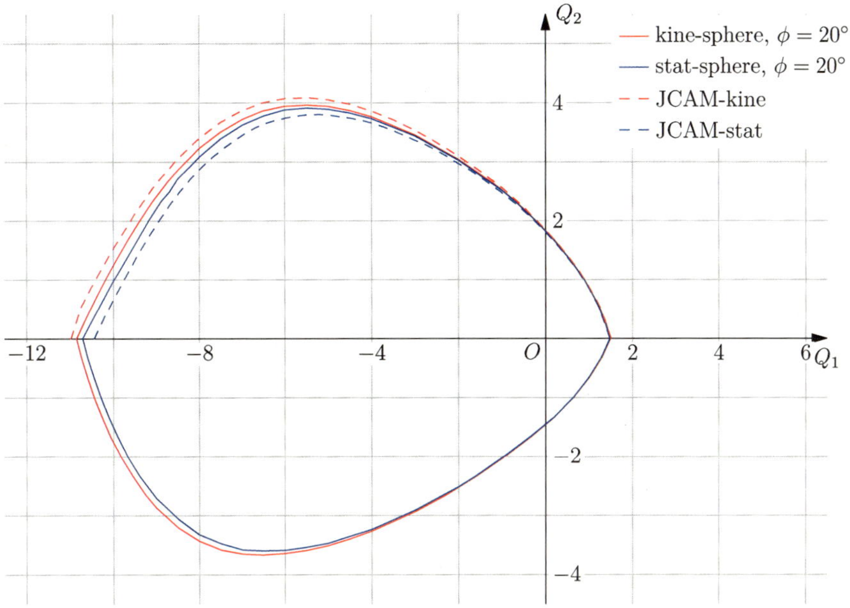

Fig. 3 Present results—Coulomb matrix—$f = 0.1$—spherical cavity—$\phi = 20°$; $c = 1$

also that the proposed mixed kinematic code has allowed similar improving of the previous results (dashed red lines) obtained with a classic kinematic approach that cannot run the refined meshes allowed by the mixed code.

The final kinematic and static criterions are very close, and always in the good order, even when they are almost indistinguishable. The presence of the singular points on the average stress axis are confirmed. The influence of the third stress invariant is also shown (recall that the second invariant is here $\Sigma_{eq} = |Q_2|$), since, owing to the isotropy of the macroscopic material, the plasticity criterion only depends on the three stress invariants. Indeed, in the Figs. 2 and 3, the Σ_m axis would be a symmetry axis of the plasticity criterion if this one does not depend on the third stress invariant. It must be emphasized that in the present case, the influence of third stress invariant comes from a combination of an effect that exists even if the matrix is not sensitive to the this third invariant, and of that which comes from the Coulomb matrix. In particular, the apexes on figure are signatures of this sensitivity of the Coulomb matrix. These points will be commented again in the following when comparing Coulomb and Drucker-Prager matrices.

6.3 Porous Coulomb Material with Oblate Voids

We consider first the case of the 0.5 aspect ratio a_1/b_1 where both axisymmetric codes are used. The case of the aspect ratio 0.2 is then investigated by using the same static and mixed code. Recalling that this problem has not been studied in the literature up to now, it should be useful to remember that the present axisymmetric bounds are also strict bounds when considering the full 3D problem, in the Q_i space.

6.3.1 Case $a_1/b_1 = 0.5$

In Figs. 4 and 5 are shown the results for an oblate cavity having an aspect ratio $a_1/b_1 = 0.5$.

First, the kinematic and static bounds, here also are very close, always remaining correctly ordered as in the case of spherical cavities. It can be seen that the graphs presents a small rotation around the frame origin, the singular point being not on the average stress axis. As expected in this case of transverse isotropy, taking into account that the axis of the anisotropy frame (x, y, z) are also loading axis, the influence of the third stress invariant is here more important.

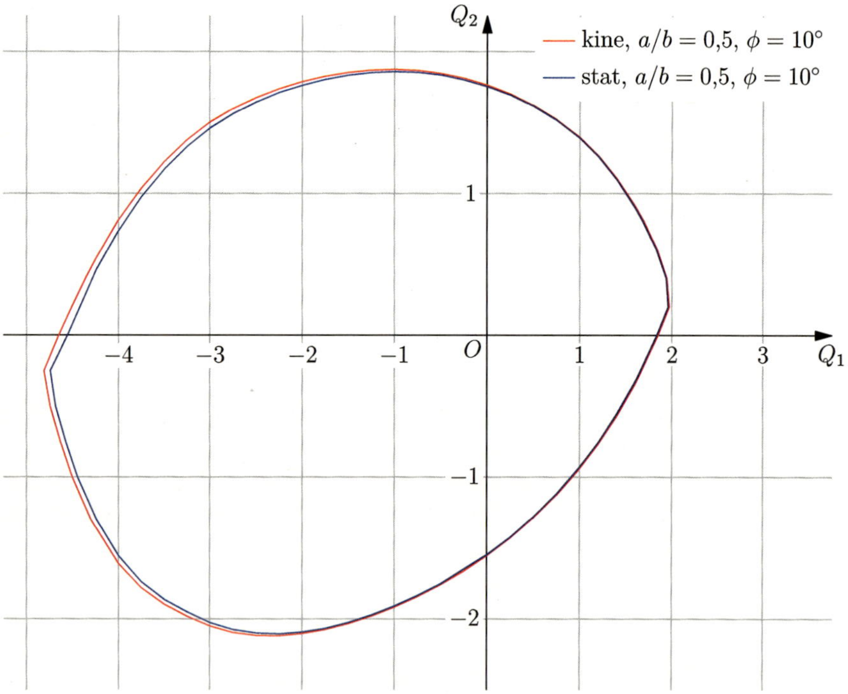

Fig. 4 Present results—Coulomb matrix—$f = 0.1$; $a_1/b_1 = 0.5$—$\phi = 10°$; $c = 1$

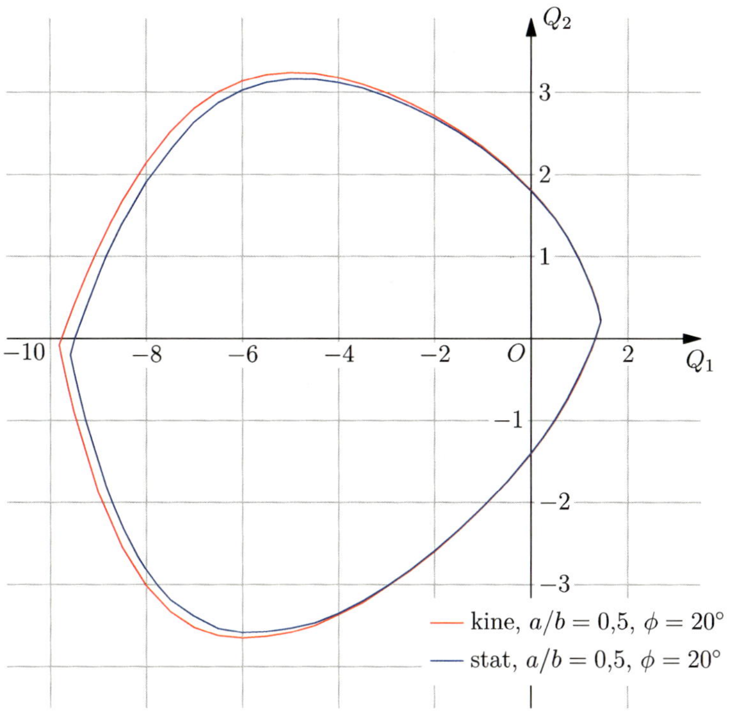

Fig. 5 Present results—Coulomb matrix—$f = 0.1$; $a_1/b_1 = 0.5$—$\phi = 20°$; $c = 1$

6.3.2 Case $a_1/b_1 = 0.2$

Figures 6 and 7 illustrate the results for an oblate cavity having this time the aspect ratio $a_1/b_1 = 0.2$. Here, the bounds are close, however in a lesser extent than above for mainly compressive loadings, and the dissymmetry with respect to the average stress axis is noticeably more important.

For comparison between the two cases with the two usual pressure-dependent matrices, we report also the graphs obtained for a Drucker-Prager matrix in the reference [18]; both graphs are reported to the cohesion c, after identification of the two criteria in plane strain from [25], in such a manner that the boundary of the domain of plastically admissible stresses for Drucker-Prager is innerly tangent to the corresponding Coulomb boundary. As expected, the criterion of the porous Drucker-Prager material remains innerly tangent to the porous Coulomb criterion, without any angular point in the Drucker-Prager case. Again, existence of such angular points is a signature of the sensitivity of the Coulomb matrix to the third invariant of the stress deviator. Since angular points might induce strong effects in strain localization phenomena, it can be seen that replacing the Coulomb criterion with an *ad hoc*

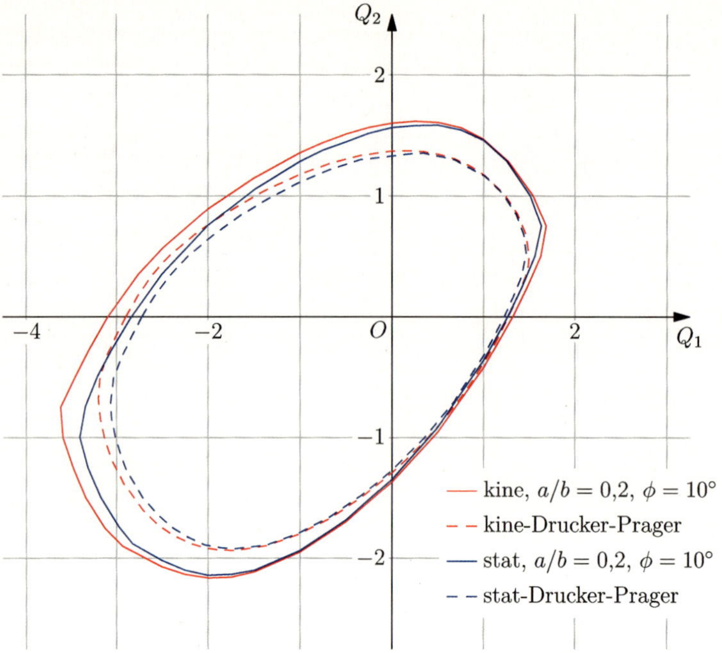

Fig. 6 Present results—Drucker-Prager matrix—$f = 0.1$; $a_1/b_1 = 0.2$—$\phi = 10°$; $c = 1$

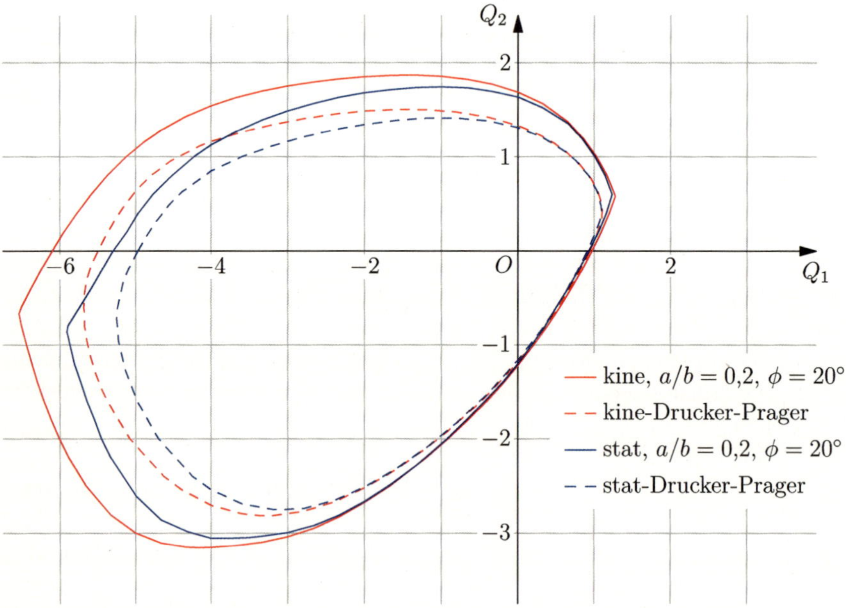

Fig. 7 Present results—Coulomb matrix—$f = 0.1$; $a_1/b_1 = 0.2$—$\phi = 20°$; $c = 1$

Drucker-Prager for elastoplastic 3D computations can be problematic, even for the present porous materials.

7 Conclusion

The first purpose of the present paper was to generalize the formulations of the limit analysis (LA) in the axisymmetric case of [24], in order to obtain robust conic formulations without losing the rigorous bound character of the solutions. The second purpose was to provide numerical, but rigorous bounds to the macroscopic criterion of a "porous Coulomb material" with spherical and oblate voids, a problem not studied in the literature up to now, at least to our knowledge. These bounds, not only allow to characterize the macroscopic plastic properties of this class of materials, but are also expected to serve as a reference results for forthcoming theoretical investigations.

To obtain accurate results for the non-spherical voids, the static approach was modified to use the conic projection algorithm of [2], and an original mixed axisymmetric approach has been elaborated which results in a second order conic programming problem. Both resulting codes have allowed remarkable performances by using specific quadratic fields together with FEM refinements of the problem unreachable before. After a validation in the case of a spherical cavity and isotropic loadings, by comparison to the known exact solution, we provide illustrations of the macroscopic plasticity criterion. This has been done for usual values of the friction angle and of the void aspect ratio. The obtained results allow to highlight, among others, the shape and the size of the plasticity criterion depending on the aspect ratio of the cavity. A comparison with the porous Drucker-Prager material is finally given in order to illustrate the noticeable differences with the present porous Coulomb material, the same geometric and loading conditions being considered. In view of all the above results, it is clear that the paper provides interesting bounding data which call for a great effort in the theoretical side in order to formulate new macroscopic criterion corresponding to the porous material with a Coulomb matrix.

References

1. Anderheggen E, Knopfel H (1972) Finite element limit analysis using linear programming. Int J Solids Struct 8:1413–1431
2. Ben-Tal A, Nemirovskii A (2001) Lectures on modern convex optimization. Analysis, algorithms and engineering applications. Series on Optimization. SIAM-MPS
3. Byer J (2000) XA, optimisation library. Sunset Software Technology. San Marino
4. Francescato P, Pastor J, Riveill-Reydet B (2004) Ductile failure of cylindrically porous materials. Part I: plane stress problem and experimental results. Eur J Mech/A Solids 23:181–190
5. Garajeu M, Suquet P (1997) Effective properties of porous ideally plastic or viscoplastic materials containing rigid particles. J Mech Phys Solids 45:873–902
6. Gologanu M, Leblond JB (1993) Approximate models for ductile metals containing non-spherical voids—case of axisymmetric prolate ellipsoidal cavities. J Mech Phys Solids 41(11):1723–1754

7. Gologanu M, Leblond JB, Perrin G, Devaux J (1994) Approximate models for ductile metals containing non-spherical voids—case of axisymmetric oblate ellipsoidal cavities. J Eng Mater Technol 116:290–297
8. Guo TF, Faleskog J, Shih CF (2008) Continuum modeling of a porous solid with pressure sensitive dilatant matrix. J Mech Phys Solids 56:2188–2212
9. Gurson AL (1977) Continuum theory of ductile rupture by void nucleation and growth—part I: yield criteria and flow rules for porous ductile media. J Eng Mater Technol 99:2–15
10. Krabbenhoft K, Lyamin AV, Hijaj M, Sloan SW (2005) A new discontinuous upper bound limit analysis formulation. Int J Numer Methods Eng 63:1069–1088
11. Lee JH, Oung J (2000) Yield functions and flow rules for porous pressure-dependent strain-hardening polymeric materials. J Appl Mech 67:288–297
12. MOSEK ApS. (2010) C/O Symbion Science Park, Fruebjergvej 3, Box 16, 2100 Copenhagen ϕ, Denmark, http://www.mosek.com
13. Makrodimopoulos A, Martin CM (2007) Upper bound limit analysis using simplex strain elements and second-order cone programming. Int J Numer Anal Methods Geomech 31(6):835–865
14. Makrodimopoulos A, Martin CM (2008) Upper bound limit analysis using discontinuous quadratic displacement fields. Commun Numer Methods Eng 24:911–927
15. Monchiet V, Charkaluk E, Kondo D (2007) An improvement of Gurson-type models of porous materials by using Eshelby-like trial velocity fields. C R Mécanique 335:32–41
16. Monchiet V, Charkaluk E, Kondo D (2013) Macroscopic yield criteria for ductile materials containing spheroidal voids: an Eshelby-like velocity fields approach. Mech Mater http://dx.doi.org/10.1016/j.mechmat.2013.05.006
17. Pastor F (2007) Résolution par des méthodes de point intérieur de problèmes de programmation convexe posés par l'analyse limite. Thèse de doctorat, Facultés universitaires Notre-Dame de la Paix, Namur
18. Pastor F, Kondo D (2014) Limit analysis and lower/upper bounds to the macroscopic criterion of Drucker-Prager materials with spheroidal voids. C R Mecanique 342:96–105
19. Pastor F, Kondo D, Pastor J (2013) 3D-FEM formulations of limit analysis methods for porous pressure-sensitive materials. Int J Numer Methods Eng 95:847–870
20. Pastor F, Loute E, Pastor J, Trillat M (2009) Mixed method and convex optimization for limit analysis of homogeneous Gurson materials: a kinematical approach. Eur J Mech-A/Solids 28:25–35
21. Pastor F, Kondo D (2013) Assessment of hollow spheroid models for ductile failure prediction by limit analysis and conic programming. Eur J Mech A/Solids 38:100–114
22. Pastor J, Turgeman S (1982) Limit analysis in axisymmetrical problems: numerical determination of complete statical solutions. Int J Mech Sci 24:95–117
23. Pastor J, Francescato P, Trillat M, Loute E, Rousselier G (2004) Ductile failure of cylindrically porous materials. Part II: other cases of symmetry. Eur J Mech/A Solids 23:191–201
24. Pastor J, Thoré Ph, Pastor F (2010) Limit analysis and numerical modeling of spherically porous solids with Coulomb and Drucker-Prager matrices. J Comput Appl Math 234:2162–2174
25. Salençon J (1983) Calcul à la rupture et analyse limite. Presses des Ponts et Chaussées, Paris
26. Sloan SW (2013) Geotechnical stability analysis. Géotechnique 63:531–572
27. Thai-The H, Francescato P, Pastor J (1998) Limit analysis of unidirectional porous media. Mech Res Commun 25:535–542
28. Thoré P, Pastor F, Pastor J (2011) Hollow sphere models, conic programming and third stress invariant. Eur J Mech/A Solids 30:63–71
29. Thoré Ph, Pastor F, Pastor J, Kondo D (2009) Closed form solutions for the hollow sphere model with Coulomb and Drucker-Prager materials under isotropic loadings. C R Mécanique Acad Sci Paris 337:260–267
30. Trillat M, Pastor J (2005) Limit analysis and Gurson's model. Eur J Mech A/Solids 24:800–819
31. Turgeman S, Pastor J (1982) Limit analysis: a linear formulation of the kinematic approach for axisymmetric mechanic problems. Int J Numer Anal Methods Geomech 6:109–128

A Direct Method Formulation for Topology Plastic Design of Continua

Zied Kammoun and Hichem Smaoui

Abstract In the present paper a method is proposed for continuous and black and white topology optimization of continuum structures subject to static and plastic admissibility conditions relative to a prescribed load. A significant feature of the continuous topology optimization problem is its outstanding similarity with the direct static formulation of the limit analysis problem that can be written as a conic programming problem. The discrete, e.i. black and white, topology optimization problem is derived by simply introducing a penalization of intermediate densities in the objective function and is solved as a sequence of conic programming problems of the same form as the continuous design problem. The proposed method is formulated in plane strain using Tresca materials and is illustrated on continuous and discrete example design problems taken from the literature.

Keywords Discrete topology · Optimization · Limit analysis

1 Introduction

Topology optimization of continuum structures has witnessed an intense research effort during the last decades [1–5] that led to remarkable developments. The numerous successful applications of topology optimization in industry [6] and the emergence of powerful dedicated topology optimization software [4] reflect the degree of maturity this discipline has reached.

It is noteworthy, however, that the work on topology optimization of continuum structures has been predominantly focused on linear elastic material behavior. Elastic design is traditionally the most common and most demanded type of design and continuum topology design is not an exception in this regard. On the other hand,

Z. Kammoun · H. Smaoui (✉)
Ecole Nationale D'Ingénieurs de Tunis, LR11ES16 Laboratoire de Matériaux, Optimisation et Energie pour la Durabilité, Université de Tunis El Manar, 1002 Tunis, Tunisie
e-mail: hismaoui@yahoo.fr

Z. Kammoun
e-mail: kammounzied@yahoo.fr

P. Fuschi et al. (eds.), *Direct Methods for Limit and Shakedown Analysis of Structures*,
DOI 10.1007/978-3-319-12928-0_3

historically the analysis and design developments in the framework of elasticity are usually followed by sustained activity extending these developments to plastic behavior. But this does not seem to be the case for topology optimization.

Following the extensive work on topology optimization of elastic structures the research in topology design has been concerned more with the incorporation of advanced materials, multiphysics and multidisciplinary applications such as MEMS and compliant mechanisms than with the extension to nonlinear material behavior.

Among other reasons, this is partly because the mathematical approaches underlying the topology design methodologies, such as the homogenization approach [7] and the Simple Isotropic Microstructure (or Material) with Penalization (SIMP) [8], strongly rely on the linear elasticity assumption.

Moreover, elastoplastic analyses that seek to determine response quantities, e.g. evolution methods, are known for their high computational demand. On the contrary, direct methods of limit analysis require lower computational effort to determine limit states in terms of either stress field or displacement/velocity field solutions. When information on the load response history is not needed and, say, only the limit stress field is of interest, direct methods present an adequate alternative for plastic collapse analysis. They are appropriate analysis tools for such applications as geotechnical works like foundations, soil reinforcement and retaining structures. In an automated design framework, where computational efficiency in the analysis is a primary factor, direct methods of limit analysis become attractive for their considerable computational saving potential. Member sizing optimization of structures with specified topologies, e.g. trusses and frames, subject to plastic design constraints has been treated in the literature using direct methods of limit analysis [9]. To the authors' knowledge, topology design of continuum media, involving direct limit analysis, is inexistent in the literature prior to the recent work presented in [10, 11]. Previous research includes some work that has dealt with continuum topology design of nonlinear elastic structures where the tools developed for the linear behavior were adapted and extended to the nonlinear case [2] and the few tentatives that were directed to design based on elastoplastic (e.g. [12]) or plastic analyses (e.g. [13]) involved evolution analysis methods.

The present work is concerned with the integration of direct methods of limit analysis precisely, into a methodology for topology design of continuum structures. It builds on the method presented in [10] and reported in [11] that generates optimum continuous, also called porous, topologies to develop a methodology that synthesizes so called black and white topologies.

The continuous design problem is formulated according to the microscopic (or material) approach [11] and expressed in terms of continuous material densities as design variables. The outstanding feature of the formulation is that it takes on a mathematical form that is largely similar to that of a direct limit analysis problem. The same code that is used in solving the analysis problem can actually be employed in finding the optimum design. Moreover, the computational demand of the topology design problem is expectedly, and has been demonstrated ([10]) to be, in the order of magnitude of that of a single limit analysis. Among the properties shared by both problems a significant one is convexity, which has desirable implications on convergence.

The continuous formulation has long been regarded as merely a prelude to the end goal of a discrete, black and white, design methodology which often incorporates continuous optimization routines as part of the discrete optimization strategy. However, the recent breakthroughs in material technology have regenerated interest in continuous topologies. State of the art technology enables the manufacturing of materials with a large variety of microstructures to suit an unlimited range of properties and property distributions, which creates demand for designs with continuous topologies.

Yet discrete topology remains a major goal from the perspectives of industrial applications and academic interest and the present work is intended to be a step forward towards optimization of black and white topologies based on direct limit analysis. After a brief presentation of the continuous formulation of the topology optimization problem a modified formulation, involving a penalization technique, is proposed to generate discrete topologies. A simple search procedure is then proposed for solving the discrete design problem as a sequence of continuous design problems. The procedure is discussed and tested on a number of example design problems and solutions are compared with topologies found in the literature.

2 The Static Method of Limit Analysis

The main design constraint considered in the present work is that a specified loading supported by the designed structure be statically and plastically admissible. The terminology adopted here is defined in [14, 15] where a stress field σ is said to be statically admissible (*SA*) if field equilibrium equations, stress vector continuity, and stress boundary conditions are satisfied. It is said to be plastically admissible if $f(\sigma) \leqslant 0$, where $f(\sigma)$ is the plasticity criterion of the material. A stress field σ that is both *SA* and plastically admissible will be said to be "admissible". A loading system $Q \in \mathbb{R}^n$ in equilibrium with a statically admissible stress field σ, $Q = Q(\sigma)$, is said to be admissible. The n components of Q are called loading parameters. The relationship $Q = Q(\sigma)$, which usually describes either field equilibrium equations, when body forces are present, or boundary conditions on the stress vector, is linear in both cases. A solution of the limit analysis problem relative to the ith loading parameter is found by solving the following optimization problem for an admissible stress field σ such that:

$$\begin{aligned} Q_{lim} &= (Q_1^d, \ldots, \lambda_0 Q_i^d, \ldots, Q_n^d) \\ \lambda_0 &= \max\{\lambda\ ,\ Q(\sigma) = (Q_1^d, \ldots, \lambda Q_i^d, \ldots, Q_n^d)\}. \end{aligned} \tag{1}$$

where Q^d is a specified admissible loading. The resulting loading $Q(\sigma)$ is a limit loading of the mechanical domain with respect to the loading component Q_i^d. This formulation defines the static, lower bound problem of limit analysis, or static limit analysis (LA) problem, as it will be dealt with in the present work. The static LA method determines the stress field at the limit state only. It generates no information on the stress field at intermediate stages of loading nor on kinematic quantities at any loading step. The payoff of this missing information is a lower computational demand which is a paramount advantage in case that information is not necessary. Another merit of the static approach is the status of rigorous lower bound of the limit load.

3 Finite Element Formulation of the Static Problem

The numerical plane strain formulation of the static, lower bound problem is described in detail in [16]. The planar mechanical domain Ω is discretized into triangular finite elements. Considering the global reference frame (x, y), the stress field is chosen to be linear in x and y within the element. Across inter element interfaces it can be discontinuous provided the stress vector acting on the interface remains continuous. The Tresca criterion is written as:

$$f(\sigma) = \sqrt{(\sigma_x - \sigma_y)^2 + (2\tau_{xy})^2} - 2s \leqslant 0, \tag{2}$$

or equivalently as:

$$S(\sigma) = \sqrt{\left(\frac{\sigma_x - \sigma_y}{2}\right)^2 + \tau_{xy}^2} \leqslant s \tag{3}$$

where s denotes the shear strength, or cohesion, of the material. The following conditions are imposed on the stress field in order to ensure its static and plastic admissibility:

- Within the element, the equilibrium equations $\sigma_{ij,j} + \gamma_i = 0$ expressed in the Cartesian reference frame, where $\gamma = \rho g$ is the specific weight vector, ρ the mass density and g the acceleration of gravity.
- Continuity conditions of the stress vector across discontinuity lines: for each discontinuity segment of normal n, the continuity of the stress vector $T_i = \sigma_{ij} n_j$ is imposed at the ends defining this discontinuity segment.
- Boundary conditions on the stress vector: $\sigma_{ij} n_j = T_i^d$ at each end of the boundary element sides where the linearly varying stress vector T^d is imposed.
- Stress field plastic admissibility at each triangle vertex. This ensures plastic admissibility over the total domain from the linear variation of the stress in a triangle and the convexity of the criterion (2).

Introducing a change of variables such that the stress vector σ is defined by the components $\frac{(\sigma_x+\sigma_y)}{2}$, $\frac{(\sigma_x-\sigma_y)}{2}$ and τ_{xy} denoted σ_+, σ_- and τ, respectively, the plasticity criterion can be written directly in the conic form $s \geqslant \sqrt{\sigma_-^2 + \tau^2}$. The numerical optimization problem, expressing the static limit analysis problem, can thus be written as a conic programming problem in the form:

$$\begin{aligned} &\lambda_0 = max\ \lambda \\ &Q(\sigma) = (Q_1^d, \ldots, \lambda Q_i^d, \ldots, Q_n^d) \\ &S(\sigma) \leq s, \\ &\sigma\ SA \end{aligned} \tag{4}$$

and can, therefore, be solved using the conic programming code MOSEK [17] as in [18–20].

4 Topology Optimization of Continua

Topology optimization of continuum structures aims at simultaneously optimizing the shape of external and internal boundaries, the number of holes and the connectivity within a specified domain Ω_0 with given boundary conditions with respect to a design objective function subject to a set of constraints. For elastic materials, in most formulations the design objective consists of a global stiffness measure or a compliance functional and the constraint is essentially a specified bound on the amount of material. Among the various approaches that have been developed, two major classes can be distinguished: the Microstructure and the Macrostructure approaches [2].

In the Microstructure approaches, also known as material approaches, the material is assumed to be distributed in some microstructural form over the design domain Ω_0. The miscrostructure may correspond to some real or fictitious composite [7] or to an isotropic porous material [SIMP]. It is customary to use a fixed, uniform finite element mesh to model the structure within the entire design domain. The design variables are assumed to be constant within each finite element. For the analysis, finite elements are applied with properties that are related to the material characteristics of the microstructures [2]. The discrete optimization consists in determining whether each element in the design domain should be solid or void. The density of material within each finite element is used as a design variable that should eventually converge to one of the limiting values 0 and 1. The continuous optimization admits designs with intermediate densities. Even if continuous topologies are not the goal, continuous optimization is often used as a step in a discrete design strategy.

In the class of Macrostructure, also called geometric, approaches the structure occupies a subset of the design domain and solid materials are considered as opposed to porous ones. Optimization of the topology is performed in conjunction with shape optimization. This implies that the finite element mesh should change with the moving boundaries of the solid domain during the design process.

Of the two abovementioned classes of approaches to topology design of continua, the material approach appears to be more straightforwardly applicable. It will be applied to the present plastic design problem based on direct limit analysis.

Topology optimization of elastic continuum structures raised difficulties that took a sustained research effort to understand and resolve. These include mainly the multiplicity of local minima, the ill-posedness of the mathematical problem, the lack of convergence with respect to the size of the finite element mesh [21] and the difficulty associated with local stress constraints [22].

In dealing with topology design based on direct limit analysis, given the distinct nature of its mathematical formulation, these difficulties will have to be addressed from a different perspective.

5 Topology Optimization Problem Formulation

A rectangular domain Ω_0 is considered with a unit thickness, made out of a Tresca material having a shear strength $\bar{s}$ and a density $\bar{\rho} = 1$, subjected to a loading system $Q = (Q_1, \ldots, Q_i, \ldots, Q_n)$. The design goal is to find the structural configuration included in the domain Ω_0 which achieves a minimum amount of material while maintaining a statically and plastically admissible stress field associated with the specified loading Q. For the discrete topology design problem the solution is a structure with a black and white configuration. That is, the domain thickness is unity or zero, respectively. The optimum design problem may be written as:

$$
\begin{aligned}
&\min \int_{\Omega_0} \rho \, d\Omega \\
&\text{s. t. } Q(\sigma) = (Q_1^d, \ldots, Q_i^d, \ldots, Q_n^d), \\
&S(\sigma) \leq \bar{s}, \\
&\sigma \ SA \\
&\rho \in \{0, 1\}
\end{aligned}
\tag{5}
$$

where $\rho(x, y)$ is a Boolean function and, in the void regions, it is understood that the stresses should vanish. Furthermore, it is important to note that, in case the self weight is taken into account, the condition σ *SA* implicitly depends on the density given the design dependence of the specific weight $\gamma = \rho g$. A necessary condition for the existence of a feasible solution to this design problem is that the specified applied load be no larger than the limit load corresponding to the fully dense domain, i.e. $\rho = \bar{\rho}$ everywhere. The problem statement (5) may express the optimization problem for both the exact (infinite dimensional) and the discretized structural problems. In the latter, obtained here through finite element modeling, the piecewise constant function ρ is interpreted as a Boolean N-vector where N is the number of finite elements spanning the domain Ω_0 and the piecewise linear field σ as a vector of nodal stress components.

Using the finite element formulation the problem (5) is an N-dimensional combinatorial optimization problem. Direct exact methods are excluded because of the exponential complexity of the problem. The numerous heuristics that have been proposed, including evolutionary methods [23, 24], suffer many limitations and hardly compete with the smooth methods such as homogenization methods and SIMP.

At the present preliminary stage of the work it is tempting to use a penalization technique for the black and white topology optimization problem given its recognized performance and its simplicity of implementation. The method consists in approaching the discrete problem solution through the solution of a sequence of continuous design problems where the intermediate densities are penalized. For this reason, the continuous design formulation is recalled as a preliminary to the discrete problem presentation.

5.1 The Continuous Design Problem Formulation

In order to model the continuous range of intermediate solutions between void and solid at a given point in the domain a relaxation scheme is applied. The "amount" of material represented by the density ρ of a fictitious material is allowed to continuously span the interval [0, 1].

For the present plastic design problem the relaxation scheme requires interpolation of the failure criterion. Assuming the intermediate material also obeys a Tresca failure criterion, it is reasonable to choose for the shear strength s as a function of density the proportional law [10]

$$s(\rho) = \rho\bar{s}. \tag{6}$$

Because the Tresca criterion is one that allows arbitrary spherical stress, another issue that arises from the relaxed formulation is the way to control the vanishing of stresses in the void regions while maintaining the smooth character of the optimum design problem. This requires a further condition on the stress field to bring the stress tensor to zero when the density vanishes, which can be achieved by bounding the trace of the stress tensor by a multiple of the density as follows:

$$-K\rho \leq \sigma_+ \leq K\rho \tag{7}$$

where K is a constant sufficiently large so that the constraint tends to be activated near zero density only. This linear constraint pair maintains the conic nature of the design problem. It should be noted, however, that it introduces an artificial truncation in the failure criterion for low densities.

The interpolation of the intermediate material having been defined, the continuous (porous) topology design problem takes the conic form:

$$\begin{aligned}
&\min \int_{\Omega_0} \rho \, d\Omega \\
&\text{s. t. } Q(\sigma) = (Q_1^d, \ldots, Q_i^d, \ldots, Q_n^d), \\
&S(\sigma) \leq \rho\bar{s}, \\
&\sigma \; SA \\
&0 \leq \rho \leq 1 \\
&-K\rho \leq \sigma_+ \leq K\rho
\end{aligned} \tag{8}$$

where the *SA* condition is understood to be design dependent if gravity loads are considered. Premultiplying the objective by the real shear strength $\bar{s}$, redefining the constant K as $K\bar{s}$ and substituting s for $\rho\bar{s}$ the optimum design problem can be rewritten in the alternative form: Eq. 9, where the shear strengths replace the densities as design variables:

$$\begin{aligned} &\min \int_{\Omega_0} s \, d\Omega \\ &\text{s. t. } Q(\sigma) = (Q_1^d, \ldots, Q_i^d, \ldots, Q_n^d), \\ &S(\sigma) \leq s, \\ &\sigma \ SA \\ &0 \leq s \leq \bar{s} \\ &-Ks \leq \sigma_+ \leq Ks \end{aligned} \tag{9}$$

The last two sets of constraints are either bounds on the variables or simple constraints. It can be seen that, aside from these constraints, the two problems differ only in that the roles of the shear strengths and the load parameter, as variables or fixed parameters or as objective variables, are interchanged. The problem (9) recalls the so-called strength reduction formulation of the static problem where the shear strength, treated as a variable, is minimized for a given loading parameter as in [18]. It reveals the close similarity between the mathematical forms of the continuous design problem and the direct limit analysis problem (4).

That the design problem be convex is significant as it should discard convergence difficulties that are commonly encountered in the, usually nonconvex, elastic and elastoplastic topology design problems, e.g. the issue of local minima. Since the number of constraints, except for the simple constraints, is the same in both problems, the computational effort has been found [10, 11], as expected, to be of the same order of magnitude. Another merit of the proposed design problem formulation combining variable densities with direct limit analysis is that no numerical difficulties arise when densities approach zero. Consequently, there is no need for imposing a finite lower bound on the density, a routine practice in continuum elastic design to avoid singularities in the stiffness matrix. Finally, it is expected that when dealing with 0-1 design the formulation based on direct limit analysis will not favor the formation of checkerboard patterns. This is because the stress field being zero in the void elements, continuity of the stress vector forces the normal and shear stresses to approach zero in the, solid, material elements. This continuity is rigorously imposed everywhere in the discretized structure. Such argument that does not apply to the case of elastic design where checkerboard patterns typically appear in optimized structures. One reason is that elastic analysis is usually formulated in displacements so that inter-element continuity of the stress vector generally does not hold. Another reason is that the checkerboard phenomenon is known to be associated with a numerical aspect of elastic analysis that is unrelated to the direct limit analysis problem. This phenomenon is attributed to finite element discretization errors that make the stiffness of checkerboard configurations artificially higher in the model than in the real structure [21].

5.2 The Discrete Problem Formulation

An effective technique to drive the densities to the limits 0 and 1 is penalization of intermediate densities. A common form of penalization is the power law used in [8] to penalize the relationship between the density and the stiffness parameter. Here, the penalization is performed by replacing the linear cost term ρ by the power law $\rho^{1/p}$ where $p > 1$, which increases the unit cost of grey elements. This leads to the following problem:

$$\begin{aligned}
&\min \int_{\Omega_0} \rho^{1/p}\, d\Omega \\
&\text{s. t. } Q(\sigma) = (Q_1, \ldots, Q_i, \ldots, Q_n), \\
&S(\sigma) \leq \rho \bar{s}, \\
&\sigma \ SA \\
&0 \leq \rho \leq 1 \\
&-K\rho \leq \sigma_+ \leq K\rho
\end{aligned} \tag{10}$$

For a given power parameter p the optimization problem is no longer convex and may not be solved using *MOSEK*. However, it belongs to the special problem class of concave minimization over a convex set which has been investigated, particularly by Jacobsen [25, 26] and appropriate deterministic search strategies have been proposed, including cutting plane methods. Focusing on the feasibility rather than the performance of the discrete design strategy, a simple iterative method is used for solving the nonconvex problem (10). It consists of solving a sequence of conic problems where the objective is a linear approximation of the power law function. In each cycle the problem to be solved is formed by exactly the same constraints as in problem (10), and the cost function can be written as $\int_{\Omega_0} c_{\rho_0} \rho \, d\Omega$ where $c_\rho = \rho^{\frac{1}{p}-1}$ and ρ_0 denotes the solution of the preceding cycle. The resulting problem differs from the problem (8) only in the cost coefficients. It is solved using the code *MOSEK* with a LINUX operating system. In the numerical examples of the next section the value $p = 2$ is chosen for the power parameter.

6 Numerical Examples

In this section, the proposed method for plastic topology optimization of continuum media is tested through a selection of examples of topology design problems that are treated in the literature in the framework of elastic material behavior. These tests are performed essentially to demonstrate the capability of the method to generate sound black and white topology designs and to compare the optimum topologies produced using the proposed plastic design method with published topologies that have been generated using design methods for elastic media. The long and medium cantilever beam problems will be presented first, followed by the benchmark Messerschmitt-Bölkow-Blohm (MBB) beam problem.

The loading Q will be restricted to a single load component denoted F. The finite element meshes are all uniform. The rectangular design domain is divided into $n_x \times n_y$ elementary rectangles, each divided into four triangular elements separated by the two diagonals. Before proceeding with the design, the limit load for the full density domain is determined, if not known, since it represents an upper bound on the specified limit loads for which the topology optimization problem is feasible.

All examples are treated with version 6 of the commercial code MOSEK using an Intel Core i5 processor (2.4 GHz). The continuous design is optimized directly using *MOSEK*. The discrete design problem formulated according to Eq. (10) is treated using the iterative procedure described in the previous section where each problem is also solved using *MOSEK*.

6.1 The Long Cantilever Beam

The problem described in Fig. 1 deals with the design of a long cantilever beam having an aspect ratio of 4 and made of a Tresca material with shear strength $\bar{s} = 1$ kPa. The beam is loaded at its free end by a transverse line load distributed along a line element of length b. Its design domain dimensions are the length denoted by L and the height by H. The height of the design domain is $H = 1$ m and the length $L = 4$ m. Applied at the free end is a centered, practically concentrated load ($b = 10$ cm).

Feasibility of the design problem requires that the applied load be no larger than the limit load corresponding to the fully dense domain. As reported in [10, 11], the limit load calculated for the domain with full density was found to be $\bar{F} = 0.1$ kN independently of the aspect ratio. This is explained by the localization of the failure zone in a small region, around the applied load, compared to the dimensions of the beam. The degree of discretization recommended to achieve a reasonable accuracy

Fig. 1 Long cantilever beam

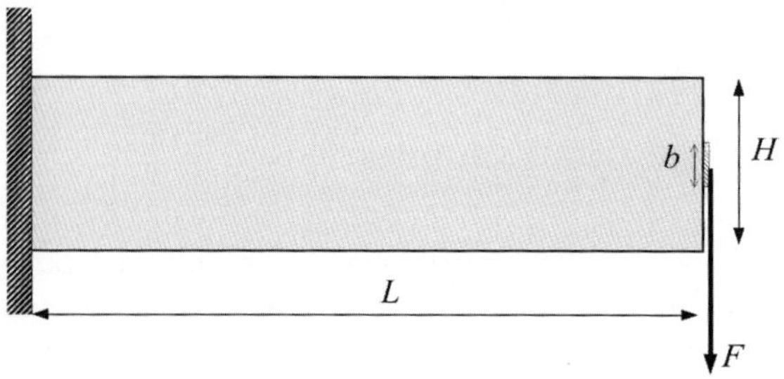

for the limit load of the long cantilever beam was determined to be in the vicinity of 3,200 finite elements. Furthermore, a series of continuous topology design problems were run with different mesh sizes and K factors. It was found that a 28,800 element mesh was appropriate. It was noted that for $K < 3$, the algorithm did not converge. For large values ($K > 40$), difficulties of convergence were also observed. Considering the criteria of optimality in terms of objective value, constraint accuracy, CPU time and minimum density, the best results were obtained for the range of K values from 3 to 10. All the results reported in the remaining of the paper are produced with $K = 10$.

6.1.1 Continuous Design

The topology design problem was solved in [10, 11] for the long beam subject to various intensities of the applied load F. As expected, the optimization CPU times were near double those spent on analysis using the same FE model. It was noted that for low loads ($F < 0.006$ kN) the density in the optimal design was less than unity everywhere and the material distribution pattern was nearly independent of the load. For higher loads the pattern of material distribution tends to be more load dependent.

The optimum objective value obtained using the Intel Core i5 processor for a 51,200 element beam model subject to the load $F = 0.04$ kN is 0.4630. The CPU time was 100 s, that is less than the 121 s required by the limit analysis based on the same finite element model. The optimum design is visualized at the top left of Fig. 2. It can be seen that the material fills the fuselage shaped domain without internal holes. The density tends to saturate at the upper and lower edges towards the supports. This solution provides an optimal material distribution and a clear definition of the external shape but little indication about the optimum layout structure.

6.1.2 Discrete Design

The same long cantilever beam, modeled with 51,200 elements, is considered now seeking discrete topologies. The iteration history for the load case $F = 0.04$ kN is

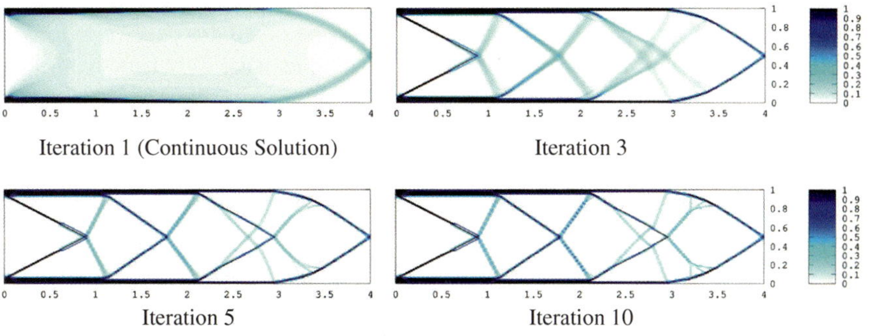

Fig. 2 Long cantilever beam designs at selected iterations

Fig. 3 Iteration history for long cantilever beam design

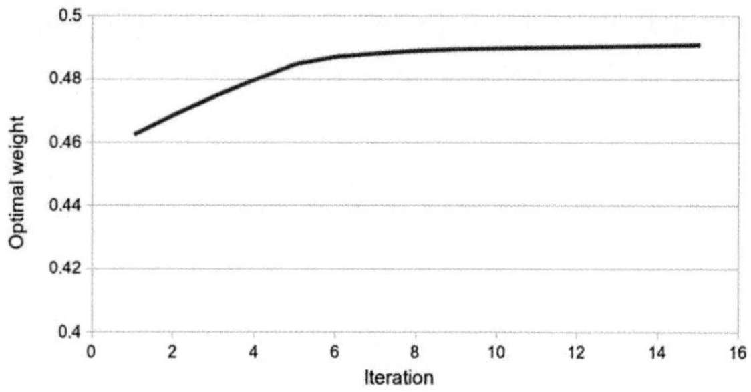

plotted in Fig. 3. The objective value increases asymptotically. The iterative process is stopped at the 20th iteration and the accuracy of the solution is expressed by the relative change in objective value between consecutive iterations. The objective values 0.4903 and 0.4924 are attained within 10 and 20 iterations with relative changes $\varepsilon_{10} = 0.06\,\%$ and $\varepsilon_{20} = 0.03\,\%$, respectively, while, by visual inspection, the design is practically unchanged beyond the 5th iteration. The topologies generated at selected iterations are illustrated in Fig. 2. The one that results from the first iteration naturally coincides with the solution of the continuous design. It can be seen that the design converges steadily towards a well defined layout, without apparent occurrence of checkerboard patterns. Nevertheless, the following two limitations are observed. The first is that the process does not accurately achieve the ultimate 0-1 solution in and close to the solid structure. This calls for an improvement in the formulation of the discrete design problem. The second is that some brace members exhibit an alternation of dark and light gray segments. Further, investigation is needed to analyze this phenomenon.

The CPU time taken by a direct problem solution using *MOSEK* does not vary significantly from an iteration to another. Therefore, the time it takes to solve a discrete problem fairly amounts to the product of the CPU time of one direct (continuous) problem solution by the number of iterations.

The above topology resulting from a plastic design procedure is compared with the optimal topology shown in Fig. 4 for an elastic cantilever beam generated using a compliance method. It can be seen that the left half of the beam has the same topology in both designs. The differences that may be noted are only variations in shape and size. The difference in topology is remarkable in the right hand half of the beam where the plastic design exhibits ramifications that are not seen in the elastic topology.

Fig. 4 Long cantilever beam optimum topology (From [27])

6.2 Medium Cantilever Beam

The medium cantilever beam has the dimensions $L = 2\,\text{m}$ and $H = 1\,\text{m}$. The load is centered and distributed along a distance $b = 10\,\text{cm}$. Since the area of the design domain is now half that of the long beam, the medium beam is modeled with only 28,800 finite elements. The CPU time required for the limit analysis of the full density beam is 18 s. As observed for the long cantilever example, the optimal topology is load dependent [10, 11]. For a load $F = 0.10\,\text{kN}$ the optimal objective 0.3676 is calculated within 28 s of CPU time. The optimal continuous configuration is visualized at the top left of Fig. 5. Except for differences in the shape of the solid zone at top and bottom edges near the support, the material distribution is visually similar to that reported in [3] and illustrated in Fig. 6a.

Solving for the black and white design, the objective value 0.3861 is obtained after 20 iterations with a relative change in objective $\varepsilon_{20} = 0.0034\,\%$. Convergence in this case is faster than observed in the long cantilever problem (Fig. 7). From visual inspection the design has nearly converged at the 5th iteration. Examination of the designs produced in the first few iterations raises the following observations. The distribution of material converges steadily to the final layout. The main braces are formed from the second and third iteration. The thinner elements evolve each from a cloud to a pair of fuzzy elements that then join into a single solid line. The thin solid lines form a regular square grillage. In the final design, the horizontal edge elements are perfectly black. The thick oblique elements are also black except that they are delimited by thin gray layers. This is apparently a discretization related problem that should not affect convergence. It is related to the difficulty to model smooth lines of arbitrary orientation using finite element sides at 0, 45 or 90°. As for the thin members, the density is uniform and gray. This could be because their thickness is so close to the finite element size that the black and white solution cannot compete with the gray one (Fig. 5).

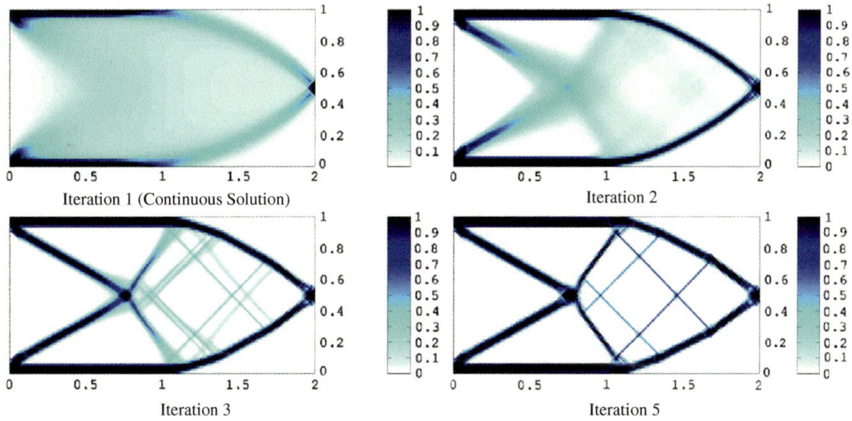

Fig. 5 Medium cantilever beam designs at selected iterations

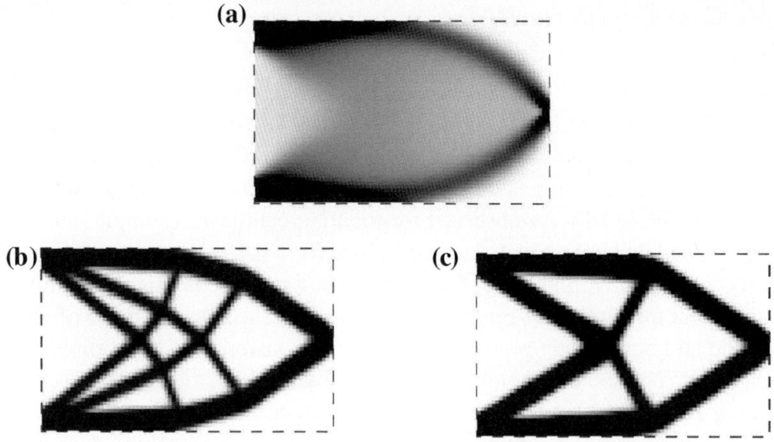

Fig. 6 Optimal topology for the medium cantilever beam. **a** Voigt upper-bound interpolation. **b** Discrete topology by SIMP (p = 2). **c** Discrete topology by H-S interpolation (From [1])

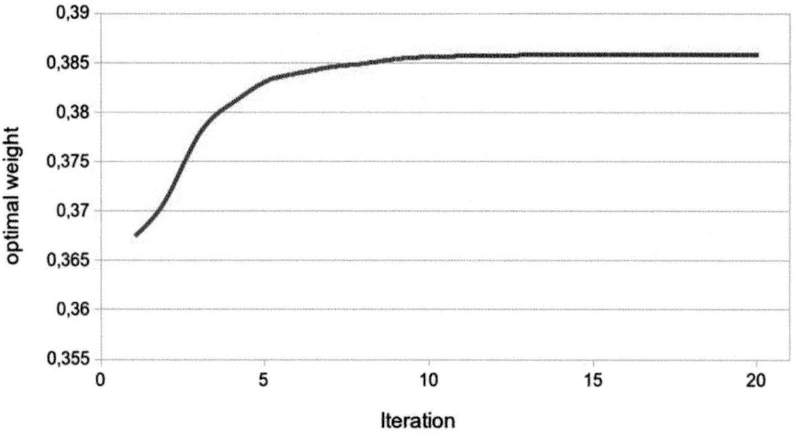

Fig. 7 Iteration history for medium cantilever beam design

Comparison will be made with two elastic topologies drawn from the literature, shown in Fig. 6. The topology given in Fig. 6b, obtained using a homogenization approach, corresponds to the same main structure (thick members) as the direct plastic design solution. The difference lies in the existence or not of the grillage in the right hand frame. On the other hand, the elastic topology in Fig. 6c, obtained by penalization of intermediate densities, is clearly different from the plastic solution but shares with it the existence of a grillage structure in the right frame, although reduced to a simple cross. Thus, in this example, the optimal plastic topology does not appear to stand out from typical optimum elastic ones.

6.3 The MBB Beam

The MBB beam problem treated in this section is illustrated in Fig. 8. The geometrical dimensions of the problem are $L = 6\,\text{m}$, $H = 1\,\text{m}$, $b = 0.05\,\text{m}$ and $a = 0.025\,\text{m}$. For a shear strength $\bar{s} = 1\,\text{kPa}$ and using a mesh with $240 \times 50 \times 4 = 48{,}000$ elements the limit analysis conducted for the full density domain yields a limit load $\bar{F} = 0.2\,\text{kN}$ in 52 s of CPU time. The continuous design problem solved for a specified applied load $F = 0.15\,\text{kN}$ takes 122 s of CPU time. The resulting design is shown at the top left of Fig. 9. It shows a significant resemblance with the reference continuous configuration depicted in Fig. 10. The outer contour and the shape of the black zones are largely similar. The differences that may be noted are only variations in size. The material distribution in the interior also exhibits common patterns with some differences in shape and size. Finally, the void in the plastic design solution is white and clear whereas it is light gray in the reference (elastic) configuration.

Pursuing the iteration process in search for a discrete topology, the optimal layout is practically reached within less than 10 cycles. At the 20th iteration the relative error in the objective value is only $\varepsilon_{20} = 0.0086\,\%$. The configurations generated at

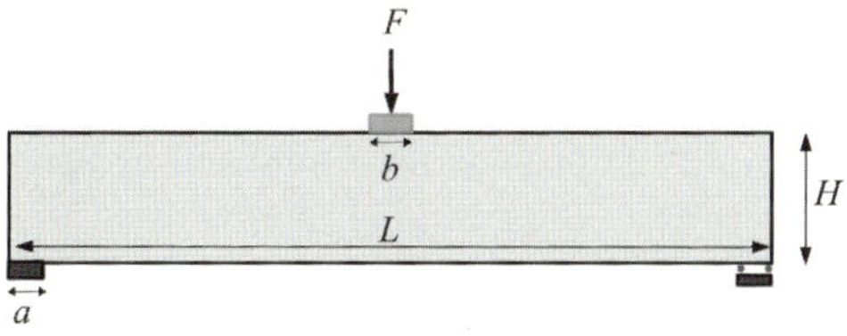

Fig. 8 The MBB beam

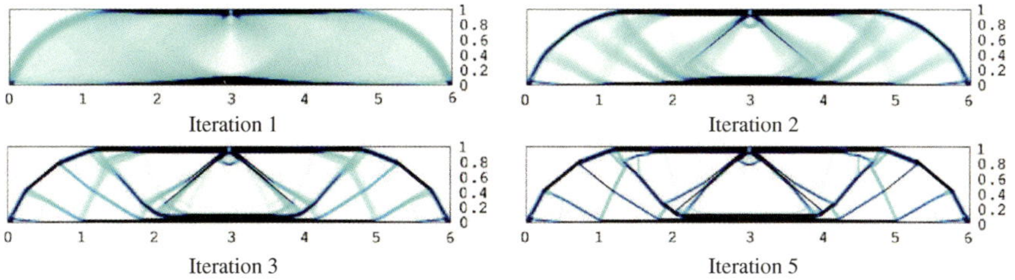

Fig. 9 MBB beam solutions at selected iterations

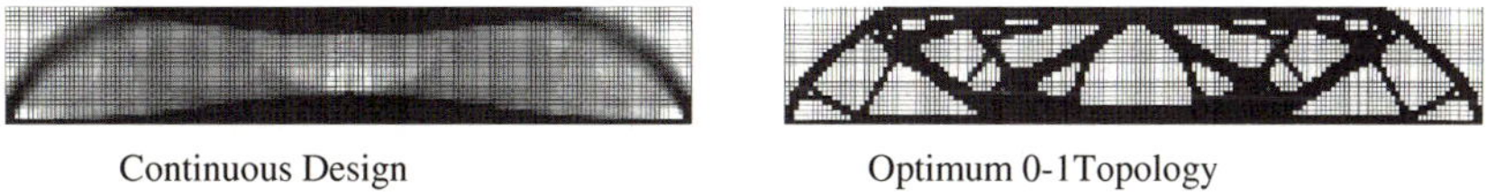

Fig. 10 Solutions for the MBB beam (From [28])

selected iterations are displayed in Fig. 9. The final topology is nearly identical to that corresponding to the 5th iteration. No checkerboard patterns appear in the design, although no measures for preventing their formation were taken. The layout lines are well defined. The void space is clear white. Comparison with the 0-1 reference solution in Fig. 10 reveals nearly the same topology. Differences are noted essentially in the shape and size of the layout details, a remarkable observation given the difference in the underlying problem natures. As noted in the previous examples, though, in the limit analysis based design, the converged solution is not perfectly a 0-1 design.

7 Conclusion

A direct method for plastic topology design recently proposed by the authors for the minimum weight design of continuum structures subject to a specified statically and plastically admissible loading is reexamined and extended to the discrete, or black and white, topology optimization. The continuous topology design integrates the optimization problem with direct limit analysis in a single mathematical programming problem exhibiting the same algebraic structure and the same order of magnitude of computational demand as the direct static limit analysis problem. The mathematical problem is a conic programming problem which presents attractive convergence properties.

The discrete topology design problem is formulated by applying a penalization scheme to the continuous one. The resulting, non convex, problem is treated by solving a converging sequence of continuous design problems. The properties of the latter translate into desirable features in the generated discrete topologies, such as the unlikely occurrence of checkerboard effects.

The proposed method is illustrated on example problems taken from the literature. Comparison of the results of the proposed method with optimal black and white topologies generated using elastic design methods shows some remarkable similarities although the material behavior and the type of analysis are largely different.

A wide range of open questions arise from the present work. Among these, it is interesting to identify example problems that highlight differences between elastic and plastic topologies.

References

1. Bendsøe MP, Sigmund O (1999) Material interpolation schemes in topology optimization. Arch Appl Mech 69(9–10):635–654
2. Eschenauer HA, Olhoff N (2001) Topology optimization of continuum structures: a review*. Appl Mech Rev 54(4):331–390
3. Allaire G (2007) Conception optimale de structures, mathématiques and applications, vol 58
4. Rozvany GIN (2009) A critical review of established methods of structural topology optimization. Struct Multidiscip Optim 37:217–237

5. Browne PA (2013) Topology optimization of linear elastic structures. PhD thesis, Department of Mathematical Sciences, University of Bath
6. Duddeck F (2011) New approaches for shape and topology optimization for crashworthiness. In: NAFEMS world congress. Ecole Centrale de Lille, France, May 2011
7. Bendsøe MP, Kikuchi N (1988) Generating optimal topologies in structural design using a homogenization method. Comput Methods Appl Mech Eng 71(2):197–224
8. Bendsøe MP (1989) Optimal shape design as a material distribution problem. Struct Optim 1(4):193–202
9. Woo TH, Schmit LA (1981) Decomposition in optimal plastic design of structures. Int J Solids Struct 17(1):39–56
10. Kammoun Z, Smaoui H (2013) A direct method formulation for topology plastic design of continua. In: 4th international workshop on direct methods—DM2013. University Mediterranea of Reggio Calabria, Italy, 1–2 October 2013
11. Kammoun Z, Smaoui H (2014) A direct approach for continuous topology optimization subject to plastically admissible loading. Comptes Rendus Mécanique, submitted April 2014
12. Maute K, Schwarz S, Ramm E (1998) Adaptive topology optimization of elastoplastic structures. Struct Optim 15(2):81–91
13. Yuge K, Kikuchi N (1995) Optimization of a frame structure subjected to a plastic deformation. Struct Optim 10(3–4):197–208
14. Salençon J (1967) Théorie des charges limites: poinçonnement d'une plaque par deux poinçons symétriques en déformation plane. Comptes Rendus, Acad Sci Paris, 265:869–872
15. Salençon J (1974) Théorie de la plasticité pour les applications à la mécanique des sols. Eyrolles, Paris
16. Pastor J (1978) Analyse limite: détermination numérique de solutions statiques complètes. Application au talus vertical. J Méca Appl (now Eur J Mech—A/Solids) 2:167–196
17. MOSEK ApS (2002) C/O Symbion Science Park, Fruebjergvej 3, Box 16, 2100 Copenhagen ϕ, Denmark. www.mosek.com
18. Kammoun Z, Pastor F, Smaoui H, Pastor J (2010) Large static problem in numerical limit analysis: a decomposition approach. Int J Num Anal Methods Geomech 34:1960–1980
19. Kammoun Z, Pastor J, Smaoui H (2013) Limit analysis of a soil reinforced by micropile group: a decomposition approach. In: Limit state of materials and structures, Springer, pp 179–195
20. Pastor F, Kammoun Z, Loute E, Pastor J, Smaoui H (2009) Large problems in numerical limit analysis: a decomposition approach. In: Limit states of materials and structures, Springer, pp 23–43
21. Sigmund O, Petersson J (1998) Numerical instabilities in topology optimization: a survey on procedures dealing with checkerboards, mesh-dependencies and local minima. Struct Optim 16(1):68–75
22. Holmberg E, Torstenfelt B, Klarbring A (2013) Stress constrained topology optimization. Struct Multidiscip Optim 48(1):33–47
23. Chapman CD, Jakiela MJ (1996) Genetic algorithm-based structural topology design with compliance and topology simplification considerations. J Mech Des 118(1):89–98
24. Garcia-Lopez NP, Sanchez-Silva M, Medaglia AL, Chateauneuf A (2013) An improved robust topology optimization approach using multiobjective evolutionary algorithms. Comput Struct 125:1–10
25. Hillestad RJ, Jacobsen SE (1980) Reverse convex programming. Appl Math Optim 6(1):63–78
26. Jacobsen SE, Moshirvaziri K (1996) Computational experience using an edge search algorithm for linear reverse convex programs. J Glob Optim 9(2):153–167
27. Zakhama R (2009) Multigrid implementation of cellular automata for topology optimisation of continuum structures with design dependent loads. PhD thesis, TUDelft, Faculteit Luchtvaart- en Ruimtevaarttechiek
28. Duysinx P (1996) Optimisation topologique: du milieu continu à la structure élastique. PhD thesis, Université de Liège

The Influence of Limited Kinematical Hardening on Shakedown of Materials and Structures

Jaan-W. Simon

Abstract One of the most important tasks in design for construction engineers is the determination of the load bearing capacity of the considered engineering structure. This can be particularly challenging when the applied thermo-mechanical loads vary with time and are high enough to exceed the material's elastic regime. In these cases, the lower bound shakedown analysis provides a convenient tool. Since accounting for the realistic material behavior is inevitable to achieve reliable results, it is highly relevant to consider limited kinematical hardening. Although there exist different formulations in the literature, in which limited kinematical hardening is incorporated, these usually do not take into account the underlying hardening law in an explicit manner. The most important question in that context is whether such formulations can cover both linear and nonlinear hardening laws. In consequence, the aim of this paper is to investigate the effect of nonlinearity of the hardening law by showing that in certain scenarios the introduction of an explicit hardening law as a subsidiary constraint is unavoidable.

1 Introduction

One of the most important tasks in design for construction engineers is the determination of the load bearing capacity of the considered engineering structure. This can be particularly challenging when the applied thermo-mechanical loads vary with time and are high enough to exceed the material's elastic regime. Then, the computation of the so-called shakedown loading factor is necessary, which is the maximum loading factor such that the system can be considered as 'safe', such that neither alternating plasticity nor spontaneous or incremental collapse occur.

J.-W. Simon (✉)
Institute of Applied Mechanics, RWTH Aachen University,
Mies-van-der-Rohe-Straße 1, 52074 Aachen, Germany
e-mail: jaan.simon@rwth-aachen.de

P. Fuschi et al. (eds.), *Direct Methods for Limit and Shakedown Analysis of Structures*,
DOI 10.1007/978-3-319-12928-0_4

In most cases, conventional step-by-step computations are performed whenever the exact stress-strain distributions are needed. In contrast, if only the material's or structure's limit state is of interest, the load bearing capacity can be conveniently determined by means of limit or shakedown analysis. In general, there exist two different approaches to shakedown analysis, which complement each other: The lower bound approach by Melan [19, 20], which is formulated in statical quantities, and the upper bound approach of Koiter [15], which makes use of the kinematical ones. From these, the lower bound approach is adopted in this work, because the formulation in terms of stresses is particularly suited for the extension to kinematical hardening.

The majority of elasto-plastic materials exhibit kinematical hardening during the evolution of plastic deformations. Therefore, this phenomenon needs to be incorporated into the procedure in order to obtain realistic results. In its original formulation the statical shakedown theorem only holds for elastic-perfectly plastic continua as well as for unlimited kinematical hardening ones. Since the unlimited kinematical hardening case does not cover incremental collapse at all, taking into account limited (or bounded) kinematical hardening is inevitable. Hence, this issue has been addressed by several authors in the field of shakedown analysis [6, 8, 10, 18, 21–27, 30, 36–40].

From those, the first explicit formulation for limited kinematical hardening materials has been given by Weichert and Groß-Weege [40] (WGW), who introduced a two-surface model. Their formulation is based on the concept of generalized standard materials [11], and thus implies an associated hardening rule, together with the assumption of limited *linear* kinematical hardening. Almost at the same time, Stein et al. [37–39] have proposed another approach based on an overlay model. The formulation presented therein has been said to be valid for limited *general nonlinear* kinematical hardening with associated flow. Noteworthy, Heitzer [12] has investigated the relation between the two different formulations. He has stated that both theorems, even though formulated differently, lead to the same optimal value for the shakedown factor, and that the only difference might appear in the corresponding residual stress fields.

More recently Pham has presented an extension of the theorem proposed by Weichert and Groß-Weege for the generally nonlinear case [23]. He has claimed, that this theorem holds for any generally nonlinear hardening law as long as the hysteresis is positive for any closed cycle of plastic deformations.

In all of these works, it turns out that the shakedown load is independent of the underlying hardening law and the according stress-strain curve. The shakedown load seems to depend only on the magnitudes of the initial yield stress σ_Y and the ultimate stress σ_H.

In contrast, independently of each other, Staat and Heitzer [36] as well as Bouby et al. [4, 5] have presented results with significant differences in the shakedown limit load between the limited linear hardening and the limited nonlinear hardening. Thus, the aim of the present paper is to investigate the effect of different hardening rules on shakedown loads.

2 Lower Bound Shakedown Analysis Accounting for Limited Kinematical Hardening

In the following, an elastic-perfectly plastic body with volume V and surface A is considered, which is subjected to: temperature loads $T(\boldsymbol{x}, t)$ in V, body forces $\boldsymbol{f}_V(\boldsymbol{x}, t)$ in V, surface loads $\boldsymbol{f}_A(\boldsymbol{x}, t)$ on $A_f \subseteq A$, and prescribed displacements $\boldsymbol{u}(\boldsymbol{x}, t)$ on $A_u \subseteq A$, such that $A = A_f \cap A_u$ and $A_f \cup A_u = \emptyset$. Only time- and temperature-independent material behavior is taken into account, while material damage and geometrical nonlinearity are neglected. The existence of a convex yield function $f[\boldsymbol{\sigma}(\boldsymbol{x}, t)]$ is assumed. Then, the elastic limit is described by a yield surface in stress space $\mathbb{S}$ as closure of the convex domain $\mathscr{C}_Y \subseteq \mathbb{S}$ of admissible states of stress with the strict interior $\mathscr{C}_Y^i$:

$$\mathscr{C}_Y^i = \left\{ \boldsymbol{\sigma} \in \mathbb{S} \,\middle|\, f\left[\boldsymbol{\sigma}(\boldsymbol{x}, t)\right] < \sigma_Y^2(\boldsymbol{x}), \quad \forall \boldsymbol{x} \in V, \quad \forall t \right\} \tag{1}$$

2.1 *Melan's Statical Shakedown Theorem for Elastic-Perfectly Plastic Materials*

As already mentioned above, this work is based on the statical shakedown theorem by Melan [19, 20], which provides a lower bound to the shakedown loading factor. To apply Melan's theorem, the total stress $\boldsymbol{\sigma}(\boldsymbol{x}, t)$ in a point $\boldsymbol{x} \in V$ within the volume V of the considered body at time t is decomposed into an elastic reference stress $\boldsymbol{\sigma}^E(\boldsymbol{x}, t)$ and a residual stress $\boldsymbol{\rho}(\boldsymbol{x}, t)$ induced by the evolution of plastic strains.

$$\boldsymbol{\sigma}(\boldsymbol{x}, t) = \boldsymbol{\sigma}^E(\boldsymbol{x}, t) + \boldsymbol{\rho}(\boldsymbol{x}, t) \tag{2}$$

The fictitious stress state $\boldsymbol{\sigma}^E(\boldsymbol{x}, t)$ is the one which would occur in a purely elastic reference body under the same conditions and loadings as the original one. Both the elastic reference stresses and the residual stresses satisfy the equilibrium constraints as well as the statical boundary conditions.

$$\text{equilibrium:} \qquad \nabla \cdot \boldsymbol{\sigma}^E = -\boldsymbol{f}_V \qquad \nabla \cdot \boldsymbol{\rho} = \boldsymbol{0} \qquad \text{in } V \tag{3}$$

$$\text{statical bc:} \qquad \boldsymbol{n} \cdot \boldsymbol{\sigma}^E = \boldsymbol{f}_A \qquad \boldsymbol{n} \cdot \boldsymbol{\rho} = \boldsymbol{0} \qquad \text{on } A_f \tag{4}$$

Then, Melan's shakedown theorem for elastic-perfectly plastic materials can be formulated as follows:

If there exist a loading factor $\alpha > 1$ and a time-independent residual stress field $\bar{\boldsymbol{\rho}}(\boldsymbol{x})$, such that the yield condition is satisfied for any loading path within the considered loading domain Ω at any time t and in any point $\boldsymbol{x}$ of the structure, then the system will shake down.

$$f\left[\alpha\, \boldsymbol{\sigma}^E(\boldsymbol{x}, t) + \bar{\boldsymbol{\rho}}(\boldsymbol{x})\right] \leq \sigma_Y^2(\boldsymbol{x}), \ \forall \boldsymbol{x} \in V, \ \forall t \tag{5}$$

It is worth to mention that the numerical procedure allows for computing values $\alpha < 1$ as long as it is positive, $\alpha > 0$. Nevertheless, shakedown only can be guaranteed if $\alpha > 1$ holds, because only then the plastic dissipative energy is guaranteed to be bounded.

2.2 *Two Surface Model for Limited Linear Kinematical Hardening by Weichert and Groß-Weege (WGW)*

The first explicit formulation of the statical shakedown theorem accounting for limited kinematical hardening has been proposed by Weichert and Groß-Weege [40] in 1988. The formulation presented therein is based on the Generalized Standard Material Model (GSMM) introduced by Halphen and Nguyen [11]. Thus, it is implied that the normality rule holds, restricting the formulation to associated hardening laws. Moreover, the proof of the theorem makes use of the assumption of limited *linear* kinematical hardening.

The kinematical hardening is considered as a translational motion of the yield surface—described by $f = \sigma_Y^2$—in stress space without change of orientation, form or size. This motion is limited by the bounding surface, $f = \sigma_H^2$, which corresponds to the ultimate stress σ_H. Further, the motion is defined by the six-dimensional vector of back-stresses $\boldsymbol{\pi}$ representing the translation of the yield surface's center, see Fig. 1. Thereby, the total stresses $\boldsymbol{\sigma}(\boldsymbol{x}, t)$ are decomposed into the back stresses $\boldsymbol{\pi}$ and the so-called reduced stresses $\boldsymbol{v}$. The latter are responsible for the occurrence of plastic strains.

$$\boldsymbol{\sigma}(\boldsymbol{x}, t) = \boldsymbol{\pi}(\boldsymbol{x}, t) + \boldsymbol{v}(\boldsymbol{x}, t) \tag{6}$$

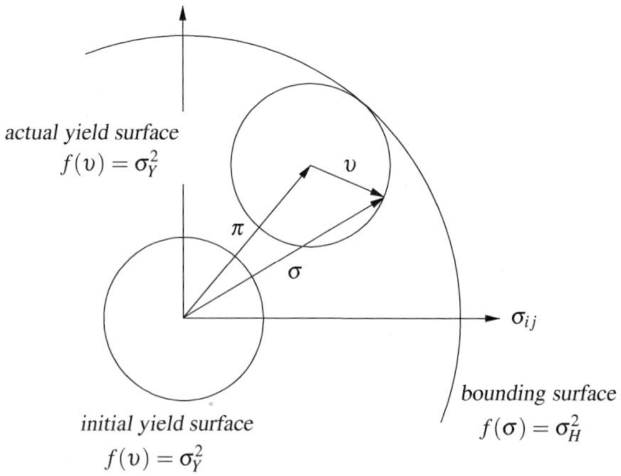

Fig. 1 Limited kinematical hardening considered as translation of the yield surface in stress space

Since the bounding surface is fixed in stress space, the back-stresses have to be time-independent, which will be indicated in the following by an overbar, $\boldsymbol{\pi} = \bar{\boldsymbol{\pi}}(\boldsymbol{x})$. Further, the decomposition of the total stresses into an elastic reference part and a residual one still holds. Consequently, the reduced stresses $\boldsymbol{v}(\boldsymbol{x}, t)$ can be expressed as follows:

$$\boldsymbol{v}(\boldsymbol{x}, t) = \boldsymbol{\sigma}(\boldsymbol{x}, t) - \bar{\boldsymbol{\pi}}(\boldsymbol{x}) = \boldsymbol{\sigma}^E(\boldsymbol{x}, t) + \bar{\boldsymbol{\rho}}(\boldsymbol{x}) - \bar{\boldsymbol{\pi}}(\boldsymbol{x}) \tag{7}$$

Finally, a shakedown theorem accounting for limited kinematical hardening can be written as:

If there exist a loading factor $\alpha > 1$, a time-independent self-equilibrated (residual) stress field $\bar{\boldsymbol{\rho}}$ and a time-independent field of back-stresses $\bar{\boldsymbol{\pi}}$,such that the yield condition and the bounding condition are satisfied for any loading path within the considered loading domain Ω at any time t and in any point $\boldsymbol{x}$ of the structure, then the system will shake down.

$$f\left[\alpha\, \boldsymbol{\sigma}^E(\boldsymbol{x}, t) + \bar{\boldsymbol{\rho}}(\boldsymbol{x}) - \bar{\boldsymbol{\pi}}(\boldsymbol{x})\right] \leq \sigma_Y^2(\boldsymbol{x}) \tag{8}$$

$$f\left[\alpha\, \boldsymbol{\sigma}^E(\boldsymbol{x}, t) + \bar{\boldsymbol{\rho}}(\boldsymbol{x})\right] \leq \sigma_H^2(\boldsymbol{x}) \tag{9}$$

It should be noted, that the consideration of specific hardening rules—such as the linear one—implies restrictions on the field of back-stresses $\bar{\boldsymbol{\pi}}$. This issue and its consequences will be discussed in Sect. 3.

2.3 Extension of WGW-Model for Limited General Nonlinear Kinematical Hardening by Pham

While the proof of the theorem given above is based on the assumption of limited linear kinematical hardening, the same formulation has been derived for generally-nonlinear hardening laws later by Pham [22, 23]. The only restriction Pham has postulated on the considered hardening is the *positive hysteresis postulate*, which states that for any closed cycle of plastic deformations ($t \in [0, \theta]$) the following condition has to hold:

$$\oint \boldsymbol{\pi} : d\boldsymbol{\varepsilon}_\pi^p = \int_0^\theta \boldsymbol{\pi} : \dot{\boldsymbol{\varepsilon}}_\pi^p dt \geq 0 \qquad \left(\boldsymbol{\varepsilon}_\pi^p(0) = \boldsymbol{\varepsilon}_\pi^p(\theta)\right) \tag{10}$$

where $\boldsymbol{\pi}$ denotes the back-stresses, and $\boldsymbol{\varepsilon}_\pi^p$ denotes the corresponding plastic deformation. In the case of a simple loading-unloading closed plastic cycle, this restriction implicates that the hysteresis loop is followed in clockwise direction, but not anti-clockwise.

2.4 Overlay-Model for Limited Nonlinear Kinematical Hardening by Stein et al.

An alternative approach to formulate a shakedown theorem taking into account limited kinematical hardening has been presented by Stein et al. [37–39]. This formulation is based on the overlay model. As in the WGW-model, the normality condition is assumed to hold. Except of this restriction the authors have stated that the formulation is valid for limited *general nonlinear* kinematical hardening.

Interestingly, the formulation of Stein et al. is rather similar to the one by Weichert and Groß-Weege. However, there is one very important difference in the way the bounding condition is expressed. Instead of the condition (9), Stein et al. derive the following one:

$$f\,[\bar{\boldsymbol{\pi}}\,(\boldsymbol{x})] \leq [\sigma_H(\boldsymbol{x}) - \sigma_Y(\boldsymbol{x})]^2 \tag{11}$$

The most severe difference between (9) and (11) is the fact that the back-stresses $\bar{\boldsymbol{\pi}}$ appear only implicitly in the first one, whereas they show up in an explicit manner in the second one. It is worth to mention, that the relation between these two different formulations has been investigated in more detail by Heitzer [12]. As a result of these investigations, Heitzer has stated that both theorems lead to the same optimal value for the shakedown factor, even though the corresponding residual stress fields might differ. As will be shown within the example in Sect. 4, this turns out not to be true in certain scenarios.

2.5 Description of the Loading Domain

In the following, the loading histories $\mathscr{H}(\boldsymbol{x}, t)$ under consideration are assumed to be describable as superposition of a finite number NL of different loading sets $P_\ell(\boldsymbol{x}, t)$. The latter can be expressed in terms of load multipliers $\mu_\ell(t)$ for any loading case ℓ and the unity load $P_0(\boldsymbol{x})$.

$$\mathscr{H}(\boldsymbol{x}, t) = \sum_{\ell=1}^{NL} P_\ell(\boldsymbol{x}, t) = \sum_{\ell=1}^{NL} \mu_\ell(t)\, P_0(\boldsymbol{x}) \tag{12}$$

As shown by König [16], it is sufficient to consider only the convex hull of the loading history, which is polyhedral with $NC = 2^{NL}$ corners. These corners are defined in the loading space by introducing bounding values μ_ℓ^+ and μ_ℓ^- for each multiplier μ_ℓ. Doing so, the set $\mathscr{U}$ is defined, which contains all possible combinations of loading sets within these bounds through merging all loading multipliers to the vector $\boldsymbol{\mu} = \mu_\ell\, \boldsymbol{e}_\ell$.

$$\mathscr{U} = \left\{ \boldsymbol{\mu} \in \mathbb{R}^{NL} \;\middle|\; \mu_\ell^- \leq \mu_\ell \leq \mu_\ell^+ \,,\; \forall \ell \in [1, NL] \right\} \tag{13}$$

Then, the loading domain Ω is described as set of all possible loading histories contained within $\mathscr{U}$.

$$\Omega = \left\{ \mathscr{H}(\boldsymbol{x}, t) \;\middle|\; \mathscr{H}(\boldsymbol{x}, t) = \sum_{\ell=1}^{NL} \mu_\ell(t)\, P_0(\boldsymbol{x}) \,,\ \forall \boldsymbol{\mu} \in \mathscr{U} \right\} \tag{14}$$

Consequently, the elastic reference stresses are split in analogy to (12).

$$\boldsymbol{\sigma}^E(\boldsymbol{x}, t) = \sum_{\ell=1}^{NL} \mu_\ell(t)\, \boldsymbol{\sigma}_\ell^E(\boldsymbol{x}) \tag{15}$$

2.6 Discretization

Using the finite element method (FEM), the stresses are approximately represented by their values in the Gaussian points, which will be referred to by the index $r \in [1, NG]$. Here NG is the total number of Gaussian points in the system. Consequently, the fictitious elastic stresses $\boldsymbol{\sigma}_{r,\ell}^E$ can be computed for any loading case ℓ by purely elastic analysis.

$$\boldsymbol{\sigma}_r^E(t) = \sum_{\ell=1}^{NL} \mu_\ell(t)\, \boldsymbol{\sigma}_{r,\ell}^E \tag{16}$$

To ensure shakedown for all possible loading paths inside of the loading domain, only its corners need to be examined. Thus, the time-dependence of $\boldsymbol{\sigma}_r^E$ can be expressed through the stress states in the corners $j \in [1, NC]$ of the loading domain. For this, the matrix $\boldsymbol{U}_{NL} \in \mathbb{R}^{NC \times NL}$ with entries $U_{j\ell}$ is introduced, where $j \in [1, NC]$ and $\ell \in [1, NL]$.

$$\boldsymbol{\sigma}_r^{E,j} = \sum_{\ell=1}^{NL} U_{j\ell} \boldsymbol{\sigma}_{r,\ell}^E \tag{17}$$

Each row of this matrix $\boldsymbol{U}_{NL}$ represents the coordinates of one corner of the loading domain in the NL-dimensional loading space, which are defined by the factors μ_ℓ^- and μ_ℓ^+ as introduced in (13). The matrix can be defined in an automatic way for arbitrary numbers of loading cases NL, as shown in [35].

Since the elastic reference stress field $\boldsymbol{\sigma}^E$ is in equilibrium with the external loading, the residual stress field $\bar{\boldsymbol{\rho}}$ has to be self-equilibrated. This fact can be expressed by means of the principle of virtual work [9],

$$\int_V \delta\boldsymbol{\varepsilon} : \bar{\boldsymbol{\rho}}\, dV = 0 \tag{18}$$

where $\delta\boldsymbol{\varepsilon}$ denotes any virtual strain field which satisfies the kinematical boundary conditions.

Using the FEM with isoparametric elements, the displacements $\boldsymbol{u}$ are approximated by shape functions and nodal displacements $\boldsymbol{u}_K$. Thus, the virtual strain field can be expressed through the nodal displacements as well: $\delta\boldsymbol{\varepsilon}(\boldsymbol{x}) = \delta\boldsymbol{u}_K \cdot \boldsymbol{B}$, where $\boldsymbol{B}(\boldsymbol{x})$ is the differentiation matrix. Furthermore, the integration is carried out numerically. Thereby (18) is approximated by a system of linear equations for the residual stresses $\bar{\boldsymbol{\rho}}_r$ in the Gaussian points.

$$\int_V \boldsymbol{B}(\boldsymbol{x}) : \bar{\boldsymbol{\rho}}\, dV =: \sum_{r=1}^{NG} \mathbb{C}_r \cdot \bar{\boldsymbol{\rho}}_r = \boldsymbol{0} \tag{19}$$

The equilibrium matrices $\mathbb{C}_r$ depend only on the geometry of the system and the applied element type as well as the kinematical boundary conditions.

2.7 Resulting Nonlinear Optimization Problem

Based on the Eqs. (17) and (19), the extended Melan's theorem for limited kinematical hardening can be expressed in terms of an optimization problem for the loading factor $\alpha > 1$:

$$(\mathscr{P}^H) \qquad \alpha_{SD} = \max_{\bar{\boldsymbol{\rho}},\, \bar{\boldsymbol{\pi}}} \alpha$$

$$\sum_{r=1}^{NG} \mathbb{C}_r \cdot \bar{\boldsymbol{\rho}}_r = \boldsymbol{0} \tag{20a}$$

$$\forall j \in [1, NC],\ \forall r \in [1, NG]:$$

$$f\left(\alpha\, \boldsymbol{\sigma}_r^{E,j} + \bar{\boldsymbol{\rho}}_r - \bar{\boldsymbol{\pi}}_r\right) \le \sigma_{Y,r}^2 \tag{20b}$$

$$f_H\left(\boldsymbol{\sigma}_r^j, \bar{\boldsymbol{\pi}}_r, \sigma_{H,r}\right) \le 0 \tag{20c}$$

Depending on which formulation is used, the according bounding condition, (9) or (11), needs to be inserted:

$$\text{WGW \& Pham}: \quad f_H\left(\boldsymbol{\sigma}_r^j, \bar{\boldsymbol{\pi}}_r, \sigma_{H,r}\right) = f\left(\alpha\, \boldsymbol{\sigma}_r^{E,j} + \bar{\boldsymbol{\rho}}_r\right) - \sigma_{H,r}^2 \tag{21a}$$

$$\text{Stein}: \quad f_H\left(\boldsymbol{\sigma}_r^j, \bar{\boldsymbol{\pi}}_r, \sigma_{H,r}\right) = f\left(\bar{\boldsymbol{\pi}}_r\right) - \left(\sigma_{H,r} - \sigma_{Y,r}\right)^2 \tag{21b}$$

The solution procedure for this nonlinear convex optimization problem is not in the scope of the current paper. The interested reader is referred to [31–34].

3 Effect of the Underlying Kinematical Hardening Law

In this section, the effect of the underlying limited kinematical hardening model on the shakedown load is investigated. As already has been mentioned, it is frequently stated in literature that shakedown limits only depend on the initial yield stress σ_Y and the ultimate stress σ_H, but not at all on the hardening behavior in between, see e.g. [1, 13, 22–24, 37, 40]. Even so, in the following it will be shown that in principle there can exist cases, in which the shakedown load is in fact influenced by the applied hardening law.

To do so, the shakedown theorems presented in the Sects. 2.2–2.4 are rewritten in a more formal way. For this, the set of all time-independent and self-equilibrated stress fields is denoted by $\mathscr{R}$. Further, the set of all time-independent and permissible fields of back-stresses is denoted by $\mathscr{B}$. Here, the definition of a permissible stress field depends on the considered theorem:

WGW: A back-stress field is permissible if it can evolve under the given loading domain following the corresponding linear hardening law with an associated flow rule.

Stein: A back-stress field is permissible if it can evolve under the given loading domain following any hardening law with an associated flow rule.

Pham: A back-stress field is permissible if it satisfies the positive hysteresis assumption.

Then, the three theorems can be written in the following way:
If there exist a scalar $\alpha > 1$ and fields $\bar{\boldsymbol{\rho}}$ and $\bar{\boldsymbol{\pi}}$, such that the following conditions hold, then the system will shake down.

$$\bar{\boldsymbol{\pi}} \in \mathscr{B} \tag{22a}$$

$$\bar{\boldsymbol{\rho}} \in \mathscr{R} \tag{22b}$$

$$\forall j \in [1, NC], \ \forall r \in [1, NG]:$$

$$f\left(\alpha\, \boldsymbol{\sigma}_r^{E,j} + \bar{\boldsymbol{\rho}}_r - \bar{\boldsymbol{\pi}}_r\right) \leq \sigma_{Y,r}^2 \tag{22c}$$

$$f_H\left(\boldsymbol{\sigma}_r^j, \bar{\boldsymbol{\pi}}_r, \sigma_{H,r}\right) \leq 0 \tag{22d}$$

If one compares this set of conditions with the subsidiary constraints of the optimization problem formulated in Sect. 2.7, one can observe that not all of the conditions (22a)–(22d) are reflected. While the yield and the bounding conditions (22c) and (22d) are represented by (20b) and (20c), respectively, the equation (20a) ensures that $\bar{\boldsymbol{\rho}}$ is self-equilibrated (22b). In contrast, the condition (22a) is not incorporated anymore.

In fact, as long as a general hardening case is considered, in which the evolution of $\bar{\boldsymbol{\pi}}$ is not restricted by any specific kind of hardening law, the optimization problem

($\mathscr{P}^H$) is equivalent to (22a–22d) without loss of generality. In this case the back-stresses can be considered as unrestricted variables of the optimization problem. Therefore, the corresponding hardening law will be called *unrestricted hardening* in the following.

Since α is maximized over $\bar{\boldsymbol{\pi}}$, the solution of the optimization problem ($\mathscr{P}^H$) involves the one particular field $\bar{\boldsymbol{\pi}}^*$, which leads to the maximum value of α. Hence, the unrestricted hardening, which allows the back-stresses $\bar{\boldsymbol{\pi}}^*$ to evolve under the given loading domain, is the most advantageous one leading to the highest shakedown loading factors amongst all possible hardening rules. In other words, the computed shakedown factor might be higher than the one, which can be obtained if a specific hardening rule is applied.

Clearly, only the yield and bounding surface need to be defined by σ_Y and σ_H, respectively, when the unrestricted hardening is considered using the optimization problem ($\mathscr{P}^H$). This is in accordance with the above mentioned references. Even so, the shakedown limit may depend on the hardening behavior in between the initial yield state and the ultimate state. If a specific hardening law is to be considered, then the feasible set of back-stresses can be restricted, such that $\bar{\boldsymbol{\pi}} \in \mathscr{B}^\circ$ and $\mathscr{B}^\circ \subset \mathscr{B}$. To ensure that the solution is admissible, this restriction for the back-stresses has to be included into the optimization problem as a separate constraint. Otherwise, a non-admissible solution might be obtained, $\bar{\boldsymbol{\pi}}^* \notin \mathscr{B}^\circ$.

If, for example, the special case of limited *linear* kinematical hardening shall be investigated, the set of feasible back-stresses $\mathscr{B}^\circ$ has to be formulated such that the resulting solution $\bar{\boldsymbol{\pi}}^\circ$ can evolve under the given loading domain following the corresponding linear hardening law, $\dot{\boldsymbol{\pi}}^\circ = C\,\dot{\boldsymbol{\varepsilon}}^p$, where C is a material parameter. However, introducing such additional constraints directly into the optimization problem can be problematic, because kinematic variables (e.g. plastic strains $\boldsymbol{\varepsilon}^p$) would have to show up at least implicitly in the statical theorem, which is formulated in stresses. Nevertheless, the generally-nonlinear hardening can be directly incorporated into the procedure by defining the restrictions on the back-stresses.

4 Sample Under Constant Tension and Alternating Torsion

To illustrate the correlation between different hardening laws, an illustrative example is presented in this section. In particular, a specimen is considered, which is subjected to a constant tension $\bar{\sigma} > 0$ and alternating torsion τ with zero mean shear stress, such that $\tau_{\min} = -\tau_{\max}$. The according loading domain consists of only two points $(\bar{\sigma}, \tau_{\max})$ and $(\bar{\sigma}, \tau_{\min})$. Noticeably, such a system has been previously examined by e.g. Portier et al. [28], where ratcheting has been investigated experimentally as well as numerically. Furthermore, numerical and analytical results of shakedown analysis accounting for different types of kinematical hardening are presented in [7, 14, 17, 36] for the plane stress state, while the plane strain state has been investigated in [4, 5].

In the following, the plane stress state is considered:

$$\boldsymbol{\sigma} = \begin{pmatrix} \bar{\sigma} & \tau & 0 \\ \tau & 0 & 0 \\ 0 & 0 & 0 \end{pmatrix} \quad \text{and} \quad \boldsymbol{\pi} = \begin{pmatrix} X & Y & 0 \\ Y & 0 & 0 \\ 0 & 0 & 0 \end{pmatrix} \tag{23}$$

In this simple problem both the stress and the strain field are uniformly distributed. Therefore, the residual stresses $\bar{\boldsymbol{\rho}}$ need not to be considered. In fact, the back-stresses $\boldsymbol{\pi}$ play the role of residual stresses in this case.

In the following, the yield criterion is expressed by the von Mises yield condition, which reads:

$$f_Y\left(\boldsymbol{\sigma} - \boldsymbol{\pi}, \sigma_Y\right) = (\bar{\sigma} - X)^2 + 3\left(\tau - Y\right)^2 - \sigma_Y^2 = 0 \tag{24}$$

Since both maxima $(\bar{\sigma}, \tau_{\text{max}})$ and $(\bar{\sigma}, -\tau_{\text{max}})$ shall be located on the yield surface which is described by (24), the following two equations have to hold:

$$(\bar{\sigma} - X)^2 + 3\left(\tau_{\text{max}} - Y\right)^2 - \sigma_Y^2 = 0 \tag{25a}$$

$$(\bar{\sigma} - X)^2 + 3\left(-\tau_{\text{max}} - Y\right)^2 - \sigma_Y^2 = 0 \tag{25b}$$

The difference between these Eq. (25a), (25b), gives:

$$(\tau_{\text{max}} - Y)^2 - (-\tau_{\text{max}} - Y)^2 = -4\,\tau_{\text{max}}\,Y = 0 \tag{26}$$

A non-trivial solution, $\tau_{\text{max}} \neq 0$, can therefore only be obtained if $Y = 0$. Thus, to reach the shakedown state, the yield surface is moved in stress space only in the direction of σ. The yield condition simplifies to:

$$f_Y\left(\boldsymbol{\sigma} - \boldsymbol{\pi}, \sigma_Y\right) = (\bar{\sigma} - X)^2 + 3\,\tau^2 - \sigma_Y^2 = 0 \tag{27}$$

The positive solution of (27) is:

$$\sqrt{3}\,\tau = \sqrt{\sigma_Y^2 - (\bar{\sigma} - X)^2} \tag{28}$$

From (28) it can be observed that the shakedown load τ can in fact depend on the according back-stress X. However, this back-stress is restricted by the applied hardening law. To illustrate the influence of different kinematical hardening rules, the following types of plastic behavior are incorporated:

1. **Perfectly plastic behavior**:
   No hardening occurs and thus no back-stress evolves, $X = 0$, leading to the following maximum value of admissible shear stress:
   $\sqrt{3}\,\tau_0 = \sqrt{\sigma_Y^2 - \bar{\sigma}^2}$ (black dashed line in Fig. 2)
   Alternating plasticity only occurs in pure shear, $\bar{\sigma} = 0$, whereas the remaining shakedown domain represents incremental collapse.

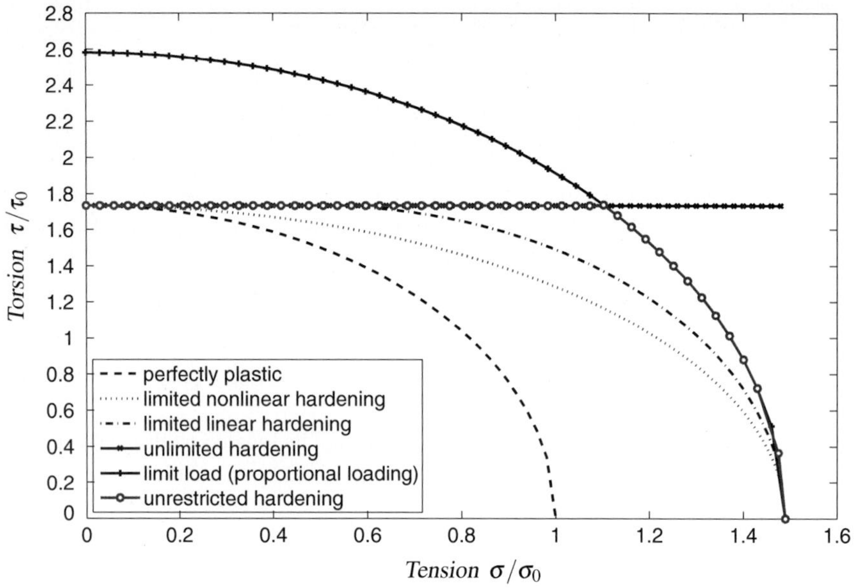

Fig. 2 Shakedown domains for a specimen under constant tension and alternating torsion with different hardening rules

2. **Limit load for proportional loading path**:
For the proportional loading path, the limit domain is similar to the one in the perfectly plastic case (case 1). The only difference is that σ_Y needs to be substituted by σ_H. Hence, the following maximum value of admissible shear stress is obtained:
$\sqrt{3}\,\tau_L = \sqrt{\sigma_H^2 - \bar{\sigma}^2}$ (black solid line with + in Fig. 2)
3. **Unlimited kinematical hardening**:
The shakedown state is defined solely by alternating plasticity. It can be obtained by any hardening rule setting $\sigma_H \rightarrow \infty$. The evolution of back-stresses is not restricted at all, and consequently $\forall \bar{\sigma} : X = \bar{\sigma}$, leading to the following maximum value of admissible shear stress:
$\sqrt{3}\,\tau_u = \sigma_Y$ (black solid line with × in Fig. 2)
4. **Limited unrestricted kinematical hardening**:
The unrestricted kinematical hardening is obtained from the solution of the optimization problem ($\mathscr{P}^H$), in which no explicit restriction is formulated for the back-stresses, since no condition in terms of (22a) is accounted for. Since the back-stress X is not restricted, $\forall \bar{\sigma} : X = \bar{\sigma}$ holds. This leads to alternating plasticity in case of $\bar{\sigma} \leq \bar{\sigma}^* = \sqrt{\sigma_H^2 - \sigma_Y^2}$, where the admissible shear stress is τ_u (see case 3).
On the other hand, if $\bar{\sigma} > \bar{\sigma}^*$, the bounding condition enforces incremental collapse independently of the back-stresses. The according bounding condition

is the one given by WGW and Pham:

$$f_H(\boldsymbol{\sigma}, \sigma_H) = \bar{\sigma}^2 + 3\,\tau^2 - \sigma_H^2 = 0 \tag{29}$$

This leads to the following maximum value of shear stresses:

$$\sqrt{3}\,\tau = \sqrt{\sigma_H^2 - \bar{\sigma}^2} \tag{30}$$

As a result, the maximum value of admissible shear stress reads:

$$\sqrt{3}\,\tau_{un} = \begin{cases} \sqrt{3}\,\tau_u = \sigma_Y & \text{if } \bar{\sigma} \leq \bar{\sigma}^* \\ \sqrt{3}\,\tau_L = \sqrt{\sigma_H^2 - \bar{\sigma}^2} & \text{if } \bar{\sigma} > \bar{\sigma}^* \end{cases} \quad \text{(blue solid line with } \circ \text{ in Fig. 2)}$$

It should be mentioned, that the yield surface is allowed to partly move outside of the bounding surface, as long as the considered stress points on the yield surface $(\bar{\sigma}, \tau_{\max})$ and $(\bar{\sigma}, \tau_{\min})$ stay inside.

5. **Limited linear kinematical hardening**:
For the limited linear kinematical hardening, the hardening rule of Prager [29] is applied: $\dot{\boldsymbol{\pi}} = C\,\dot{\boldsymbol{\varepsilon}}^p$, where C denotes the kinematical hardening modulus and $\dot{\boldsymbol{\varepsilon}}^p$ denotes the plastic strain rate. As shown in [14, 38], the back-stresses are restricted by:

$$f(\boldsymbol{\pi}, \sigma_H) = X^2 - (\sigma_H - \sigma_Y)^2 = 0 \tag{31}$$

Hence, the back-stress $X \leq \bar{\sigma}^\circ$ cannot exceed the value $\bar{\sigma}^\circ = \sigma_H - \sigma_Y$. In consequence, alternating plasticity can only occur if $\bar{\sigma} \leq \bar{\sigma}^\circ$, because then $X = \bar{\sigma}$ is possible.
On the contrary, for $\bar{\sigma} > \bar{\sigma}^\circ$, the restriction of the back-stresses leads to $X = \bar{\sigma}^\circ$, which enforces incremental collapse. The resulting maximum value of admissible shear stress for limited linear hardening is:

$$\sqrt{3}\,\tau_P = \begin{cases} \sqrt{3}\,\tau_u = \sigma_Y & \text{if } \bar{\sigma} \leq \bar{\sigma}^\circ \\ \sqrt{\sigma_Y^2 - (\bar{\sigma} - \bar{\sigma}^\circ)^2} & \text{if } \bar{\sigma} > \bar{\sigma}^\circ \end{cases} \quad \text{(dotted line in Fig. 2)}$$

It is worth to mention, that this is in agreement with the solution presented in [3] for a specific nonlinear Prager's rule: $\dot{\boldsymbol{\pi}} = C\,\dot{\boldsymbol{\varepsilon}}^p - (\gamma/C)^2\,X_{eq}^2\,\dot{\boldsymbol{\varepsilon}}^p$. The back-stress corresponding to the stabilized cycle has only to be replaced by $\bar{\sigma}^\circ = C/\gamma$.
Further, it should be mentioned that this is exactly the result which can be obtained by using the theorem by Stein, because (31) obviously corresponds to (21b). This could be expected, since the considered hardening law is based on an associated flow rule.

6. **Limited nonlinear kinematical hardening**:
Finally, for the limited nonlinear kinematical hardening the hardening rule of Armstrong and Frederick [2] is used:

$$\dot{\boldsymbol{\pi}} = \frac{2}{3}\,C\,\dot{\boldsymbol{\varepsilon}}^p - C\,\frac{\boldsymbol{\pi}}{X_\infty}\,\dot{p} \tag{32}$$

where $\dot{p} = \sqrt{\frac{2}{3}\,\dot{\boldsymbol{\varepsilon}}^p : \dot{\boldsymbol{\varepsilon}}^p}$ and $X_\infty = \bar{\sigma}^\circ = \sigma_H - \sigma_Y$.

For this hardening law, the considered example has been intensively investigated by several authors. The first solution has been proposed by Lemaitre and Chaboche [17], followed by Saxcé and coworkers [7], where an analytical solution is derived and verified by an alternative theoretical calculation on the basis of the bipotential approach. In [4, 5] an analytical solution is presented for the plane strain state, which can be transferred to the plane stress state considered here simply by setting $\nu = 0$. Moreover, a numerical implementation is given, which is in perfect agreement with the analytical solution. Finally, Staat and Heitzer [36] obtained lower bound results using a finite element computation with basis reduction. In all these references, it turns out that the back-stresses are restricted even more than in the linear hardening case:

$$X = \bar{\sigma}\,\frac{\sigma_H - \sigma_Y}{\sigma_H} \tag{33}$$

Since $\sigma_H > \sigma_Y$, this leads to $\forall\bar{\sigma} : X < \bar{\sigma}$. Noteworthy, this restriction enforces incremental collapse in the whole loading domain, and alternating plasticity only can occur in case of pure shear, $\bar{\sigma} = 0$. Consequently, a significant influence of the restriction on the back-stresses can be observed, which contradicts some statements presented in literature, see e.g. [1]. The according maximum value of admissible shear stress reads:

$$\sqrt{3}\,\tau_{AF} = \frac{\sigma_Y}{\sigma_H}\sqrt{\sigma_H^2 - \bar{\sigma}^2} \qquad \text{(dash-dot line in Fig. 2)}$$

The results for all of the above mentioned cases are shown in Fig. 2 for an arbitrarily chosen value $\sigma_H/\sigma_Y = 3/2$, where both axes of the plot are scaled to the according shakedown value in the perfectly plastic case, σ_0 and τ_0, respectively.

Noteworthy, the limited linear hardening model [29] predicts a higher shakedown load than the limited nonlinear one [2]. Furthermore, the nonlinear model only predicts failure due to incremental collapse, whereas in the linear case two different regions exist, one of which represents the incremental collapse and the other one represents alternating plasticity. However, both models give a lower value for the shakedown load than the unrestricted one resulting from the optimization problem ($\mathscr{P}^H$), as expected according to the discussion above.

5 Conclusions

Concluding, it could be shown that the lower bound shakedown analysis is well suited to determine the limit states of materials or structures. In particular, formulations have been investigated which addressed: perfectly plastic behavior, unlimited kinematical hardening, limited linear kinematical hardening, and limited nonlinear kinematical hardening. The most important result of these investigations is the fact that an unrestricted hardening law is implicitly applied whenever the back-stresses are not restricted by an according subsidiary constraint in the optimization problem.

Even more, the result of such unrestricted problems gives the highest shakedown factor, which can be obtained by any hardening law. Nonetheless, specific hardening laws can be incorporated as indicated above by formulating additional constraints in the optimization problem. This is still covered by the shakedown theorem and its proof. In that sense, the general character of the theorem itself is not curtailed by the discussion above.

References

1. Abdel-Karim M (2005) Shakedown of complex structures according to various hardening rules. Int J Press Vessel Pip 82(5):427–458
2. Armstrong P-J, Frederick C-O (1966) A mathematical representation of the multiaxial Bauschinger effect. Technical report, C.E.G.B. Report No. RD/B/N907
3. Bodovillé G, De Saxcé GD (2001) Plasticity with nonlinear kinematic hardening: modelling and shakedown analysis by the bipotential approach. Eur J Mech A/Solids 20:99–112
4. Bouby C, De Saxcé G, Tritsch J-B (2006) A comparison between analytical calculations of the shakedown load by the bipotential approach and step-by-step computations for elastoplastic materials with nonlinear kinematic hardening. Int J Solids Struct 43:2670–2692
5. Bouby C, De Saxcé G, Tritsch J-B (2009) Shakedown analysis: comparison between models with the linear unlimited, linear limited and non-linear kinematic hardening. Mech Res Commun 36:556–562
6. Corigliano A, Maier G, Pycko S (1995) Kinematic criteria of dynamic shakedown extended to nonassociate constitutive laws with saturation nonlinear hardening. Redic Accademia Lincei IX 6:55–64
7. De Saxcé GD, Tritsch J-B, Hjiaj M (2000) Shakedown of elastic-plastic structures with nonlinear kinematic hardening by the bipotential approach. In: Weichert D, Maier G (eds) Inelastic analysis of structures under variable loads. Kluwer Academic Publishers, Dordrecht
8. Fuschi P (1999) Structural shakedown for elastic-plastic materials with hardening saturation surface. Int J Solids Struct 36:219–240
9. Groß-Weege J (1997) On the numerical assessment of the safety factor of elastic-plastic structures under variable loading. Int J Mech Sci 39(4):417–433
10. Groß-Weege J, Weichert D (1992) Elastic-plastic shells under variable mechanical and thermal loads. Int J Mech Sci 34:863–880
11. Halphen B, Nguyen QS (1975) Sur les materiaux standards generalises. J Mech 14:39–63
12. Heitzer M (1999) Traglast- und Einspielanalyse zur Bewertung der Sicherheit passiver Komponenten, PhD thesis, Forschungszentrum Jülich, RWTH Aachen, Germany
13. Heitzer M, Pop G, Staat M (2000) Basis reduction for the shakedown problem for bounded kinematical hardening material. J Glob Optim 17:185–200
14. Heitzer M, Staat M, Reiners H, Schubert F (2003) Shakedown and ratcheting under tension-torsion loadings: analysis and experiments. Nucl Eng Des 225:11–26
15. Koiter WT (1960) General theorems for elastic-plastic solids. In: Sneddon IN, Hill R (eds) Progress in solid mechanics. North-Holland Publisher, Amsterdam, pp 165–221
16. König JA (1987) Shakedown of elastic-plastic structures. Elsevier, Amsterdam
17. Lemaitre J, Chaboche J-L (1990) Mechanics of solid materials. Cambridge University Press, Cambridge
18. Mandel J (1976) Adaptation d'une structure plastique ecrouissable et approximations. Mech Res Commun 3:483–488
19. Melan E (1938) Der Spannungszustand eines "Mises-Hencky'schen" Kontinuums bei veränderlicher Belastung. Sitzungsber Akad Wiss Wien, Math-Nat Kl, Abt IIa 147:73–87
20. Melan E (1938) Zur Plastizität des räumlichen Kontinuums. Ing-Arch 9:116–126

21. Nguyen Q-S (2003) On shakedown analysis in hardening plasticity. J Mech Phys Solids 51:101–125
22. Pham DC (2007) Shakedown theory for elastic plastic kinematic hardening bodies. Int J Plast 23:1240–1259
23. Pham DC (2008) On shakedown theory for elastic-plastic materials and extensions. J Mech Phys Solids 56:1905–1915
24. Pham DC, Weichert D (2001) Shakedown analysis for elastic-plastic bodies with limited kinematical hardening. Proc R Soc Lond A 457:1097–1110
25. Pham PT, Vu DK, Tran TN, Staat M (2010) An upper bound algorithm for shakedown analysis of elastic-plastic bounded linearly kinematic hardening bodies. In: Proceedings of the ECCM 2010
26. Polizzotto Castrenze (1986) A convergent bounding principle for a class of elastoplastic strain-hardening solids. Int J Plast 2(4):359–370
27. Polizzotto Castrenze (2010) Shakedown analysis for a class of strengthening materials within the framework of gradient plasticity. Int J Plast 26(7):1050–1069
28. Portier L, Challoch S, Marquis D, Geyer P (2000) Ratchetting under tension-torsion loadings: experiments and modelling. Int J Plast 16:303–335
29. Prager W (1959) An introduction to plasticity. Addison-Wesley, London
30. Pycko S, Maier G (1995) Shakedown theorems for some classes of nonassociative hardening elastic-plastic material models. Int J Plast 11(4):367–395
31. Simon J-W (2013) Direct evaluation of the limit states of engineering structures exhibiting limited, nonlinear kinematical hardening. Int J Plast 42:141–167
32. Simon J-W, Weichert D (2011) Numerical lower bound shakedown analysis of engineering structures. Comput Methods Appl Mech Eng 200:2828–2839
33. Simon J-W, Weichert D (2012) Interior-point method for lower bound shakedown analysis of von Mises-type materials. In: de Saxcé G, Oueslati A, Charkaluk E, Tritsch J-B (eds) Limit states of materials and structures—direct methods 2. Springer, Berlin, pp 103–128
34. Simon J-W, Weichert D (2012) Shakedown analysis of engineering structures with limited kinematical hardening. Int J Solids Struct 49(4):2177–2186
35. Simon J-W, Weichert D (2012) Shakedown analysis with multidimensional loading spaces. Comput Mech 49(4):477–485
36. Staat M, Heitzer M (2002) The restricted influence of kinematical hardening on shakedown loads. In: Proceedings of the WCCM V
37. Stein E, Zhang G, König JA (1992) Shakedown with nonlinear strain-hardening including structural computation using finite element method. Int J Plast 8(1):1–31
38. Stein E, Zhang G, Huang Y (1993) Modeling and computation of shakedown problems for nonlinear hardening materials. Comput Methods Appl Mech Eng 103(1–2):247–272
39. Stein E, Zhang G, Mahnken R, König JA (1990) Micromechanical modelling and computation of shakedown with nonlinear kinematic hardening including examples for 2-D problems. In: Axelard DR, Muschik W (eds) Recent developments of micromechanics. Springer, Berlin
40. Weichert D, Groß-Weege J (1988) The numerical assessment of elastic-plastic sheets under variable mechanical and thermal loads using a simplified two-surface yield condition. Int J Mech Sci 30(10):757–767

Theoretical Basis and a Finite Element Formula for the Direct Calculation of Steady Plastic States

Denis A. Tereshin and Oleg F. Cherniavsky

Abstract Direct methods discussed in the literature address mainly the determination of an elastic shakedown boundary, and insufficient attention has been paid to the practically important case of inelastic steady cyclic structural response. To fill this gap, this paper extends a systematic direct approach to safety factor and cyclic state assessment beyond elastic shakedown, which is applicable for any kind of steady cyclic elastic-plastic deformation. In order to construct the limit state boundaries, the basic theorems of elastic and inelastic shakedown accounting for material hardening are formulated uniformly for all the possible combinations of ratcheting and alternating plasticity. A special emphasis is put on practical material description for each kind of limit state. After the cyclic state under prescribed loading is classified, a direct approach created on the same theoretical framework is employed to determine the evolution of the strain and stress fields over the cycle. Mathematically, this results in a convex constrained optimization problem, which is formulated making use of finite element discretization. The proposed optimization approach is validated through the Bree problem, with the results agreeing well with the analytical solution to the problem.

1 Introduction

Structural low cycle deformation has three stages: the first one is an initial stage with material hardening or softening, the second one is a steady cycle stage and the third one is a tertiary stage leading to crack initiation. The time span of the initial stage and the amount of the correspondent damage strongly depend on the fabrication technique, with material property data being usually unavailable, and the material models being complicated and inaccurate. Therefore, the damage

D.A. Tereshin (✉) · O.F. Cherniavsky
South-Ural State University, Chelyabinsk, Russia
e-mail: denisat75@gmail.com

O.F. Cherniavsky
Engineering Research Centre of the Ural Department of the Russian Academy of Sciences, Yekaterinburg, Russia

P. Fuschi et al. (eds.), *Direct Methods for Limit and Shakedown Analysis of Structures*,
DOI 10.1007/978-3-319-12928-0_5

gained over this stage is usually neglected. At a tertiary stage, material softening caused by the increasing damage prevails. Material behaviour at this stage has been insufficiently investigated, and the unstable deformation processes peculiar to material softening are difficult to be evaluated and analyzed with the existing computational techniques. Practically, the damage yielded at the last stage is usually assessed and predicted using the data obtained by monitoring and calculations. The calculations are conducted under the assumption of the stability of deformation.

In a steady state, material hardening is balanced by softening. Since this stage occupies the most lifetime, the whole structural life can be often predicted correctly knowing the steady cycle parameters. In the framework of the steady cycle model, the stress cycle is supposed to be closed, and plastic strain rates—to repeat exactly each cycle. There are four main types of steady cyclic states [1], with material behaviour and properties being particular to each type. Each type of steady deformation leads to a certain structural limit state and failure mode. The deformation type is determined by the combination of the plastic strain increment $\Delta\varepsilon''$:

$$\Delta\varepsilon'' = \int_0^T \dot{\varepsilon}'' d\tau = \varepsilon''(\tau + T) - \varepsilon''(\tau), \tag{1a}$$

(where $\dot{\varepsilon}''$ denotes the plastic strain rate, τ is time, T is the cycle period) and by the plastic strain range $\delta\varepsilon''$:

$$\delta\varepsilon'' = \max_\tau \varepsilon'' - \min_\tau \varepsilon'' - \left|\Delta\varepsilon''\right| \tag{1b}$$

where τ ranges over one cycle. These states are conveniently classified with a load interaction diagram also known as the Bree diagram, shown in Fig. 1 for a pipe under constant internal pressure and through-wall linear thermal transients [2]:

- elastic shakedown ($\delta\varepsilon'' = 0$, $\Delta\varepsilon'' = 0$) occurs in regions *I* and *II* of the diagram;
- alternating (reversed) plasticity ($\delta\varepsilon'' \neq 0$, $\Delta\varepsilon'' = 0$)—in region *III*;
- incremental plasticity collapse (ratcheting) ($\delta\varepsilon'' = 0$, $\Delta\varepsilon'' \neq 0$)—in region *V*;
- a combination of the last two ($\delta\varepsilon'' \neq 0$, $\Delta\varepsilon'' \neq 0$)—in region *IV*.

It is worth noting that whereas material deformation and damage behaviour for the first three states can be described correctly by the engineering material models, the study to the combined deformation material behaviour is only at the beginning [3]. Another complication is that the engineering material models are usually specific for each deformation type. In particular, the elastic-perfectly plastic material model can represent material behaviour for pure ratcheting, but it does not reflect the essential deformation features in the case of non-infinitesimal strains in alternating plasticity or in combined deformation. So a priory the engineer has to provide the computational software with an appropriate material model before understanding the real structural response, though the response is dependent on loading and can rarely be predicted without computations. To solve this issue, the interaction diagram is useful as its boundaries give the estimates of limit loads and allow determining

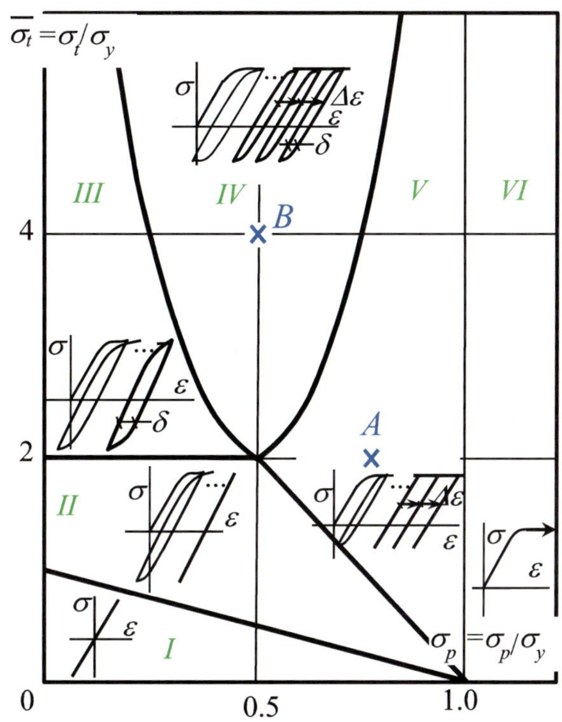

Fig. 1 Load interaction diagram

the type of component deformation. Thus, the problem of steady cycle assessment is twofold: firstly, it is necessary to construct the interaction diagram to find the loading corresponding to the limit states of the structure, and, secondly, the steady structural response under prescribed repeated loading should be computed.

For the first time the general theorems of the existence and uniqueness of cyclic steady state were proved by Frederick and Armstrong [4] using Drucker's postulate and the assumption of material stability. They argue that physical convergence to a steady state is asymptotic and obeys the exponent law, so it was inferred that for steady cyclic state assessment, a computation over only a few first cycles is sufficient. However, the engineering practice evidences that this postulate does not generally hold.

In reality, the deformation convergence to a steady state in excess of shakedown can proceed quite slowly over many tens of cycles. The computations over loading history show that cyclic deformation converges especially slowly in severe alternating flow combined with incremental collapse [1] (zone *IV* in Fig. 1). In such a case the whole stress-strain history is arduous to compute using the step-by-step approach, as not the amount of the degree of freedom but a long loading history becomes the main obstacle. The problem of a slow convergence to a steady state is caused by the fact that the parameters of the exponential law evolve during deformation, and a small change in the cycle parameters between subsequent cycles does not necessarily implies that the state is close to a steady one. Consequently, extrapolation to a period greater than an order of the computed cycle number becomes meaningless especially in combined deformation.

This issue necessitates the development of an approach capable of finding the steady cycle parameters (strain range and increment) directly. Although a variety of efficient procedures for elastic shakedown have already been proposed and implemented employing low and upper bound theorem as well as primal-dual theory [5–8], there are only few approaches being developed for steady plastic response. Among them are the algorithm by Chen and Ponter [9], which captures the plastic shakedown boundary between zones *III* and *IV*, and the technique by Spiliopoulos and Panagiotou [10] employing the residual stress decomposition method to determine structural response for any steady cyclic state.

To find the solution to a steady cycle problem, the latter two approaches iteratively adjust either the elasticity parameters or the residual stress field. Alternatively, the problem of determining a steady plastic response under prescribed cyclic loading can be stated as a convex constrained optimization problem [11] by the unified approach used in this study to derive the shakedown theorem. So the aim of the present study is to extend a mathematical optimization approach combined with finite element discretization to the steady cycle problem so as to create an approach for obtaining the cycle parameters for any point of the interaction diagram.

This work begins with stating the governing equations for a solid plastic body in Sect. 2; then Sect. 3 outlines the rules for material description in steady cyclic states. The individual theorems for separate elastic shakedown boundary determination are formulated in a unified way in Sect. 4, and a similar formula for an arbitrary steady cycle under given loading is deduced in Sect. 5. Finally, the approach is tested with a model problem for different steady cycles in Sect. 6.

2 Governing Equations for a Plastic Body Under Cyclic Loading

Direct steady cycle computations are based on the classical plasticity theory relations complemented by a stress cycle closure condition.

The total stress in a steady cycle is constituted by the fictitious elastic $\boldsymbol{\sigma}^{(e)}$ and residual $\boldsymbol{\rho}$ stresses; the latter includes the initial residual stress $\boldsymbol{\rho}^{\mathrm{o}}$ at the beginning of a cycle and the residual stress increment accumulated since then over time τ:

$$\boldsymbol{\sigma} = \boldsymbol{\sigma}^{(e)} + \boldsymbol{\rho} = \boldsymbol{\sigma}^{(e)} + \boldsymbol{\rho}^{\circ} + \int_{0}^{\tau} \dot{\boldsymbol{\rho}}\, d\tau \tag{2}$$

The total strain consists of the elastic strain $\boldsymbol{\varepsilon}'$, which is caused by the fictitious elastic $\boldsymbol{\sigma}^{(e)}$ and the residual $\boldsymbol{\rho}$ stresses, the plastic $\boldsymbol{\varepsilon}''$ and thermal $\boldsymbol{\varepsilon}'''$ strains:

$$\boldsymbol{\varepsilon} = \boldsymbol{\varepsilon}' + \boldsymbol{\varepsilon}'' + \boldsymbol{\varepsilon}''' = \boldsymbol{\varepsilon}^{(e)} + \mathbf{E}^{-1} : \boldsymbol{\rho} + \boldsymbol{\varepsilon}'' + \boldsymbol{\varepsilon}''', \tag{3}$$

$$\dot{\boldsymbol{\varepsilon}} = \dot{\boldsymbol{\varepsilon}}' + \dot{\boldsymbol{\varepsilon}}'' + \dot{\boldsymbol{\varepsilon}}''' = \dot{\boldsymbol{\varepsilon}}^{(e)} + \mathbf{E}^{-1} : \dot{\boldsymbol{\rho}} + \dot{\boldsymbol{\varepsilon}}'' + \dot{\boldsymbol{\varepsilon}}''' \tag{4}$$

For the sake of simplicity, the elasticity tensor **E** in Eqs. (3) and (4) is assumed to be time independent.

The equilibrium for the fictitious elastic stresses $\boldsymbol{\sigma}^{(e)}$ holds naturally. For the residual stress field $\boldsymbol{\rho}$, equilibrium over a body and the boundary conditions at the part of its surface S_p with prescribed tractions can be stated as follows:

$$\nabla \cdot \boldsymbol{\rho}^{\circ} = \mathbf{0} \ \text{ in } \ V, \qquad \boldsymbol{\rho}^{\circ} \cdot \bar{n} = \mathbf{0} \ \text{ at } S_p, \tag{5}$$

$$\nabla \cdot \dot{\boldsymbol{\rho}} = \mathbf{0} \ \text{ in } V, \qquad \dot{\boldsymbol{\rho}} \cdot \bar{n} = \mathbf{0} \ \text{ at } S_p \tag{6}$$

Since the fields of the total strain rate $\dot{\boldsymbol{\varepsilon}}$ and sum of the elastic fictitious and thermal strain rates in Eq. (4) are compatible, the plastic strain rate $\dot{\boldsymbol{\varepsilon}}''$ field and the strain rate field caused by the residual stresses $\dot{\boldsymbol{\rho}}$ are also compatible to a displacement rate $\dot{\mathbf{u}}$ field at any time. For small strains and displacements this leads to

$$\dot{\boldsymbol{\varepsilon}}'' + \mathbf{E}^{-1} : \dot{\boldsymbol{\rho}} = \frac{1}{2}(\nabla \dot{\mathbf{u}} + (\nabla \dot{\mathbf{u}})^T) \tag{7}$$

For simplicity let us omit the boundary conditions at the body surface where displacements are given.

To account for kinematic and isotropic hardening through the back stress tensor **S** and parameter k respectively, the stress $\boldsymbol{\sigma}^*$ on the yield surface obeys [12]

$$f\left(\boldsymbol{\sigma}^* - \mathbf{S}\right) - k^2 = 0 \tag{8}$$

The actual stress lies inside or on the yield surface:

$$f\left(\boldsymbol{\sigma} - \mathbf{S}\right) - k^2 \le 0 \tag{9}$$

and is associated with the plastic strain by the flow rule:

$$\dot{\boldsymbol{\varepsilon}}'' = \alpha \nabla_{\sigma} f(\boldsymbol{\sigma} - \mathbf{S}), \ \alpha \ \ge \ 0, \tag{10}$$

where plastic multipliers α are determined by the hardening law. The plastic strains can be nonzero only if the stress $\boldsymbol{\sigma}$ is on the yield surface, i.e.

$$\alpha[f\left(\boldsymbol{\sigma} - \mathbf{S}\right) - k^2] = 0 \tag{11}$$

Even if $f(\boldsymbol{\sigma} - \mathbf{S}) - k^2 = 0$ is satisfied, these strains are zero in unloading:

$$\alpha = 0 \ \text{ if } f(\boldsymbol{\sigma} - \mathbf{S}) - k^2 = 0 \ \text{ and } \dot{f}\left(\boldsymbol{\sigma} - \mathbf{S}\right) \le 0 \tag{12}$$

For the determination of $\Delta\varepsilon''$ and $\delta\varepsilon''$, condition (12) is usually not used in the analysis as yield surface is a mathematical model of a complicated phenomenon, and, practically, the parameters of a cycle are expressed through finite increments.

Nonlinear constraints (11) are satisfied at every material point over all times and can be substituted by the requirement that the functional

$$I = \int_{0}^{T} d\tau \int_{V} \left(\boldsymbol{\sigma}^{*} - \boldsymbol{\sigma}\right) : \dot{\boldsymbol{\varepsilon}}'' dv \tag{13}$$

must be zero:

$$I = 0 \tag{14}$$

in which the stress $\boldsymbol{\sigma}^{*}$ satisfies Eq. (8) and relates to the strain $\dot{\boldsymbol{\varepsilon}}''$ through Eq. (10).

If the actual stresses $\boldsymbol{\sigma}$ satisfying (9) were not equal to $\boldsymbol{\sigma}^{*}$ associated with $\dot{\boldsymbol{\varepsilon}}''$ by Eq. (10), it follows from Drucker's postulate that the functional I would take on a positive value. At a solution, within plastic zones, stress histories are uniquely defined, and equality holds at least for those components of $\boldsymbol{\sigma}$ and $\boldsymbol{\sigma}^{*}$ which are associated with $\dot{\boldsymbol{\varepsilon}}''$, so the functional I yields the minimum equal to zero.

The closure condition for a steady cycle may be enforced either in terms of residual stresses by forcing them to repeat every cycle:

$$\Delta\boldsymbol{\rho}(T) = \int_{0}^{T} \dot{\boldsymbol{\rho}} d\tau = 0, \tag{15}$$

or, equivalently, in terms of strains by ensuring the plastic strain field increment $\Delta\boldsymbol{\varepsilon}''$ over a cycle to be compatible with the displacement field increment $\Delta\,\mathbf{u}$:

$$\Delta\boldsymbol{\varepsilon}''(T) = \frac{1}{2}(\nabla(\Delta\mathbf{u}) + (\nabla(\Delta\mathbf{u}))^{T}) \tag{16}$$

Note that making use of Eqs. (5)–(7), (15), functional (13) can be simplified for steady cycles as [11]:

$$I = \int_{0}^{T} d\tau \int_{V} \left(\boldsymbol{\sigma}^{*} - \boldsymbol{\sigma}\right) : \dot{\boldsymbol{\varepsilon}}'' dv = \int_{0}^{T} d\tau \int_{V} \left(\boldsymbol{\sigma}^{*} - \boldsymbol{\sigma}^{(e)}\right) : \dot{\boldsymbol{\varepsilon}}'' dv \tag{17}$$

Thus, the system defining a steady cyclic state is constituted by Eqs. (5)–(10) and (14) complemented by cycle closure condition (15) or (16).

3 Material Behaviour Description

In engineering strength and service life assessments it is important to properly prescribe the relation between the stresses $\boldsymbol{\sigma}^{*}$ (and, consequently, all the stress-strain history in steady cycling) and plastic strains through Eqs. (8) and (10). Although the

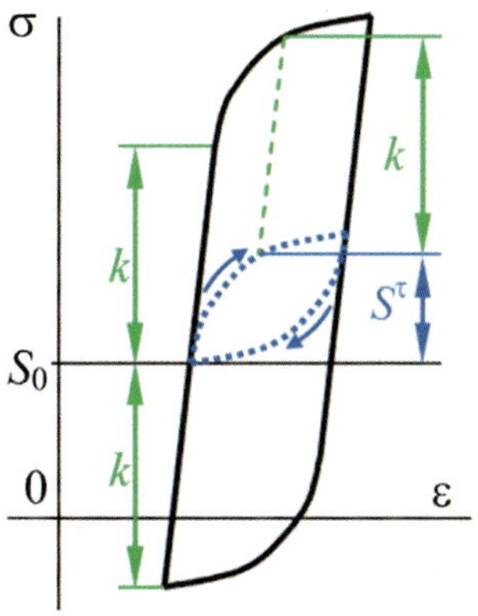

Fig. 2 Isotropic and kinematic hardening description

stresses $\boldsymbol{\sigma}^*$ depend on material hardening at an initial unsteady stage, practically, the material hardening properties can often be supplied in spite of the lack of the data for this stage. This section discusses both the main rules for material description and the tests necessary for material data acquisition. These tests should be conducted in the same environment, temperature range and with the same cyclic pre-conditioning as the structural component.

In cyclic loading, the deviator back stress tensor $\mathbf{S}$ in Eqs. (8)–(12) is conveniently presented as the sum: $\mathbf{S} = \mathbf{S}^0 + \mathbf{S}^\tau$. The first component $\mathbf{S}^0$ determines the origin of a cycle as Fig. 2 shows; it depends on the preceding load history and does not vary in a cycle. The cycle origin corresponds to the center of the yield surface (8), so the parameter k signifies the yield stress. The second term $\mathbf{S}^\tau$ accounts for kinematic hardening during a cycle. The parameters $\mathbf{S}^\tau$ and k depend on such conditions as temperature, environment and the strain history from the beginning of a cycle. Like $\mathbf{S}^0$, the parameters $\mathbf{S}^\tau$ and k can also depend on the loading history preceding the cycle. However in a steady cycle, $\mathbf{S}^0$, $\mathbf{S}^\tau$ and k become independent from the cycle count.

The functions $\mathbf{S}$ and k and the domains of their validity are determined by tests, with the experiments being set for a particular deformation mode and for the purpose of computation.

Any deformation assessment necessitates yield criterion (8) to be particularized for the working conditions experienced by the component. If alternating plasticity is expected, the material properties are determined from strain-controlled cyclic tests (Fig. 3a). The material model should account for the Bauschinger effect, the cyclic hardening or softening gained over the initial unsteady stage and finite (nonzero) tangent stiffness at the tips of the stress-strain loop. Material parameters for an alternating plasticity steady cycle are often independent from the mean stress and the strain accumulated, i.e. from $\mathbf{S}^0$.

In incremental plasticity, material behaves as it does in a steady loading combined with elastic reverse loadings (Fig. 3b). A steady strain accumulation can take place only without strain-hardening in a cycle ($\mathbf{S}^\tau = \mathbf{0}$).

Material properties in combined loading depend on both cyclic and static damages, and, therefore, the material behaviour is affected by interacting alternating plasticity

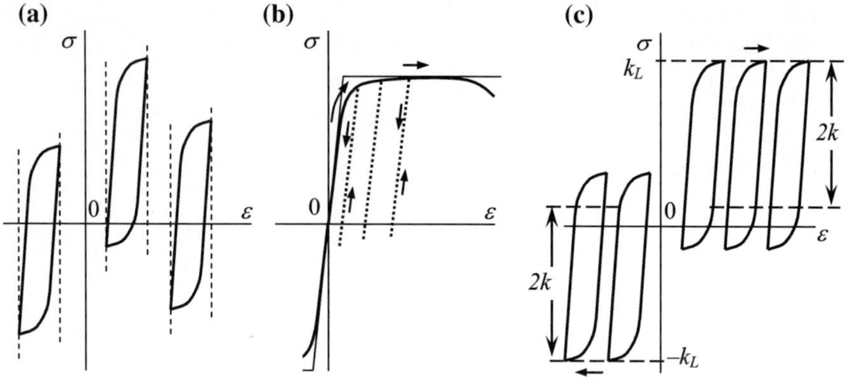

Fig. 3 Deformation behaviour in alternating plasticity (**a**), incremental collapse (**b**) and a combined deformation (**c**)

and ratcheting. In a steady cyclic state this complicated phenomenon can be described in a simpler way through hysteresis loop parameters and the loop tip locus curve defined through parameter k_L as shown in Fig. 3c.

The experimental studies conducted for a number of steels and alloys [13] show that if the static damage is less than the cyclic one, ratchet strain influence on a steady cyclic state can often be neglected. This fact supports the hypothesis that under some restrictions the function $\mathbf{S}^{\tau}$ does not depend on the strain accumulated in the cycle and over the previous deformation. However, the horizontal locus line of loop tips does depend on cyclic hardening and softening, with the line location being impossible to be determined solely from monotonous loading tests except for cyclically stable materials.

The material model for a particular cyclic limit state created according to the above reasoning should be supplied individually to the correspondent formula deduced in the following sections.

4 Cyclic Limit State Theorems

The individual formulae for different cyclic limit states derived in this section provide theoretical ground for constructing an interaction diagram. With the interaction diagram one can infer the type of the steady cyclic state under given loading and assess the safety factors against the limit states. In the sequel, the cyclic limit state theorems are uniformly derived from the complete set of governing relations (Sect. 2), with the formulae making use of the material models according to Sect. 3.

Let us parameterize each load with its own parameter. These parameters define the combined load for any instant and specify the path to a limit state. As the parameters are introduced for convenience, let us define the parameter n_i for the ith load as a factor for the corresponding elastic fictitious stress field. In computations for all the limit state boundaries, without loss of generality one can suppose that all the factors are specified except for one denoted as n_p.

4.1 Individual Theorems for the Elastic Shakedown Boundaries

Let us start from the formulae for elastic shakedown boundary delimiting region *II* from regions *III–V* in the interaction diagram (Fig. 1). These formulae are deduced individually for reversed plasticity and incremental collapse.

4.1.1 The Formulae for Alternating Plasticity

The boundary between elastic shakedown region *II* and alternating plasticity one *III* in Fig. 1 can be found either from the condition that n_p attains maximum without alternating plasticity, or that n_p is minimal subject to the strain range $\delta\boldsymbol{\varepsilon}''$ is nonzero. In both statements there must be no strain increment over the cycle. The first argument leads to the lower bound statement, and the second—to the upper bound:

$$n_p^- = \max n_p \quad \text{if } \delta\boldsymbol{\varepsilon}'' = 0,\ \Delta\boldsymbol{\varepsilon}'' = 0,\ \text{i.e. } \dot{\boldsymbol{\varepsilon}}'' \equiv 0, \tag{18}$$

$$n_p^+ = \min n_p \quad \text{subject to } \delta\boldsymbol{\varepsilon}'' \neq 0,\ \Delta\boldsymbol{\varepsilon}'' = 0 \tag{19}$$

Let us eliminate the redundant relations form the system of Eqs. (5)–(10), (14), and (15) or (16) for a steady cycle.

As according to Eq. (18) there is no plastic deformation, Eqs. (6), (7), (10), (14) and (16) are satisfied trivially, and, therefore, can be excluded from the system of constraints. Plastic deformation in the limit state of alternating plasticity is local: only the most stressed point or some equally stressed points of the body undergo plasticity, whereas the other body part forms an elastic core which controls the deformation. According to the solution uniqueness statement [4], the residual stresses are unique only in plastic zones. Therefore, one can always find a residual stress field that satisfies Eq. (5). Consequently, this constraint is also redundant. The plastic strain range of a loop vanishes, so deformation hardening is negligible and $\mathbf{S}^{\tau}$ becomes zero as well. Denoting the result of combining $\boldsymbol{\rho}^0$ and a time-independent part of the fictitious stresses as $\boldsymbol{\sigma}^{\circ}$, the only nontrivial constraint (9) for problem (18) can be rewritten as:

$$f\left(\boldsymbol{\sigma}_{\tau}^{(e)} + \boldsymbol{\sigma}^{\circ} - \mathbf{S}^0\right) - k^2 \leq 0 \tag{20}$$

where $\boldsymbol{\sigma}_{\tau}^{(e)}$ is the time-dependent part of $\boldsymbol{\sigma}^{(e)}$, k is the proportional elastic limit (in an initial assessment can be taken as the yield stress) obtained from strain controlled cyclic tests (Fig. 3a). Thus, *steady alternating plasticity cannot take place if there exists any permanent stress field* $\boldsymbol{\sigma}^{\circ}$, *the sum of which with the time dependent elastic fictitious stresses* $\boldsymbol{\sigma}_{\tau}^{(e)}$ *yields a plastically admissible stress state, i.e. Eq.* (20) *is satisfied with the appropriate* $\mathbf{S}^0$ *and k for any instant.*

The isotropic hardening gained over an initial stage is usually small except for the case of special hardening technologies. If the kinematic hardening parameter $\mathbf{S}^0$ does not affect the isotropic hardening parameter k, which is common, e.g. for isothermal deformation in a certain strain range or for a small strain-hardening, the back stress $\mathbf{S}^0$ in Eq. (20) can also be combined with $\boldsymbol{\sigma}^{\mathrm{o}}$ so as not to be accounted for as a separate variable. In such a case the initial stage gives rise to a residual stress field that makes the loop at the most stressed points of the body symmetric.

Let us extend the upper bound given by Eq. (19) by considering along with the given structure (which has the yield stress k and the stress on the yield surface $\boldsymbol{\sigma}^*$) the hardened structures, which have greater yield stresses nk ($n \geq 1$) and the correspondent stresses on the yield surface $n\boldsymbol{\sigma}^*$, with n being allowed to be distributed nonuniformly over the body. The solution to such relaxed problem is obviously the same as for the original one, with the problem extension allowing the elimination of some constraints. Since one can form the extended problem so that Eqs. (5)–(7), (9) are always satisfied, and do not affect the actual solution, these constraints are not essential for the formula and can be excluded. The associate flow rule by Eqs. (8) and (10) does not change. Taking into account that plastic flow in the limit state of alternating plasticity is instantaneous and local in the body, cycle closure condition (16) and functional (17) in Eq. (14) can be rewritten as the sums:

$$\Delta\boldsymbol{\varepsilon} = \sum_{l=1}^{m} \dot{\boldsymbol{\varepsilon}}''_l = 0, \tag{21}$$

$$\sum_{l=1}^{m} \left(n\boldsymbol{\sigma}^*_l - \boldsymbol{\sigma}^{(e)}_l \right) : \dot{\boldsymbol{\varepsilon}}''_l = 0, \tag{22}$$

where m stands for the number of time instants yielding occurs in a cycle. As Drucker's postulate leads to the inequality $(n-1)\,\boldsymbol{\sigma}^*_l : \dot{\boldsymbol{\varepsilon}}''_l \geq 0$, condition (22) finally takes the following form:

$$\sum_{l=1}^{m} \left(\boldsymbol{\sigma}^*_l - \boldsymbol{\sigma}^{(e)}_l \right) : \dot{\boldsymbol{\varepsilon}}''_l \leq 0 \tag{23}$$

Thus, the constraint system of upper bound (19) includes only Eqs. (8), (10), (21) and (23), which means that *a steady alternating flow will happen if there exists a plastic mechanism satisfying these constraints.*

Alternating plasticity limit state formula given by Eqs. (18)–(23) leads to a primal-dual problem of quadratic programming. This algorithm is implemented in the software described in [1], which can be used jointly with the conventional FEM software capable of conducting elastic analysis.

It is worth noting that a disproportional change of the stress components in a cycle leading to a curvilinear stress path and yielding more than twice per cycle can seriously influence the onset of alternating flow. So against the widespread opinion, alternating yielding can start if the largest elastic fictitious stress component has the range only of 1.5 yield stress in plain stress state and of 1 yield stress in triaxial stress state [1].

4.1.2 The Formulae for Incremental Collapse

Let us make similar formulations for the boundary between an elastic shakedown region (area *II* in Fig. 1) and a steady incremental collapse region (area *V*) employing the same approach:

$$n_p^- = \max n_p \quad \text{if } \delta\boldsymbol{\varepsilon}'' = 0,\ \Delta\boldsymbol{\varepsilon}'' = 0,\ \text{i.e. } \dot{\boldsymbol{\varepsilon}}'' \equiv 0, \tag{24}$$

$$n_p^+ = \min n_p \quad \text{subject to } \delta\boldsymbol{\varepsilon}'' = 0,\ \Delta\boldsymbol{\varepsilon}'' \neq 0 \tag{25}$$

For a correct estimate, the yield stress k_L in computations should be taken as the ultimate stress of the material [1, 11]. Lower bound statement (24) differs from problem (18) because displacements are accumulated in the process of nonlocal plasticity; therefore, Eq. (5) must be retained in the constraint system. This leads to the static formulation: *incremental collapse cannot take place if there exists any time-independent self-equilibrated stress field* $\boldsymbol{\rho}^o$ *the superposition of which with the elastic fictitious stress field* $\boldsymbol{\sigma}^{(e)}$ *yields a plastically admissible stress state for any instant.*

In kinematical counterpart (25) cycle closure condition (16) is retained, and Eq. (17) is transformed like (22) to

$$\sum_{l=1}^{m} \int_V \left(\boldsymbol{\sigma}_l^* - \boldsymbol{\sigma}_l^{(e)}\right) : \dot{\boldsymbol{\varepsilon}}_l'' dv \leq 0 \tag{26}$$

So the upper bound statement reads: *a steady incremental collapse will happen if there exists a plastic mechanism satisfying* Eqs. (8), (10), (16), (25) *and* (26).

Had the limit states not been differentiated, the results obtained would coincide with the classical Melan and Koiter theorems [14] and be true only for an elastic-perfectly plastic material with strain-independent yield stress.

4.2 Individual Theorems for Inelastic Shakedown Boundaries

The upper boundaries of pure alternating plasticity or incremental collapse regions delimit these regions from the region of combined steady deformation (region *IV* in Fig. 1). The formulation deduction in this section is made by reasoning similar to that of Sect. 4.1

4.2.1 The Formulae for the Upper Boundary of Alternating Plasticity

The upper boundary of reversed plasticity delimits regions *III* and *IV* in Fig. 1 and corresponds to the onset of ratcheting against the background of alternating plasticity.

Material behaviour is described by means of the functions $\mathbf{S}^{\tau}$ and k obtained from tests under different strain ranges. In addition, the horizontal locus line of the loop tips is defined by the parameter k_L dependent on both cyclic and static damages [13] as Fig. 3c shows.

Assume that alternating plasticity computations for material with a certain yield stress k have been performed for any loading by enforcing stain loop to be closed (see Fig. 3a), i.e. excluding the possibility of ratchet strain, to obtain the *reversed plasticity stresses* $\boldsymbol{\sigma}^{(a)}$. The stresses $\boldsymbol{\sigma}^{(a)}$ are composed of the fictitious elastic stresses $\boldsymbol{\sigma}^{(e)}$ and the residual stresses $\boldsymbol{\rho}^{(a)}$:

$$\boldsymbol{\sigma}^{(a)} = \boldsymbol{\sigma}^{(e)} + \boldsymbol{\rho}^{(a)} \tag{27}$$

With this decomposition for a limit cycle, in which ratcheting begins by the existing alternating plasticity, the optimization formulae differ from problems (24) and (25) only by the additional time-dependent self-equilibrated stress $\boldsymbol{\rho}^{(a)}$ term. So one has the lower bound formulation:

$$\begin{aligned} n_p^- &= \max n_p \\ \text{s.t.} &\left| \begin{array}{l} f(\boldsymbol{\sigma}^{(e)} + \boldsymbol{\rho}^{\circ} + \boldsymbol{\rho}^{(a)} - \mathbf{S}^0) - k_L^2 \le 0 \\ \nabla \cdot \boldsymbol{\rho}^{\circ} = \mathbf{0} \ \text{ in } V, \qquad \boldsymbol{\rho}^{\circ} \cdot \overline{n} = \mathbf{0} \ \text{ at } S_p \end{array} \right. \end{aligned} \tag{28}$$

which leads to the statement: *incremental collapse cannot take place if there exists a time-independent self-equilibrated stress field* $\boldsymbol{\rho}^{o}$*, the sum of which with the reversed plasticity stresses* $\boldsymbol{\sigma}^{(a)}$ *yields an admissible stress state for any instant.*

The upper bound formula is similar to (25) except for the fact that alternating plasticity is allowed:

$$\begin{aligned} n_p^+ &= \min n_p \\ \text{s.t.} &\sum_{l=1}^{m} \int_V \left(\boldsymbol{\sigma}_l^* - \boldsymbol{\sigma}_l^{(a)} \right) : \dot{\boldsymbol{\varepsilon}}_l'' dv \le 0 \end{aligned} \tag{29}$$

also subject to Eqs. (8), (10), and (16), in which k_L is used instead of k. This results in that *a steady incremental collapse will happen if there exists a plastic mechanism satisfying all the constraints mentioned.*

4.2.2 The Formulae for the Upper Boundary of Incremental Collapse

The upper boundary of incremental collapse separates regions *IV* and *V* in Fig. 1. Similarly to the case when alternating plasticity shakes down, if alternating plasticity is at the onset in a state of pronounced ratchet straining, the material does not usually cyclically strain-harden or soften noticeably over an initial deformation stage before a cyclic steady deformation.

At first, computations for incremental collapse should be made for a range of loads for the determination of the *incremental plasticity stresses* $\boldsymbol{\sigma}^{(p)}$. In these computations, as Fig. 3b shows, yielding at reverse half-cycles is suppressed by making k great enough as compared to k_L (see Fig. 3c), with k_L being taken equal to the ultimate stress [11]. Denoting the residual part of the stresses $\boldsymbol{\sigma}^{(p)}$ in these solutions as $\boldsymbol{\rho}^{(p)}$, the lower bound formulation reads: *alternating plasticity cannot take place if there exists such self-equilibrated stress field* $\boldsymbol{\rho}^{\text{o}}$ *that its sum with the incremental plasticity stress* $\boldsymbol{\sigma}^{(p)}$ *represents an admissible stress state for any instant*:

$$
\begin{aligned}
n_p^- &= \max n_p \\
\text{s.t.} &\left| \begin{array}{l} f(\boldsymbol{\sigma}^{(e)} + \boldsymbol{\rho}^\circ + \boldsymbol{\rho}^{(p)} - \mathbf{S}^0) - k^2 \leq 0 \\ \nabla \cdot \boldsymbol{\rho}^\circ = \mathbf{0} \ \text{ in } V, \qquad \boldsymbol{\rho}^\circ \cdot \overline{n} = \mathbf{0} \ \text{ at } S_p \end{array} \right.
\end{aligned}
\tag{30}
$$

As in the determination of the alternating plasticity bound (Sect. 4.1.1), the parameter k in Eq. (30) is determined by the proportionality limit, or approximated by the yield stress, obtained in a steady loading test. As opposed to the elastic fictitious stresses $\boldsymbol{\sigma}^{(e)}$, the stresses $\boldsymbol{\sigma}^{(p)}$ in problem (30) similarly to the stresses $\boldsymbol{\sigma}^{(a)}$ in formulae (28), (29) are nonlinearly dependent on the loads.

The upper bound formula takes the form:

$$
\begin{aligned}
n_p^+ &= \min n_p \\
\text{s.t.} &\left| \begin{array}{l} \sum\limits_{l=1}^{m} \left(\boldsymbol{\sigma}_l^* - \boldsymbol{\sigma}_l^{(p)} \right) : \dot{\boldsymbol{\varepsilon}}_l'' \leq 0 \\ \Delta \boldsymbol{\varepsilon}'' = 0 \end{array} \right.
\end{aligned}
\tag{31}
$$

also subject to Eqs. (8) and (10). It argues that *an alternating plasticity will happen if there exists a plastic mechanism satisfying all the constraint system of* Eq. (31).

5 Formula for a Steady Cyclic State Under Prescribed Loading

In the unified approach to cyclic strength and service life assessment, a direct technique for a steady cycle can serve as a tool, firstly, for pre-computation $\boldsymbol{\sigma}^{(a)}$ and $\boldsymbol{\sigma}^{(p)}$ necessary for bound construction in Sects. 4.2.1 and 4.2.2, and, secondly, for actual structural steady cyclic response determination under given cyclic loading.

A formula for a steady cyclic state can be derived by substitution for one of the steady cycle relations listed in Sect. 2 by a specific functional, which recasts the problem in an optimization form. There are different ways to write the steady cycle problem in an optimization form making use of extremum principles in mechanics [14]. One possibility is in the substitution of equality (14) by the condition of minimization of the same functional (17). The functional is subject to all the other steady cycle relations unchanged and attains zero at the exact minimizer:

$$I^+ = \min I \tag{32}$$

subject to Eqs. (5)–(10) and (15) or (16).

The mechanical sense of such relaxation is that by rejecting Eq. (11) the plastic strain rates $\dot{\boldsymbol{\varepsilon}}''$ are allowed to be nonzero even for the stresses $\boldsymbol{\sigma}$ being not on the yield surface. Condition (11) is satisfied at the exact minimizer of problem (32). A simple technique employing this formula and finite element discretization for numerical estimations of a steady cycle is considered hereafter in Sect. 6.

As the solutions to the model problems of trusses and frames have demonstrated and the test problem in Sect. 6 confirms, one of the advantages of such approaches is that the cycle parameters $\delta\ \varepsilon''$, $\Delta\ \varepsilon''$ are accurately estimated with only 4–5 time points per cycle, even though the strain rate history may deviate notably.

It is worth to stress that cyclic plastic deformation can sometimes be unstable and lead either to structural buckling [1], or to non-uniform deformation and material redistribution. These issues are not addressed in this study.

6 Direct Finite Element Computations of a Steady Cycle

This section validates formula for a steady cycle (32) by considering the classical problem [2] of a pressurized thin-walled cylinder subject to cyclic through-wall thermal transients making use of the analytical solution to this problem.

The resulting convex constrained optimization problem is stated using finite element discretization. In this simple optimization approach the equality constraints are enforced by means of quadratic penalties; and the inequality ones by logarithmic barriers. The influence of time discretization step on the result accuracy is studied.

6.1 Temporal and Spatial Discretization

Let us make time discretization as Fig. 4 shows, where time instant l is in the range from the beginning of a cycle ($l = 0$) to its end ($l = m$), with both coinciding by the fact the cycle is closed. Hereafter, the plastic strain $\boldsymbol{\varepsilon}''_l$, displacement $\mathbf{u}_l$, residual stress $\boldsymbol{\rho}_l$ ($l = \overline{0, m}$) have the meaning of increments, whereas $\boldsymbol{\rho}_0$ is the initial residual stress, and the elastic stress $\boldsymbol{\sigma}_l^{(e)}$ corresponds to the lth instant.

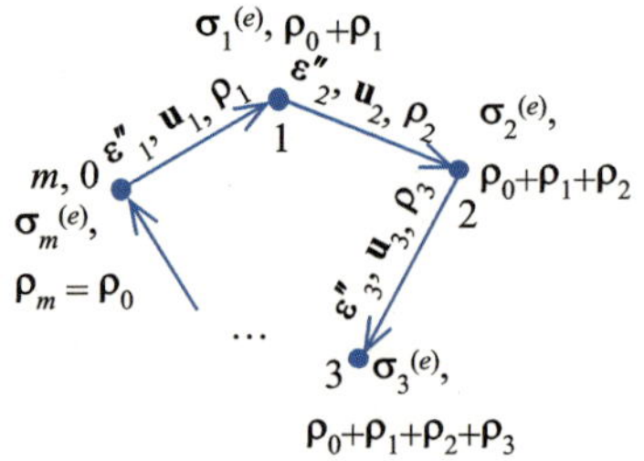

Fig. 4 Time discretization

Employing finite element spatial discretization with the deformation matrix **B**, the self-equilibrium condition (5), (6) for the initial residual stress is formulated as:

$$\mathbf{B}^T \boldsymbol{\rho}_0 \equiv \sum_{i=1}^{NG} w_i \mathbf{B}_i^T \boldsymbol{\rho}_{0i} = \mathbf{0}, \tag{33}$$

where NG is the total number of the Gauss integration points; w_i is the integration weight of point i. From now on all vectors and matrices indexed by i or j imply the local ones for the ith or jth point, whereas the matrices not indexed—the global ones. For example, Hook's relation between the stress and strain for point i at time instant l is $\boldsymbol{\sigma}_{il}^{(e)} = \mathbf{E}_i \boldsymbol{\varepsilon}_{il}^{(e)}$, with $\boldsymbol{\varepsilon}_{il}^T = \{\varepsilon_x^{il} \quad \varepsilon_y^{il} \quad \varepsilon_z^{il} \quad 2\varepsilon_{xy}^{il} \quad 2\varepsilon_{yz}^{il} \quad 2\varepsilon_{zx}^{il}\}$, but the global relation is $\boldsymbol{\sigma}_l^{(e)} = \mathbf{E}\boldsymbol{\varepsilon}_l^{(e)}$.

In place of Eq. (7), one finds the displacements $\mathbf{u}_l$ induced by the plastic strain increment $\boldsymbol{\varepsilon}_l''$ by solving the FE problem with the global stiffness matrix **K**:

$$\mathbf{u}_l = \mathbf{K}^{-1} \sum_{i=1}^{NG} \left(w_i \mathbf{B}_i^T \mathbf{E}_i \boldsymbol{\varepsilon}_{il}'' \right) \tag{34}$$

and then restores the residual stresses:

$$\boldsymbol{\rho}_l = \mathbf{E}\left(\mathbf{B}\mathbf{u}_l - \boldsymbol{\varepsilon}_l''\right) \tag{35}$$

The total stress evolves through the cycle points discretely as Fig. 4 illustrates:

$$\boldsymbol{\sigma}_{il} = \boldsymbol{\sigma}_{il}^{(e)} + \boldsymbol{\rho}_{i0} + \sum_{h=1}^{l} \boldsymbol{\rho}_{ih} \tag{36}$$

The cycle closure condition (15) takes the form of

$$\sum_{l=1}^{m} \boldsymbol{\rho}_{il} = \mathbf{0} \tag{37}$$

and the stress admissibility condition (9) is required to be satisfied:

$$f(\boldsymbol{\sigma}_{il}) \le k^2, \quad l = \overline{1, m}, \quad i = \overline{1, NG} \tag{38}$$

Accepting the von Mises yield criterion the functional I in Eq. (32) is written as:

$$I = \sum_{l=1}^{m} \sum_{i=1}^{NG} \left(\sqrt{2/3} w_i k \sqrt{\boldsymbol{\varepsilon}_{il}^{''T} \mathbf{D}_i \boldsymbol{\varepsilon}_{il}''} - w_i \boldsymbol{\sigma}_{il}^{eT} \boldsymbol{\varepsilon}_{il}'' \right) \tag{39}$$

where $\mathbf{D}_i = Diag\{1 \quad 1 \quad 1 \quad 1/2 \quad 1/2 \quad 1/2\}$.

Thus, the mathematical optimization problem is stated as: $\min_{\boldsymbol{\varepsilon}'', \boldsymbol{\rho}, \mathbf{u}} I$ subject to Eqs. (33)–(38) for $l = \overline{1, m}$, $i = \overline{1, NG}$, in addition to which, one must impose the incompressibility constraints on the plastic strains by making the strain deviatoric:

$$\mathbf{D}_{Vi}\boldsymbol{\varepsilon}''_{il} = \mathbf{0}, \quad l = \overline{1,m}, \quad i = \overline{1,NG} \tag{40}$$

where $\mathbf{D}_{Vi}$ is the diad projecting to tensor's spherical part: $\mathbf{D}_{Vi} = 1/3\,(1\ 1\ 1\ 0\ 0\ 0)^T$ $(1\ \ 1\ \ 1\ \ 0\ \ 0\ \ 0)$.

For eliminating displacements, by substitution Eqs.(34)–(35) one gets to:

$$\boldsymbol{\rho}_l = -\mathbf{E}\mathbf{P}\boldsymbol{\varepsilon}''_l, \quad l = \overline{1,m}, \tag{41}$$

with the matrix $\mathbf{P} = \mathbf{I} - \mathbf{B}\mathbf{K}^{-1}\int_V \mathbf{B}^T\mathbf{E}dV$ projecting an arbitrary strain field to the subspace of self-equilibrated strain fields.

For convenience, let us accept the notation from Vu and Yan et al. [7], which simplifies energy relations: $\mathbf{e}_{il} = w_i\mathbf{D}^{1/2}\boldsymbol{\varepsilon}''_{il}$—for strain increment vector; $\hat{\mathbf{B}}_i = w_i\mathbf{D}_i^{1/2}\mathbf{B}_i$—for deformation matrix, so that $\sum_{l=1}^{m}\mathbf{e}_{il} = \hat{\mathbf{B}}_i\mathbf{u}$; $\mathbf{t}_{il} = \mathbf{D}_i^{-1/2}\boldsymbol{\sigma}_{il}^{(e)}$ and $\boldsymbol{\beta}_{il} = \mathbf{D}_i^{-1/2}\boldsymbol{\rho}_{il}$—for stress vectors. In these definitions $\mathbf{D}_i^{-1/2}$ and $\mathbf{D}_i^{1/2}$ are diagonal symmetric matrices such that: $\mathbf{D}_i^{-1/2} = (\mathbf{D}_i^{1/2})^{-1}$ and $\mathbf{D}_i = \mathbf{D}_i^{1/2}\mathbf{D}_i^{1/2}$.

Von Mises stress admissibility condition (38) for point i and instant l can be presented in terms of the Euclidean norm:

$$\left\| (\mathbf{I}_i - \mathbf{D}_{Vi})\left(\mathbf{t}_{il} + \boldsymbol{\beta}_{i0} + \sum_{h=1}^{l}\boldsymbol{\beta}_{ih}\right)\right\| \le \sqrt{2/3}k \tag{42}$$

Let us define the matrix $\mathbf{G}$, a component $\mathbf{G}_{ij}$ of which relates plastic strain at point j to the induced residual stress at point i:

$$\boldsymbol{\beta}_{il} = \mathbf{G}_{ij}\mathbf{e}_{jl}, \quad l = \overline{1,m}, \quad i,j = \overline{1,NG} \tag{43}$$

where $\mathbf{G}_{ij} = \mathbf{D}_i^{-1}\mathbf{E}_{\_wi}\hat{\mathbf{B}}_i\mathbf{K}^{-1}\hat{\mathbf{B}}_j^T\mathbf{E}_{\_wj}\mathbf{D}_j^{-1} - \mathbf{D}_i^{-1/2}\mathbf{E}_{\_wi}\mathbf{D}_j^{-1/2}\delta_{ij}$ ($\delta_{ij} = 1$ if $i = j$, $\delta_{ij} = 0$ if $i \ne j$; $\mathbf{E}_{\_w_i} = (1/w_i)\mathbf{E}_i$). Then the problem formula simplifies to:

$$\begin{aligned}
&I^- = \min_{\mathbf{e},\,\boldsymbol{\beta}_0} \sum_{i=1}^{NG}\sum_{l=1}^{m}\left(\sqrt{2/3}k\sqrt{\mathbf{e}_{il}^T\mathbf{e}_{il} + \varepsilon_0^2} - \mathbf{t}_{il}^T\mathbf{e}_{il}\right) && \text{(a)}\\
&\text{s.t.}\ \left\|\begin{aligned}
&\sum_{i=1}^{NG}\hat{\mathbf{B}}_i^T\boldsymbol{\beta}_{i0} = \mathbf{0} && \text{(b)}\\
&\sum_{j=1}^{NG}\mathbf{G}_{ij}\sum_{l=1}^{m}\mathbf{e}_{jl} = \mathbf{0} && \text{(c)}\\
&\mathbf{D}_{Vi}\mathbf{e}_{il} = \mathbf{0} && \text{(d)}\\
&\bar{f}_{il} = \left(\boldsymbol{\beta}_{i0} + \sum_{j=1}^{NG}\mathbf{G}_{ij}\sum_{h=1}^{l}\mathbf{e}_{jh} + \mathbf{t}_{il}\right)^T(\mathbf{I}_i - \mathbf{D}_{Vi}) && \text{(e)}\\
&\quad\times\left(\boldsymbol{\beta}_{i0} + \sum_{j=1}^{NG}\mathbf{G}_{ij}\sum_{h=1}^{l}\mathbf{e}_{jh} + \mathbf{t}_{il}\right) - (2/3)\,k^2 \le 0
\end{aligned}\right.
\end{aligned} \tag{44}$$

where all the constraints are satisfied for $i = \overline{1, NG}$, $l = \overline{1, m}$, and the first constraint ensures initial residual stress self-balance, the second—cycle closure, the third—plastic incompressibility conditions, and the last one replaces Eq. (42).

According to Vu and Yan et al. [7] a small regularization parameter ε_0 was introduced to Eq. (44) in order to make it differentiable at $\mathbf{e}_{il} = \mathbf{0}$.

6.2 Simple Computational Approach

The conditions of plastic incompressibility (44d), cycle closure (44c) and initial residual stress self-equilibrium (44b) are enforced by means of quadratic penalties, which with Eq. (44a) make up an extended object function:

$$
\begin{aligned}
f_0 = & \sum_{i=1}^{NG}\sum_{l=1}^{m}\left(\sqrt{2/3}k\sqrt{\mathbf{e}_{il}^T\mathbf{e}_{il}+\varepsilon_0^2}-\mathbf{t}_{il}^T\mathbf{e}_{il}\right)+\tfrac{c_1}{2}E\sum_{i=1}^{NG}\sum_{l=1}^{m}\mathbf{e}_{il}^T\mathbf{D}_{Vi}\mathbf{e}_{il} \\
& +\frac{c_2}{2E}\sum_{i=1}^{NG}\sum_{j=1}^{NG}\left(\sum_{l=1}^{m}(\mathbf{e}_{il})^T\left(\sum_{g=1}^{NG}\mathbf{G}_{gi}^T\mathbf{G}_{gj}\right)\sum_{l=1}^{m}(\mathbf{e}_{jl})\right) \\
& +\frac{c_3}{2EL_e}\sum_{i=1}^{NG}\sum_{j=1}^{NG}\left(\boldsymbol{\beta}_{i0}^T\hat{\mathbf{B}}_i\hat{\mathbf{B}}_j^T\boldsymbol{\beta}_{j0}\right)
\end{aligned} \quad , \tag{45}
$$

where c_1, c_2, c_3 are penalty coefficients, L_e—the typical element size, E—Young modulus.

An unconstrained mathematical optimization problem is formulated using logarithmic barrier functions for inequality constraints (44e):

$$
\min_{\mathbf{e},\ \boldsymbol{\beta}_0} F_0, \tag{46}
$$

in which $F_0 = f_0 - (1/t)\sum_{i=1}^{NG}\sum_{l=1}^{m}\log(-\bar{f}_{il})$.

Having denoted the vector collecting all the inequality constraint functions $\bar{f}_{il}$ as $\bar{\mathbf{f}}$, one can write the stationarity (minimum) condition of F_0 as a central path condition:

$$
\begin{aligned}
& \nabla f_0 + \sum_{i=1}^{NG}\sum_{l=1}^{m}\lambda_{il}\nabla\bar{f}_{il} = \mathbf{0} \\
& -\mathbf{diag}(\boldsymbol{\lambda})\bar{\mathbf{f}} - (1/t)\mathbf{I} = \mathbf{0}
\end{aligned} \tag{47}
$$

which from another viewpoint is the modified Karush-Khun-Tucker equations of the problem of functional (45) minimization subject to Eq. (44e) [15], with the gradients taken with respect to the primal ($\mathbf{e}$, $\boldsymbol{\beta}_0$) and dual $\boldsymbol{\lambda}$ variables. So, solving system (47) one applies a simple primal-dual interior-point approach; and the greater the parameter t, the closer the solution of Eq. (47) to the solution to the original problem given by Eq. (44).

However, before solving Eq. (47), one has to find by means of a phase I method a feasible point $(\mathbf{e}, \boldsymbol{\beta}_0)$ satisfying all the inequality constraints (44e), e.g.:

$$\begin{aligned} &\min_{\mathbf{e},\,\boldsymbol{\beta}_0} S \\ &\text{s.t. } \bar{f}_{il}(\mathbf{e}_{il}, \boldsymbol{\beta}_{i0}) \le S, \quad i = \overline{1, NG}, \quad l = \overline{1, m}, \end{aligned} \tag{48}$$

The solution process is initialized by setting S to make all the constraints in Eq. (48) strictly satisfied and proceeds till the parameter S is reduced below zero, which means all the inequality constraints (44e) are strictly satisfied.

Equation (47) is solved by means of Newton's method using the analytical expressions of the first and second derivatives to obtain the primal $(\Delta\ \mathbf{e}, \Delta\boldsymbol{\beta}_0)$ and dual $\Delta\boldsymbol{\lambda}$ variable increments. At the second stage, exact line search was employed, since without line search, Newton's method experiences convergence difficulties when $||\mathbf{e}_{il}|| \ \geq \ \varepsilon_0$. The line search procedure uses the extended object function F_0 as a merit function and ensures that $\boldsymbol{\lambda} \ \succ \ \mathbf{0}$ and the incremented value of $(\mathbf{e}, \boldsymbol{\beta}_0, \boldsymbol{\lambda})$ is feasible. Basing on the current value of the surrogate duality gap $\eta = \bar{\mathbf{f}}^{\mathrm{T}}\boldsymbol{\lambda}$ the parameter t is determined in each iteration as: $t = \mu \cdot NG \cdot m/\eta$; where the greater parameter μ is, the more aggressively t increases.

6.3 Test Problem

The analytical solution to the problem of pressurized thin-walled tube made of elastic-perfectly plastic material under repeated thermal loading [2] was used to prove the optimization formulation (44) is capable of yielding the solution. If a Tresca yield condition is implied, this problem becomes equivalent to the plain stress problem corresponding to a cross section of the tube under the same time-varying thermal stress σ_t and the constant stress σ_p equal to the hoop stress caused by pressure in the tube. The structure was discretized with rectangular quadratic finite elements as Fig. 5a shows (0y points along a radial direction, 0x—in a hoop one), each element having nine integration points. The U_x displacement at the left edge is restrained, and u_x of all the nodes at the right edge is identical. In the beginning of a cycle temperature varies from uniform distribution at the initial time point to a linear distribution with the maximal gradient along 0y axis (see Fig. 5b), and then goes back in the end of a cycle.

For validation, comparisons between the analytical and numerical solutions were made for two loading conditions corresponding to point A, which lies in the incremental collapse region V in Fig. 1, and for point B located in the combined deformation region IV.

For initialization, S was taken such as to strictly satisfy all the inequality constraints. Computations were started with $\varepsilon_0 \approx 10^{-2}$ and small coefficients c_i of about 10^{-5} so as not to pay much attention to the equality constraints of the original problem (Eq. (44b–d)) at this stage. The phase I method (48) converged rapidly since inequality constraint functions expressed by Eq. (44e) are quadratic, and the algorithm based on Newton's method performed immediately at quadratically convergent stage.

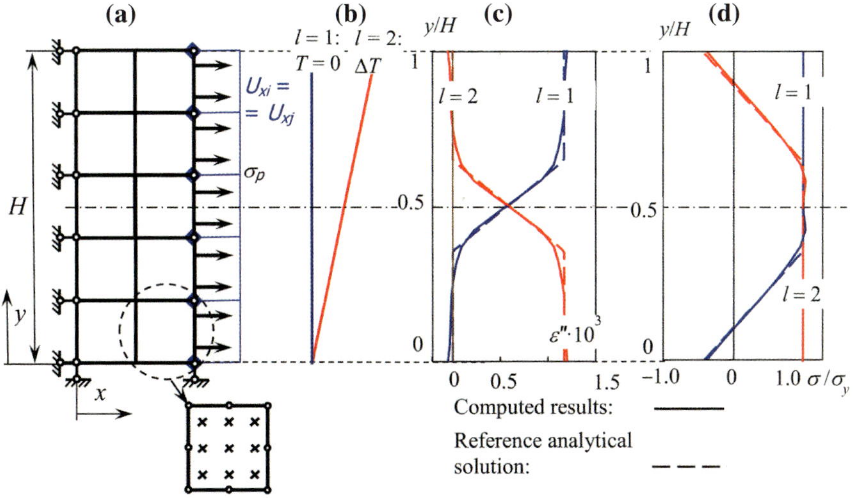

Fig. 5 Problem sketch (**a**), temperature distributions (**b**), comparison of computed and analytical plastic strains (**c**) and total stresses (**d**)

In the main minimization phase, convergence difficulties entailed by the sum of the Euclidean norms term in the object function F_0 (46) were remedied by the proper choice of μ value, combined with adjusting the regularization parameter ε_0 and the penalty coefficients c_i. In general, the object function reduced successfully when μ was relatively small of about 1–5, and all of the four terms in Eq. (45) had nearly the same order of magnitude. However, approaching the boundary defined by inequality constraints Eq. (44e), the method can drastically slow down [16]. This obstacle was dealt with by a temporal reduction of μ for several iterations to about 0.05, after which the point became repelled enough form the boundary for the object function to be reduced further.

At the end of the minimization process, ε_0 was adjusted to be 10^{-5}–10^{-6} (2–3 orders of magnitude less than the maximal $||\mathbf{e}_{il}||$), and the coefficients c_i amounted to 10^5–10^{10} for the constraints given by Eq. (44b–d) to be fulfilled.

6.3.1 Incremental Deformation

Let us first discuss the results for point A in the interaction diagram (Fig. 1), at which pure incremental collapse is expected. The constant stress σ_p amounts to 0.75 σ_y, the maximal thermal stress $\sigma_t = 2\sigma_y$.

Three mesh patterns were used to solve the problem: 5 elements in column and 2 in row, as Fig. 5a depicts, 10 × 1 elements, and 5 × 1 elements. Only two instances were considered over cycle period ($m = 2$) corresponding to no temperature applied and to the full thermal load (see Fig. 5b).

The convergence process and results obtained for different mesh patterns are nearly the same. It should be admitted that in spite of the fact that the penalty coefficients c_i were increased gradually, and the regularization coefficient ε_0 appearing in the sum of norms term was also reduced gradually, the convergence during main optimization phase was neither fast nor stable since Newton's method is naturally not suited for the sum of norms problem. So it required about a hundred of iterations to converge.

Nevertheless, even though the cycle was discretized over time only by two time points corresponding to the extremes of thermal load, one can see a good correspondence between the numerical solution (solid lines) and the reference analytical one (dashed lines) shown in Fig. 5c, d for the plastic strain increments ε_x and total stress σ_x distributions along 0y axis. Figure 5 shows the results only for 5×2 finite element mesh, as for the other meshes there is no distinguishable difference. It is worth noting that due to object function (44a) regularization with ε_0, the solution fields occurred to be almost uniform in 0x direction and showed no deformation instability intrinsic to a non-hardening material.

Figure 5c shows that the strain increment over the cycle is compatible; the checks confirm that the residual stress $\boldsymbol{\beta}_0$ is self-equilibrated, and the plastic strain is deviatoric, which means the constraints presented by Eq. (44b–d) are satisfied.

6.3.2 Incremental Deformation with Alternating Plasticity

The performance of the direct method at point B in the Bree diagram ($\sigma_p = 0.5\sigma_y$, $\sigma_t = 4.0\sigma_y$) is of special interest as it corresponds to a combined deformation, which makes physical convergence to a steady cycle quite slow. The mesh for this test was created as a 5×1 pattern (Fig. 6a).

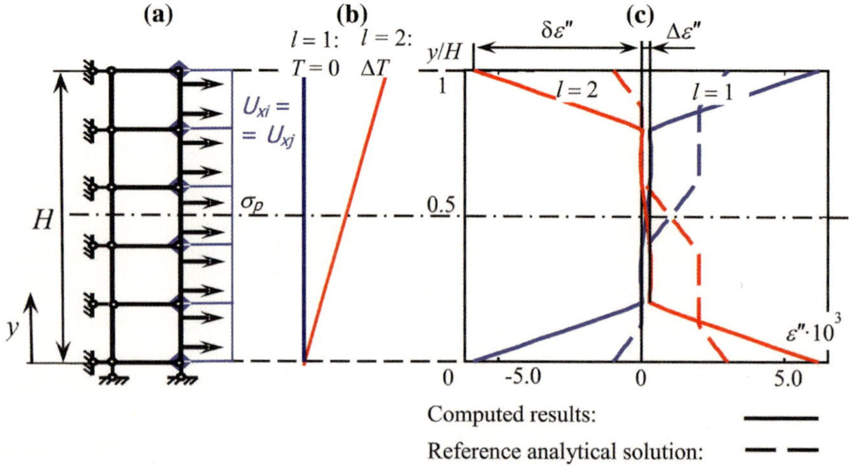

Fig. 6 Problem sketch (**a**), temperature distributions (**b**), comparison of computed and analytical plastic strains (**c**)

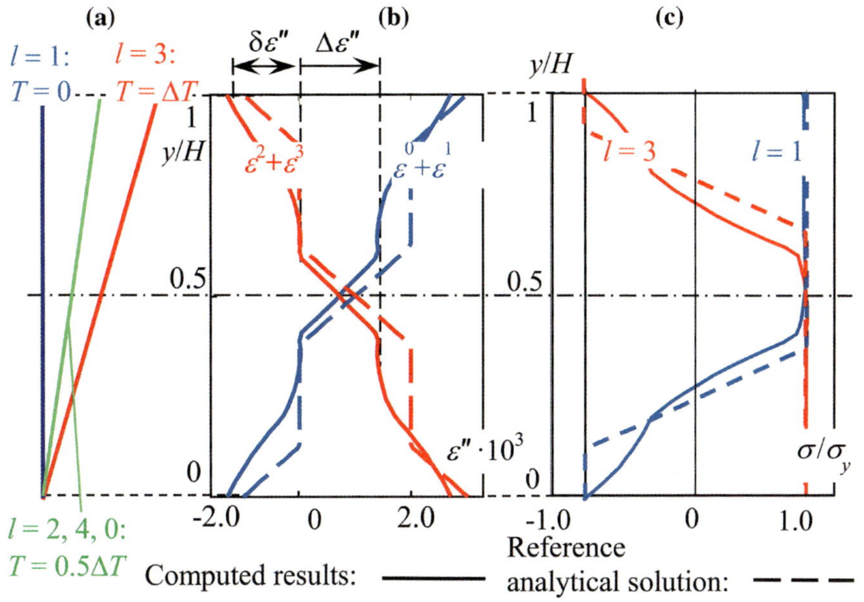

Fig. 7 Temperature distributions (**a**), comparison of computed and analytical plastic strains increments over half-cycle (**b**) and total stress (**c**)

The first trial was made, as in the previous case, with only two time instants per cycle. Figure 6 shows a considerable deviation from the exact solution both in the strain increment $\Delta\varepsilon_x$, which is underestimated, and the strain range $\delta\varepsilon_x$, which is, on the contrary, overestimated.

However, having introduced two intermediate time instances as Fig. 7a shows ($m = 4$), the results become much closer to the analytical solution, and the strain (Fig. 7b) and stress (Fig. 7c) distributions match to the exact ones much better. Therefore, though a more complicated deformation history generally requires more refined time discretization, by increasing the parameter m to a value of less than 10, one can expect a satisfactory result improvement.

7 Conclusions

This study attempts to develop a unified direct method based approach for the evaluation of an elastic-plastic steady cyclic structural response. For this aim, the theorems of elastic and inelastic shakedown accounting for material hardening are formulated in a unified way for all the four cyclic limit states. These theorems form a theoretical basis for the construction of a full interaction diagram for a structural component. This diagram allows to classify the deformation of the component and, then, to use an appropriate material model in the direct computation method for cyclic steady state assessment.

The direct formula for an arbitrary steady cycle (in incremental collapse, alternating plasticity or in combination of the both) under prescribed cyclic loading is deduced in the same way as the theorems and takes on a similar mathematical form.

The applicability of such approach has been demonstrated by the solutions to the classical Bree problem. In these solutions the resulting convex equality and inequality constrained optimization problem is stated making use of finite elements. The solution results for two types of cyclic state agree well with the analytical solutions to the problem. Thus, this formulation has been proved to be able, in principle, to capture the proper results for all the parameters of a steady cycle such as plastic strains and stresses, even though the simple computational approach is only for demonstration.

Thus, it has been shown that this formula is equally suitable for any kind of deformation including a combined one, for which the conventional step-by-step approaches are often excessively time expensive. As the computational tests show, another advantage of such direct technique is that the necessary number of time increments per cycle for reasonable strain range $\delta\varepsilon''$ and increment $\Delta\varepsilon''$ estimates seems to be small, e.g. in the case of pure incremental collapse, only two instances per cycle resulted in a good stress and strain estimate.

References

1. Abramov AV, Chernyavsky AO, Cherniavsky OF et al (2011) Low-cycle deformation and fracture of structures. Eng J 11(176). doi:10.14489/issn.0203-347X
2. Bree J (1967) Elastic-plastic behavior of thin tubes subjected to internal pressure and intermittent high-heat fluxes with application to fast nuclear reactor fuel elements. J Strain Anal 2(3):226–238
3. Rebyakov YN, Chernyavskii OF (2011) Deformation and strength properties of steels at low-cycle combined deformation. Nat Sci: 84–88
4. Frederic CO, Armstrong PJ (1966) Convergent internal stresses and steady cyclic states of stress. J Strain Anal 2:154–159
5. Simon J-W, Weichert D (2013) Interior-point method for lower bound shakedown analysis of von Mises-type materials. In: de Saxcé G, Oueslati A et al (eds) Limit state of materials and structures. Direct methods 2. Springer, Berlin, pp 103–128
6. Chen H (2009) Lower and upper bound shakedown analysis of structures with temperature-dependent yield stress. J Press Vessel Technol 132(1):011202
7. Vu DK, Yan AM, Nguyen-Dang H (2004) A primal-dual algorithm for shakedown analysis of structures. Comput Methods Appl Mech Eng 193:4663–4674
8. Garcea G, Leonetti L (2011) A unified mathematical programming formulation of strain driven and interior point algorithms for shakedown and limit analysis. Int J Numer Methods Eng 88:1085–1111
9. Chen H, Ponter ARS (2001) A method for the evaluation of a ratchet limit and the amplitude of plastic strain for bodies subjected to cyclic loading. Eur J Mech A/Solids 20:555–571
10. Spiliopoulos KV, Panagiotou KD (2012) A direct method to predict cyclic steady states of elastoplastic structures. Comput Methods Appl Mech Eng 223–224:186–198
11. Gokhfeld DA, Cherniavsky OF (1980) Limit analysis of structures at thermal cycling. Sijthoff and Noordhoff. International Publishers, The Netherlands

12. Rabotnov YN (1979) Mechanics of deformable solids. Nauka, Moscow
13. Makhutov NA, Gadenin MM, Evropin SV, Chernyavskii OF, Chernyavskii AO, Rebyakov YN (2014) Specific features of elastoplastic deformation and fracture of steels at complicated thermomechanical stress paths. Inorg Mater 50(15):1495–1505. doi:10.1134/S0020168514150096
14. Koiter W (1960) General problems for elastic-plastic solids. In: Sneddon H (ed) Progress in solid mechanics, vol 4. North-Holland, Amsterdam, pp 165–221
15. Boyd S, Vandenberghe L (2004) Convex optimization. Cambridge University Press, Cambridge
16. Wright SJ (1997) Primal-dual interior-point methods. SIAM

On the Statistical Determination of Yield Strength, Ultimate Strength, and Endurance Limit of a Particle Reinforced Metal Matrix Composite (PRMMC)

Geng Chen, Utku Ahmet Ozden, Alexander Bezold, Christoph Broeckmann and Dieter Weichert

Abstract In this paper we present a numerical methodology to determine the load bearing capacity of a random heterogeneous material. It is applied to a particulate reinforced metal matrix composite (PRMMC), WC-30 Wt.% Co, to predict its strength against both monotonic and cyclic loads. In this approach, limit and shakedown analysis based on Melan's static theorem [30] is performed on representative volume element (RVE) models generated from real material microstructure and the obtained results are converted to macroscopic load domains through homogenization. To evaluate microstructure's impact on the overall material strength, the relationship between strength and composite structure is investigated by means of statistics. Meanwhile, several numerical issues, e.g. the impact of RVE's size, mesh density, as well as the discrepancy between 2D and 3D models, are studied.

Keywords PRMMC · Shakedown · Statistical RVE · Homogenization · Melan's theorem

1 Introduction

In civil and mechanical engineering, determining the capability of structures to support different types of loadings plays a central role. The recent trend for increasing application of composite material and the gradual replacement of conventional metallic materials in structural components, set forth new challenges for this classic problem. In the context of heterogeneous materials, it is known that the strength depends on their microstructural features [17, 40]. Thus to fully exploit the potential of such materials, it requires to understand decisive factors influencing composite's strength, especially the contribution of material's microstructure.

G. Chen (✉) · U.A. Ozden · A. Bezold · C. Broeckmann
Institute of Materials Application in Mechanical Engineering,
RWTH Aachen University, Augustinerbach 4, 52062 Aachen, Germany
e-mail: G.Chen@iwm.rwth-aachen.de

D. Weichert
Institute of General Mechanics, RWTH Aachen University, Templergraben 62,
52062 Aachen, Germany
e-mail: weichert@iam.rwth-aachen.de

P. Fuschi et al. (eds.), *Direct Methods for Limit and Shakedown Analysis of Structures*,
DOI 10.1007/978-3-319-12928-0_6

Particulate Reinforced Metal Matrix Composites (PRMMC) consist of the ductile binder and brittle inclusion particles. This combination constitutes a large class of existing composite structures. Compared to fiber reinforced composites PRMMCs generally show higher isotropy and demand less effort for manufacturing [20]. However, the randomness in the distribution of reinforcement grains entails considerable spatial variation of material behavior. This leads to two consequences: first, material's response to external loading is quite localized; second, representative volume elements (RVE) do not exist in a strict sense—it means not a singular model of finite size is capable to completely reflect the overall material behavior [39]. The randomness of microstructure also introduces difficulties in the numerical representation of the material. Two prominent problems are the modeling technique of the microstructural geometry and the size effect of RVE investigated in many studies [5, 14, 21, 32, 36, 46].

For PRMMC with high binder content, the matrix failure is a major cause of overall material failure [38] and the strength of the composite is highly depends on matrix phase's capability to resist plastic failure. Similar to single phase material, strength of a globally elasto-plastic composite can be characterized by three strengths: yield strength, ultimate strength and endurance limit. They correspond to elastic limit, plastic limit and shakedown limit of the material, respectively.

The link between endurance limit and shakedown on micro scale has already been noted by Drucker [13]. The idea has further been studied by Dang Van who developed his well known criterion [12], which calculates the condition of elastic shakedown on grain level and allows determining the endurance of material under complex multi-axial loading situations. Nevertheless, since often serious stress and strain localization will be observed within the heterogeneous material, this criterion is not fully satisfying for composites. To take into account the contribution of the composite structure, one has to consider numerical approaches [7, 16].

Applying direct methods (DM) in the Melan-Koiter path-independent spirit [23, 30] to study the problem has some advantages. On the one hand, it allows to consider the composite structure and on the other hand it avoids performing cumbersome step-by-step calculation. This method has become increasingly popular in recent decades: as example, Weichert et el. [42], Schwabe [35], Maier [28], Magoariec et al. [27], Zhang et al. [45], You et al. [44], Chen et al. [10, 11] solve the problem by a static approach. In contrast, Carvelli [6], Chen and Ponter [9], Li [26], and Barrera [3] deal the problem by the kinematic approach.

The implementation of DM according to Melan's theorem involves solving a constrained extremal problem, and conventional mathematical programming methods cause serious difficulties when model size grows. This is the main reason that existing studies, to the authors' knowledge, are restricted to idealized microstructures. To study more complex microstuctures, one has to rely on highly efficient optimization algorithms. The interior point method introduced by Karmarkar [22] has demonstrated high efficiency in solving linear programming (LP) problems and it attracted interests from different disciplines [15, 41]. More recently, the algorithm has been introduced to solve nonlinear programming (NP) problems. An additional advantage of the interior point method is its capability in dealing with large scale problems.

The number of iterations it requires to converge is much smaller than the polynomial upper bound and is almost independent upon the problem size [43]. Therefore, many researchers have incorporate this algorithm with DM. This includes problem-tailored codes [25, 33, 37, 47], and reformulating problems in such a way that they can be efficiently solved by general-purpose solvers [4, 18, 31].

In the present study, WC-70 Wt.% Co is used as a typical PRMMCs to study both monotonic and cyclic limit loads. The technique developed in a previous study [8] is applied in order to overcome the aforementioned main obstacles. Using a series of scanning electron microscope (SEM) images obtained from the material finite element (FE) models are built. Then first, static limit and shakedown problems are constructed on each RVE and solved by an interior-point method based solver. In the second step, the load domains obtained in the previous step are converted to macroscopic strength through the homogenization technique. To estimate the dependence of the strength on structure, the obtained results are evaluated statistically. In order to identify factors that influence the quality of numerical results, several characteristics of FE models, e.g. the impact of RVE's size, mesh density, as well as the difference between 2D and 3D model, are investigated and discussed in detail.

2 Limit and Shakedown Analysis of Random PRMMC Material

2.1 Micromechanical Homogenization of Elasto-Plastic Materials

Homogenization theory links physical fields in two well-separated scales, i.e. the microscale $\boldsymbol{y}$ in which structural details of RVE are distinguishable and the macro scale $\boldsymbol{x}$ in which a RVE is recognized as a macroscopic point. For a heterogeneous composite, once exposed to external loading, its microscopic stress field $\boldsymbol{\Sigma}$ in $\boldsymbol{y}$ and the macroscopic counterpart in $\boldsymbol{x}$ satisfy the relationship:

$$\boldsymbol{\Sigma} = \frac{1}{\Omega} \int_{\Omega} \boldsymbol{\sigma}(y) dv. \qquad (1)$$

Here Ω indicates the RVE domain. Analogously, macroscopic strain $\boldsymbol{E}$ can be defined as

$$\boldsymbol{E} = \frac{1}{\Omega} \int_{\Omega} \boldsymbol{\varepsilon}(y) dv. \qquad (2)$$

When all constituents of the composite are elastic, $\boldsymbol{\Sigma}$ and $\boldsymbol{E}$ are correlated by the effective elastic tensor $\overline{\boldsymbol{C}}$:

$$\boldsymbol{\Sigma} = \overline{\boldsymbol{C}} : \boldsymbol{E}. \qquad (3)$$

For a globally isotropic composite, $\boldsymbol{C}$ depends only on the effective Young's modulus $\overline{E}$ and effective Poisson's ratio $\overline{\upsilon}$.

A characteristic of PRMMCs is that stress and strain are strongly non-uniform due to heterogenity, thus the onset of local plasticity can take place even when $\boldsymbol{\Sigma}$ is fairly low. To assess the plastic strain in the macro scale, Suquet [38] proposed a measure: the effective plastic strain $\boldsymbol{E^p}$ which is a work conjugate to $\boldsymbol{\Sigma}$ defined as:

$$\boldsymbol{E^p} = \boldsymbol{E} - \overline{\boldsymbol{C}}^{-1} : \boldsymbol{\Sigma}. \tag{4}$$

This way, the global yield strength of a composite $\boldsymbol{\Sigma}^Y$ can be set as the stress leading to 0.1 % global strain:

$$\boldsymbol{\Sigma}^Y = \arg(E^p_{eq}(\Sigma_{eq}) = 0.1\,\%). \tag{5}$$

Here, E^p_{eq} and Σ_{eq} represent equivalent macroscopic plastic strain and stress, respectively. In the present study, all phases are assumded to obey the von Mises yield criterion. Thus E^p_{eq} and Σ_{eq} can be formulated as:

$$E^p_{eq} = \sqrt{\frac{2}{3}(\boldsymbol{E^p})' : (\boldsymbol{E^p})'}, \tag{6a}$$

$$\Sigma_{eq} = \sqrt{\frac{3}{2}\boldsymbol{\Sigma}' : \boldsymbol{\Sigma}'}. \tag{6b}$$

The apostrophe in (6a, 6b) indicates the deviatoric part of a tensor. It is important to note that an individual RVE is not isotropic, despite macroscopic isotropy of WC-Co in the elastic as well as the plastic range. For this reason, we take the average of strength in two orthogonal directions Σ^Y_{11} and Σ^Y_{22} as the measure of composite's global strength.

2.2 Static Theorem and Its Numerical Reformulation

The static shakedown criterion for a elastic-perfectly plastic material can be formulated as follows: shakedown occurs if there exist a safety factor $\alpha > 1$ and a time-independent residual stresses field $\overline{\boldsymbol{\rho}}$, whose superposition with the purely elastic stresses $\boldsymbol{\sigma}^E$ does not exceed the yield condition F at any time $t > 0$:

$$F\left(\alpha\boldsymbol{\sigma}^E(\boldsymbol{y}, t) + \bar{\boldsymbol{\rho}}, \sigma_Y\right) \leq 0, \tag{7a}$$

$$\nabla \cdot \bar{\boldsymbol{\rho}} = \boldsymbol{0} \text{ in } \Omega, \tag{7b}$$

$$\boldsymbol{\sigma}^E(y, t) \cdot \boldsymbol{n} = \boldsymbol{t} \text{ on } \partial\Omega_t, \tag{7c}$$

$$\bar{\boldsymbol{\rho}} \cdot \boldsymbol{n} = \boldsymbol{0} \text{ on } \partial\Omega_t. \tag{7d}$$

Here Ω_t denotes the part of the boundary where tractions $\boldsymbol{t}$ are prescribed. In case of von Mises yield criterion, the function F becomes:

$$F(\boldsymbol{\sigma}, \sigma^Y) = \sqrt{\frac{3}{2}\boldsymbol{\sigma}' : \boldsymbol{\sigma}'} - \sigma^Y. \tag{8}$$

By employing (7a–7d) to each RVE, its respective admissible load domains will be identified. Since plastic limit can be regarded as a special case of shakedown where $\boldsymbol{\sigma}^E$ evolves monotonically through t, this condition holds as well for the limit analysis.

When $\boldsymbol{\sigma}^E$ is entailed by a set of NL independent loads $\boldsymbol{P}_i$, due to linearity following relationship holds:

$$\boldsymbol{\sigma}^E\left(\sum_{i=1}^{NL} \boldsymbol{P}_i(y)\right) = \sum_{i=1}^{NL} \boldsymbol{\sigma}^E\left(\boldsymbol{P}_i(y)\right). \tag{9}$$

Here, an individual load $\boldsymbol{P}_i$ can be separated into two parts including a varying magnitude scalar μ_i and a time invariant base vector $\hat{\boldsymbol{P}}_i$:

$$\boldsymbol{P}_i(y, t) = \mu_i(t)\hat{\boldsymbol{P}}_i(y). \tag{10}$$

Since each particular load μ_i varies within the interval $[\mu_i^-, \mu_i^+]$, a loading profile can be considered as a trajectory in space $\mathscr{L}$ spanned by $\{\hat{\boldsymbol{P}}_i\}$. As shown by König [24], it is sufficient to only consider the convex hull of the loading history, which is defined by the $\mathrm{NC} = 2^{\mathrm{NL}}$ corners of the polyhedral loading domain. This way, (7a) can be simplified to a time independent form:

$$F\left(\alpha\boldsymbol{\sigma}^E(\hat{\boldsymbol{P}}_i) + \bar{\boldsymbol{\rho}}, \sigma_Y\right) \le 0, \quad i \in [1 \ldots \mathrm{NC}] \tag{11}$$

By FE discretization and replacing (7a) by (11), the shakedown condition (7a–7d) becomes:

$$(\mathscr{P}_{\mathrm{ORI}}) \max_{\bar{\boldsymbol{\rho}}} . \alpha, \tag{12a}$$

$$s.t. : [\boldsymbol{C}]\{\bar{\boldsymbol{\rho}}\} = \{\boldsymbol{0}\}, \tag{12b}$$

$$F(\alpha\boldsymbol{\sigma}_i^E(\hat{\mathbf{P}}_k) + \bar{\boldsymbol{\rho}}_i, \sigma_i^Y) \le 0, i \in [1, \mathrm{NGS}], k \in [1, \mathrm{NC}]. \tag{12c}$$

Here, NGS is the number of total Gaussain points in a model, and matrix $\boldsymbol{C}$ the self-equilibrium matrix defined as:

$$[\boldsymbol{C}] = \sum_{m=1}^{\mathrm{NGS}} w_m\, |J_m| [\boldsymbol{B}_m]^T. \tag{13}$$

Matrix $\boldsymbol{B}$ consists of spatial derivatives of shape functions and maps displacements into strains; w is the weight factor of integration points; J is the determinant of the

Jacobian matrix. For FE models having NK nodes and NDoF degrees of freedom for each node, one obtains $\boldsymbol{C} \in \mathbb{R}^{\mathrm{NDoF \cdot NK} \times 6\mathrm{NGS}}$ where NDoF equals 3 for 3D case, and 2 for 2D case.

In the current study, ultimate strength $\boldsymbol{\Sigma}^U$ and endurance limit $\boldsymbol{\Sigma}^\infty$ are considered for the non-reverse axial loading which correspond to 1 load vertex and 2 load vertices in $\mathscr{L}$, respectively. To exclude anisotropy, strengths will be evaluated on various directions and their average value will be taken. The simplest way to achieve this is following the approach illustrated in Fig. 1: arbitrary orthogonal stresses $\boldsymbol{\Sigma}^E_{11}$ and $\boldsymbol{\Sigma}^E_{22}$ are prescribed alternately on a purely elastic reference RVE, and entailed microscopic stress fields $\boldsymbol{\sigma}^E_{11}(\boldsymbol{y})$ and $\boldsymbol{\sigma}^E_{22}(\boldsymbol{y})$ are calculated, respectively. By introducing an angle θ, a combined loading $\hat{\boldsymbol{P}}_1$ can be formed as a joint effect of $\boldsymbol{\Sigma}^E_{11}$ and $\boldsymbol{\Sigma}^E_{22}$. Therefore, to calculate $\boldsymbol{\Sigma}^U$, the shakedown condition is required to be satisfied at vertex $\hat{\boldsymbol{P}}_1$. Analogously, to calculate $\boldsymbol{\Sigma}^\infty$, the same condition should hold simultaneously at vertices $\hat{\boldsymbol{P}}_1$ and $\hat{\boldsymbol{P}}_2$.

Some efficiency issues associated with the implementation of shakedown theorem (12a–12c) has been noticed concerning inequality constraints (12c). For pragmatic reasons, this condition should be reformulated to improve the efficiency of optimization algorithm. Akoa et al. [2] have suggested to convert convex quadratic constraints into Euclidean ball constraints. Several key steps of this approach are briefly introduced here.

For the original shakedown problem $\mathscr{P}_{\mathrm{ORI}}$ in (12a–12c), primal variables are components of residual stress at every Gaussian point. More specifically:

1. $\bar{\boldsymbol{\rho}} = \{\bar{\rho}_{11}, \bar{\rho}_{22}, \bar{\rho}_{33}, \bar{\rho}_{12}, \bar{\rho}_{13}, \bar{\rho}_{23}\}^T$ for 3D case,
2. $\bar{\boldsymbol{\rho}} = \{\bar{\rho}_{11}, \bar{\rho}_{22}, \bar{\rho}_{12}\}^T$ for 2D plane stress case,
3. $\bar{\boldsymbol{\rho}} = \{\bar{\rho}_{11}, \bar{\rho}_{22}, \bar{\rho}_{33}, \bar{\rho}_{12}\}^T$ for 2D plane strain case.

The reformulation introduced in [2] changes the primal variables to a linear transformation of total stress $\boldsymbol{\sigma}$ where $\boldsymbol{\sigma} = \alpha\boldsymbol{\sigma}^E + \bar{\boldsymbol{\rho}}$. The relationship between the new

Fig. 1 Superposition of elastic stress

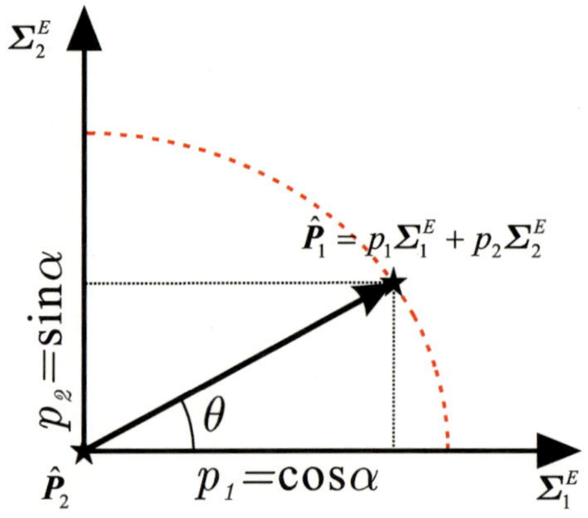

primary variable $\{\boldsymbol{u}, v\}$ and $\boldsymbol{\sigma}$ in the 3D case follows:

$$\begin{Bmatrix} u_1 \\ u_2 \\ u_3 \\ u_4 \\ u_5 \\ v \end{Bmatrix} = \frac{1}{\sigma^Y} \begin{bmatrix} 1 & -1/2 & -1/2 & 0 & 0 & 0 \\ 0 & \sqrt{3}/2 & -\sqrt{3}/2 & 0 & 0 & 0 \\ 0 & 0 & 0 & \sqrt{3} & 0 & 0 \\ 0 & 0 & 0 & 0 & \sqrt{3} & 0 \\ 0 & 0 & 0 & 0 & 0 & \sqrt{3} \\ 1 & 0 & 1 & 0 & 0 & 0 \end{bmatrix} \begin{Bmatrix} \sigma_{11} \\ \sigma_{22} \\ \sigma_{33} \\ \sigma_{12} \\ \sigma_{23} \\ \sigma_{13} \end{Bmatrix}, \tag{14}$$

and in 2D plane strain case:

$$\begin{Bmatrix} u_1 \\ u_2 \\ u_3 \\ v \end{Bmatrix} = \frac{1}{\sigma^Y} \begin{bmatrix} 1 & -1/2 & -1/2 & 0 \\ 0 & \sqrt{3}/2 & -\sqrt{3}/2 & 0 \\ 0 & 0 & 0 & \sqrt{3} \\ 1 & 0 & 1 & 0 \end{bmatrix} \begin{Bmatrix} \sigma_{11} \\ \sigma_{22} \\ \sigma_{33} \\ \sigma_{12} \end{Bmatrix}, \tag{15}$$

and in 2D plane stress case:

$$\begin{Bmatrix} u_1 \\ u_2 \\ u_3 \end{Bmatrix} = \frac{1}{\sigma^Y} \begin{bmatrix} 1 & -1/2 & 0 \\ 0 & \sqrt{3}/2 & 0 \\ 0 & 0 & \sqrt{3} \end{bmatrix} \begin{Bmatrix} \sigma_{11} \\ \sigma_{22} \\ \sigma_{12} \end{Bmatrix}. \tag{16}$$

This transformation matrix denoted by $\boldsymbol{U}$ is applied to each $\boldsymbol{\sigma}$, and v retained from all Gaussian points are put collectively into a global vector $\boldsymbol{v} = \{v_1, v_2, \ldots, v_m\}^T$ where $m = \text{NGS}$ and $\boldsymbol{v} \in \mathbb{R}^{\text{NGS}}$. Applying the new primal variable, shakedown problem $\mathscr{P}_{\text{ORI}}$ in (12a–12c) can be reformulated into an equivalent form:

$$(\mathscr{P}_{\text{Reform}}) \max . \alpha, \tag{17a}$$

$$s.t. : \sum_{r=1}^{\text{NGS}} [\boldsymbol{A}_r] \left\{\boldsymbol{u}_r^1\right\} + [\boldsymbol{B}] \left\{\boldsymbol{v}^1\right\} - \alpha \left\{\boldsymbol{w}^1\right\} = 0, \tag{17b}$$

$$\text{Where: } \left\{\boldsymbol{w}^1\right\} = [\boldsymbol{C}]\{\boldsymbol{\sigma}^E(\hat{\boldsymbol{P}}_1)\} \tag{17c}$$

$$\left\{\boldsymbol{u}_r^2\right\} - \left\{\boldsymbol{u}_r^1\right\} = \alpha\, [U] \left\{\boldsymbol{\sigma}_r^{2,E} - \boldsymbol{\sigma}_r^{1,E}\right\} \tag{17d}$$

$$\left\|\boldsymbol{u}_r^{1,2}\right\| \leq 1. \tag{17e}$$

In (17a–17e), number within superscript indicates load vertex. For all 2D and 3D models, $[\boldsymbol{A}_r]$ is defined as:

$$[\boldsymbol{A}_r] = \sqrt{2}\sigma^Y [\boldsymbol{D}_r][\boldsymbol{L}^{-T}] \tag{18}$$

The other variables of (18) in 3D case are defined by:

$$\boldsymbol{L}^{3\text{D}} = \begin{bmatrix} \sqrt{2} & 0 & 0\,0\,0 \\ \sqrt{2}/2 & \sqrt{3}/\sqrt{2} & 0\,0\,0 \\ 0 & 0 & 1\,0\,0 \\ 0 & 0 & 0\,1\,0 \\ 0 & 0 & 0\,0\,1 \end{bmatrix} \tag{19a}$$

$$\boldsymbol{D}_r^{3\text{D}} = \left[(\boldsymbol{C}_r\boldsymbol{T})^1 \; (\boldsymbol{C}_r\boldsymbol{T})^2 \; (\boldsymbol{C}_r\boldsymbol{T})^4 \; (\boldsymbol{C}_r\boldsymbol{T})^5 \; (\boldsymbol{C}_r\boldsymbol{T})^6 \right] \tag{19b}$$

$$\boldsymbol{B}^{3\text{D}} = \left[(\boldsymbol{C}_1\boldsymbol{T})^3 \; (\boldsymbol{C}_2\boldsymbol{T})^3 \; \cdots \; (\boldsymbol{C}_{\text{NGS}}\boldsymbol{T})^3 \right] \tag{19c}$$

$$\boldsymbol{T}^{3\text{D}} = \begin{bmatrix} 1/2 & 1/2 & 1/2 & 0 & 0 & 0 \\ -1/2 & 1/2 & 1/2 & 0 & 0 & 0 \\ -1/2 & -1/2 & 1/2 & 0 & 0 & 0 \\ 0 & 0 & 0 & 1/\sqrt{6} & 0 & 0 \\ 0 & 0 & 0 & 0 & 1/\sqrt{6} & 0 \\ 0 & 0 & 0 & 0 & 0 & 1/\sqrt{6} \end{bmatrix} \tag{19d}$$

while in 2D plane strain case:

$$\boldsymbol{L}^{\text{P. Strain}} = \begin{bmatrix} \sqrt{2} & 0 & 0 \\ \sqrt{2}/2 & \sqrt{3}/\sqrt{2} & 0 \\ 0 & 0 & 1 \end{bmatrix} \tag{20a}$$

$$\boldsymbol{D}_r^{\text{P. Strain}} = \left[(\boldsymbol{C}_r\boldsymbol{T})^1 \; (\boldsymbol{C}_r\boldsymbol{T})^2 \; (\boldsymbol{C}_r\boldsymbol{T})^4 \right] \tag{20b}$$

$$\boldsymbol{B}^{\text{P. Strain}} = \left[(\boldsymbol{C}_1\boldsymbol{T})^3 \; (\boldsymbol{C}_2\boldsymbol{T})^3 \; \cdots \; (\boldsymbol{C}_{\text{NGS}}\boldsymbol{T})^3 \right] \tag{20c}$$

$$\boldsymbol{T}^{\text{P. Strain}} = \begin{bmatrix} 1/2 & 1/2 & 1/2 & 0 \\ -1/2 & 1/2 & 1/2 & 0 \\ -1/2 & -1/2 & 1/2 & 0 \\ 0 & 0 & 0 & 1/\sqrt{6} \end{bmatrix} \tag{20d}$$

and in 2D plane stress case:

$$\boldsymbol{L}^{\text{P. Stress}} = \begin{bmatrix} 1 & -1/2 & 0 \\ 0 & \sqrt{3}/2 & 0 \\ 0 & 0 & \sqrt{3} \end{bmatrix} \tag{21a}$$

$$\boldsymbol{D}_r^{\text{P. Stress}} = \boldsymbol{C}_r \tag{21b}$$

Here $(\boldsymbol{X})^i$ represents the ith column of matrix $\boldsymbol{X}$. We note that, according to (16), the converted variable does not contain v for the *plane stress* case. Thus the term $[\boldsymbol{B}]\{\boldsymbol{v}^1\}$ does not exist in this case as well.

When the numerical scheme (17a–17e) is used for limit analysis, due to the absence of $\hat{\boldsymbol{P}}_2$, equality constraints (17d) is removed.

2.3 Solving the Quadratically Constrained Programming (QCP) by Primal-Dual Interior Point Method

In order to obtain the shakedown factor α, the optimization problem (17a–17e) has to be solved. This is a typical QCP problem because inequality constraints (17e) are quadratic functions. By introducing the slack variable $\boldsymbol{s}$ to convert inequality to equality constraints, this problem can be written in a general form:

$$(\mathscr{P}_{\text{Standard}})\ \min.\ -\boldsymbol{\alpha} = f(\boldsymbol{x}), \tag{22a}$$

$$s.t.: \boldsymbol{c}_E(\boldsymbol{x}) = \boldsymbol{0}, \tag{22b}$$

$$\boldsymbol{c}_I(\boldsymbol{x}) - \boldsymbol{s} = \boldsymbol{0}, \tag{22c}$$

$$\boldsymbol{s} \geq \boldsymbol{0}. \tag{22d}$$

The vector $\boldsymbol{x}$ consists of $\boldsymbol{u}_r$ and $\boldsymbol{v}$ for all elements at all $\hat{\boldsymbol{P}}_i$ as well as the loading factor α. By denoting the dimension of the converted variable in each Gaussian point as DCV, then depending on if model is in 3D, plane strain or plane stress, DCV equals 6, 4, or 3, respectively. Within the given QCP problem, primal variables are $\boldsymbol{x}$ and $\boldsymbol{s}$. In 3D or plane strain case $\boldsymbol{x} \in \mathbb{R}^{[\text{NC}\cdot(\text{DCV}-1)+1]\cdot\text{NGS}+1}$, while in plane stress case $\boldsymbol{x} \in \mathbb{R}^{\text{NC}\cdot\text{DCV}\cdot\text{NGS}+1}$ due to the absence of $\boldsymbol{v}^1$. Meanwhile, for all three cases, $\boldsymbol{s}$ is constantly a vector in $\mathbb{R}^{\text{NC}\cdot\text{NGS}}$. The objective function (23a) is linear with negative loading factor as the function value. Equality constraints (23b) are obtained from self-equilibrium condition (17b) and time-independence condition (17d). The quadratic inequality constraints (23c) represent the von Mises yield criterion.

To avoid the complication of direct dealing with (22d) a barrier problem is constructed:

$$\min_{\boldsymbol{x},\boldsymbol{s}} f(\boldsymbol{x}) - \mu \sum_{i=1}^{m} \log s_i, \tag{23a}$$

$$s.t.: \boldsymbol{c}_E(\boldsymbol{x}) = \boldsymbol{0}, \tag{23b}$$

$$\boldsymbol{c}_I(\boldsymbol{x}) - \boldsymbol{s} = \boldsymbol{0}. \tag{23c}$$

Here μ is a positive barrier parameter.

The first-order Karush-Kuhn-Tucker (KKT) conditions of (23a–23c) write:

$$\nabla f(\boldsymbol{x}) - \mathscr{A}_E^T(\boldsymbol{x})\boldsymbol{y} - \mathscr{A}_I^T(\boldsymbol{x})\boldsymbol{z} = \boldsymbol{0}, \tag{24a}$$

$$-\mu \boldsymbol{S}^{-1}\boldsymbol{e} + \boldsymbol{z} = \boldsymbol{0}, \tag{24b}$$

$$\boldsymbol{c}_E(\boldsymbol{x}) = \boldsymbol{0}, \tag{24c}$$

$$\boldsymbol{c}_I(\boldsymbol{x}) - \boldsymbol{s} = \boldsymbol{0}. \tag{24d}$$

In (24a–24d), dual variables $\boldsymbol{y}$ and $\boldsymbol{z}$ are lagrangian multipliers to equality constraints and inequality constraints, respectively. $\mathscr{A}_E$ consists of gradients of equality constraints where $\mathscr{A}_E = [\nabla c_{E,1}, \nabla c_{E,2}, \ldots, \nabla c_{E,n}]$. Because the equality constraints are linear, their gradients become constant values. In 3D and plane strain models $\mathscr{A}_E$ has (NK · NDoF) + (NC − 1) · (DCV − 1) · NGS rows and [NC · (DCV − 1) + 1] · NGS columns, while in plane stress models, $\mathscr{A}_E$ has (NK · NDoF) + (NC − 1) · DCV · NGS rows and NC · DCV · NGS columns. Similar to $\mathscr{A}_E$, $\mathscr{A}_I$ is defined as $[\nabla c_{I,1}, \nabla c_{I,2}, \ldots, \nabla c_{I,m}]$. Because c_I are quadratic, their gradients ∇c_I are linear functions. The number of rows in $\mathscr{A}_I$ is same as $\mathscr{A}_E$, but the number of its columns is NC · NGS, which is independent on the model type. Matrix $\boldsymbol{S}$ in (24b) is defined by $\boldsymbol{S} = \text{diag}(\boldsymbol{s})$, and $\boldsymbol{e}$ is a unit vector. The term $\mu \boldsymbol{S}^{-1}\boldsymbol{e}$ in (24b) is yielded from $\nabla_s(\mu \sum_{i=1}^{m} \log s_i)$.

The nonlinear system (24a–24d) can be solved with Newton's method by given numerical scheme:

$$\begin{bmatrix} \nabla^2_{xx}\mathscr{L} & \mathbf{0} & -\mathscr{A}_E^T(\boldsymbol{x}) & -\mathscr{A}_I^T(\boldsymbol{x}) \\ \mathbf{0} & \boldsymbol{Z} & \mathbf{0} & \boldsymbol{S} \\ \mathscr{A}_E^T(\boldsymbol{x}) & \mathbf{0} & \mathbf{0} & \mathbf{0} \\ \mathscr{A}_I^T(\boldsymbol{x}) & -\boldsymbol{I} & \mathbf{0} & \mathbf{0} \end{bmatrix} \begin{bmatrix} \boldsymbol{p}_x \\ \boldsymbol{p}_s \\ \boldsymbol{p}_y \\ \boldsymbol{p}_z \end{bmatrix} = - \begin{bmatrix} \nabla_x f(\boldsymbol{x}) - \mathscr{A}_E^T(\boldsymbol{x})\boldsymbol{y} - \mathscr{A}_I^T(\boldsymbol{x})\boldsymbol{z} \\ \boldsymbol{S}\boldsymbol{z} - \mu\boldsymbol{e} \\ \boldsymbol{c}_E(\boldsymbol{x}) \\ \boldsymbol{c}_I(\boldsymbol{x}) - \boldsymbol{s} \end{bmatrix}. \tag{25}$$

In (25), $\mathscr{L}$ represents the lagrangian for the barrier problem (23a–23c):

$$\mathscr{L}(\boldsymbol{x}, \boldsymbol{s}, \boldsymbol{y}, \boldsymbol{z}) = f(\boldsymbol{x}) - \boldsymbol{y}^T \mathscr{A}_E(\boldsymbol{x}) - \boldsymbol{z}^T(\mathscr{A}_I(\boldsymbol{x}) - \boldsymbol{s}). \tag{26}$$

whereas $\boldsymbol{Z}$, analogous to $\boldsymbol{S}$, is defined as $\boldsymbol{Z} = \text{diag}(\boldsymbol{z})$; $\boldsymbol{I}$ is the unit matrix.

To implement the primal-dual interior point method, one starts with a predefined barrier parameter μ and a feasible initial solution $\{\boldsymbol{x}_0, \boldsymbol{s}_0, \boldsymbol{y}_0, \boldsymbol{z}_0\}^T$. By solving the system (25) a Newton's step can be calculated. This step will be corrected with respect to the *fraction to boundary rule*, and the corrected step will be taken to update both primal and dual variables. This procedure will be repeated, and once the error function $\mathscr{E}$ defined as:

$$\mathscr{E} = \max\{\|\nabla_x f(\boldsymbol{x}) - \mathscr{A}_E^T(\boldsymbol{x})\boldsymbol{y} - \mathscr{A}_I^T(\boldsymbol{x})\boldsymbol{z}\|, \|\boldsymbol{S}\boldsymbol{z} - \mu\boldsymbol{e}\|, \|\boldsymbol{c}_E(\boldsymbol{x})\|, \|\boldsymbol{c}_I(\boldsymbol{x}) - \boldsymbol{s}\|\} \tag{27}$$

drops below a predefined threshold, μ will be updated and solution to the evolved barrier problem will be calculated by the same iterative scheme. One can prove that with $\mu \downarrow 0$, solution to the barrier problem is exactly the same as the one to original problem.

3 Numerical Results

3.1 Finite Element Models for Statistical Analysis

20 RVE models with different microstructure have been prepared. We distinguish two groups: Group 1 consists of 10, each 30 μm-by-30 μm, RVEs numbered consecutively from 1 to 10. The models in Group 2, however, have a size 40 μm-by-40 μm

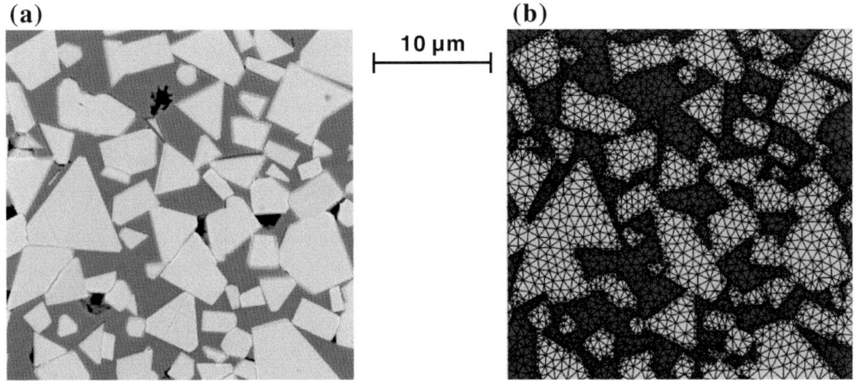

Fig. 2 **a** SEM image **b** RVE model converted from SEM image

with numbers from 11 to 20. All models are based on real WC-Co microstructures obtained from a scanning electron microscope (SEM) observation using a backscattering detector (Fig. 2). In these images the dark grey area is the Co phase, while the bright areas are the WC grains with their characteristic prismatic shape. The average grain size is $d_{WC} = 2.35\,\mu\text{m}$. To convert these SEM images to the corresponding FE models an automatic technique developed in [8] is employed. This technique is capable of generating a triangular element based adaptive finite element mesh. It adopts a denser mesh in vicinity to phase interfaces and a coarser mesh elsewhere (Fig. 2).

Both plane stress and plane strain models are investigated. It should be noted that in previous studies plane strain models are observed to give more accurate prediction on the elasto-plastic behavior of given composite in small strain regime [8, 34]. To check the error caused by finite element type, the current study considers also a special typed 3D model which consists of a thin layered wedge elements obtained by extracting the 2D model in the 3rd direction for 1 μm. Intuitively, we refer these 3D models to *2.5D* models, and they are regarded as compromise between plane stress and plane strain, which correspond to two extremes about the stiffness in the 3rd direction.

For simplicity, the materials for both constituents are considered as elastic-perfect plastic materials with parameters illustrated in Table 1. It is worthy to note that current study is restricted to small deformation and assumes plastic failure as the only failure mechanism.

The procedure of numerical limit and shakedown analysis can be summarized as follows: first, RVE models are built in commercial finite element software ABAQUS 6.12 [1]. By using a self-developed Python script, these models are prescribed with global loading as explained in Sect. 2.2 and reference elastic stress

Table 1 Material properties of both phases

	E (GPa)	μ [−]	σ^Y (MPa)
WC	700	0.24	2,000
Co	210	0.30	279

fields are calculated. In the second step, the geometrical setup of models and their associated stress results are output to MATLAB [29] to build the optimization problem (17a–17e). In the subsequent step, the constructed QCP problem is submitted to an interior point method solver, Gurobi [19], to calculate loading factor α. By taking a series of $\theta \in [0, \pi]$ strengths in many different directions are calculated and they together form an entire feasible load domain. By projecting this domain to π-plane and fit the projection to a perfect semi-circle, a direction-independent strength value is obtained which best characterizes the overall strength of a RVE. In the final step, strength of each RVE retrieved from the best fit are collected and interpreted by a statistical analysis. In such analysis three measures of a random variable x are evaluated. These measures include mean value $\bar{x}$, standard deviation x_{SD}, and coefficient of variance defined as:

$$C_x = \frac{x_{SD}}{\bar{x}}. \tag{28}$$

3.2 Comparative Study Between Plane Stress, Plane Strain and 2.5D Models

In [8] it has been shown that the discrepancy between plane stress and plane strain models becomes more obvious when global plasticity accumulates. Also, once global plasticity has reached a critical level, the mechanical behavior reflected by these two model variances are fundamentally different. In the present study, we further studied how these two element types influence the strength prediction.

It should be noticed, even for a FE model with fixed mesh pattern, the scale of its associated optimization problem still greatly depends on the element type: a 2D RVE model is arbitrarily picked to illustrate this difference. The model is taken from Group 2 and consists of 17,739 elements and 17,915 nodes. The scale of the optimization problem related to this model is given in Table 2. The table shows that the problem scale increases significantly when a 2D model is extended to 2.5D. A comparison of plane stress and 2.5D model pointed out that the number of primal variables has increased for around 6 times, while dual variables for around 2.5 times.

As has been stressed, plane stress and plane strain correspond to two extreme cases about strength in the 3rd direction and the original intention to introduce 2.5D model is to overcome such obstacle. However, beside such advantage 2.5D model also, unexpectedly, exhibits advantage in mesh insensitivity. For a 30 μm-by-30 μm RVE, two

Table 2 Scale of optimization problem for a FE mesh with different element types

	Limit analysis		Shakedown analysis	
	P. Stress	P. Strain	2.5D	
Num. Var	35,478	70,956	212,868	390,258
Num. EQ	17,914	17,914	53,742	142,437
Num. InEQ	17,739	17,739	35,478	70,956

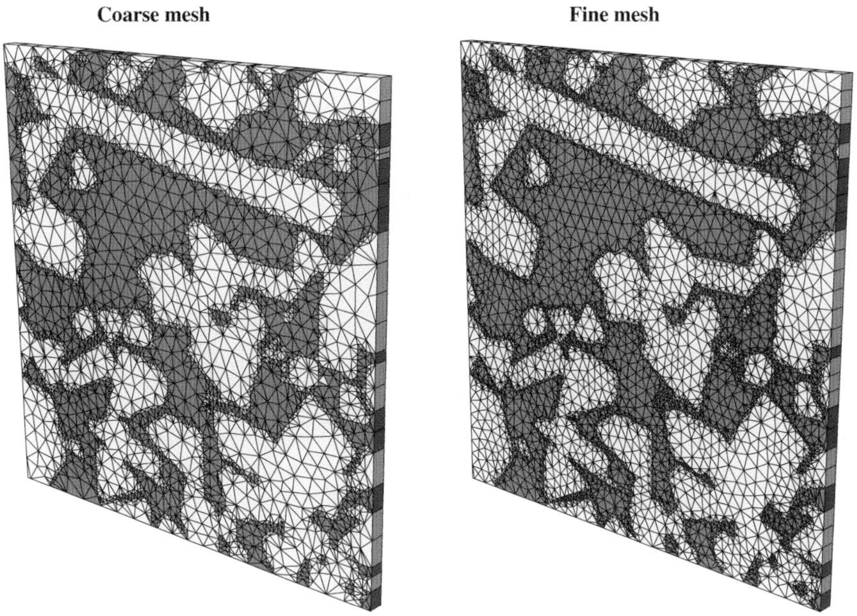

Fig. 3 RVE model with different mesh density

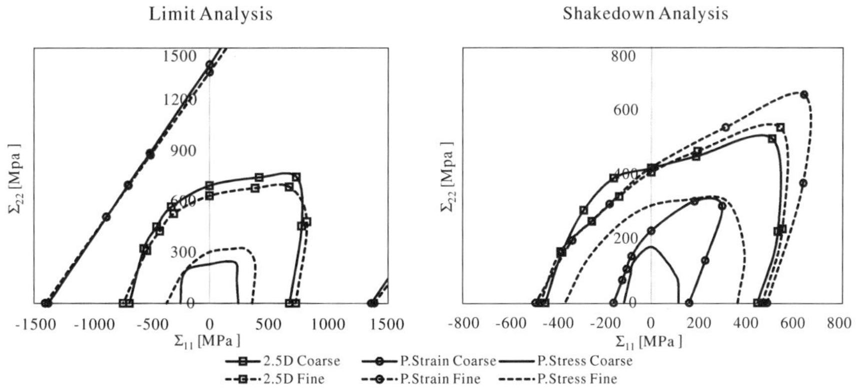

Fig. 4 Limit and shakedown domains for RVE Nr. 2 meshed with different element types

element densities are adopted. As can be observed from Fig. 3, the model with coarse mesh contains only 5,784 elements, whereas the number of elements in fine mesh model is almost doubled which renders 9,940. Limit and shakedown domains for both models are shown in Fig. 4. The domains obtained in limit analysis enlarge along with the increase of out-of-plane strength. Meanwhile, it is also evident that plane strain models have extremely high strength around $\theta = \pi/4$. This can be explained as follows: for plane strain, $\theta = \pi/4$ corresponds to the loading condition of globally hydrostatic stress. Since hydrostatic stress does not contribute to the von-Mises yield condition, material under this load can sustain exceptionally high global stress.

Unlike limit domains that are insensitive to discretization, shakedown domains appear to have more dependence on the mesh density, which means the local behavior has more influence on the overall performance of material. Comparing shakedown domains of different model types, it is manifest that 2.5D model is least sensitive to mesh density. For this reason, 2.5D models are used in the remaining part of our study.

3.3 Statistical Analysis and Study on the Size Effect

To prepare finite element models for statistical analysis, a uniform mesh setting is adopted to all involved models. For each individual microstructure, elements covering the non-critical regions were assigned with a global size of 0.8 μm; and near the phase boundaries a finer mesh is used with an edge size of 0.2 μm. Under this configuration, the number of elements for 30 μm-by-30 μm RVEs varies roughly between 6,000–9,000, and 13,000–18,000 for 40 μm-by-40 μm RVEs. The size of the optimization problem can be estimated by referring to Table 2.

As Table 3 indicates results are dispersively distributed. Because RVEs inside the same group are fixed in size and constituents, this reflects the contribution of the microstructure. Results shown in Table 3 can be interpreted as follows: The microstructure has a considerable impact on different global material properties which is in general strong for the nonlinear than the linear ones. For example, the normalized variance of Poisson's ratio for Group 1, $C_{\bar{\upsilon}}^{\text{Group2}}$ is less than 0.013. But the same indicator of endurance limit, $C_{\Sigma^U}^{\text{Group2}}$, becomes 0.193. A transverse comparison has been made between the distribution of Σ^U and Σ^∞ (Fig. 5). According to Fig. 5, the scatter of Σ^U is more pronounced than Σ^∞, for the reason outlined above. This indicates the structure to have more influence on the former parameter.

Normally larger RVEs will retain smaller C_x compared to smaller RVEs, but exceptions can occasionally be observed: e.g. $C_{\Sigma^U}^{\text{Group1}}$ is 0.176, but $C_{\Sigma^U}^{\text{Group2}}$ increases

Table 3 Statistics of selected global material properties

Model group	Parameters	$\bar{x}$	x_{SD}	C_x
Group 1	$\bar{E}$ (GPa)	394.41	33.98	0.086
(30 μm-by-30 μm)	$\bar{\upsilon}$ [-]	0.273	0.007	0.027
	Σ^Y (MPa)	440.45	72.74	0.165
	Σ^U (MPa)	504.22	88.87	0.176
	Σ^∞ (MPa)	373.93	30.64	0.082
Group 2	$\bar{E}$ (GPa)	388.78	21.18	0.055
(40 μm-by-40 μm)	$\bar{\upsilon}$ [-]	0.276	0.003	0.013
	Σ^Y (MPa)	416.83	40.65	0.097
	Σ^U (MPa)	527.65	102.37	0.193
	Σ^∞ (MPa)	369.59	27.08	0.073

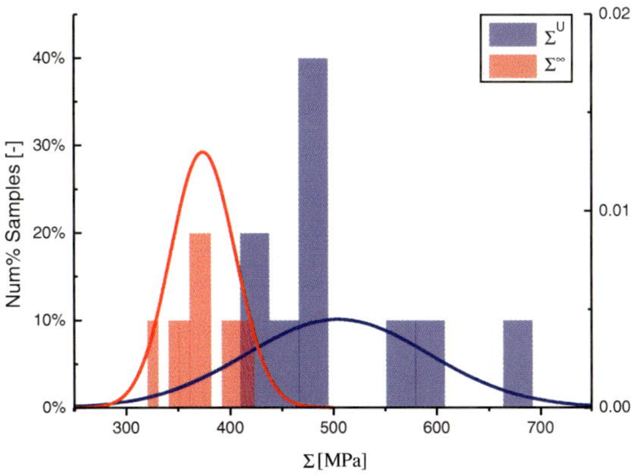

Fig. 5 Distribution of Σ^U and Σ^∞ associated to models in Group 1

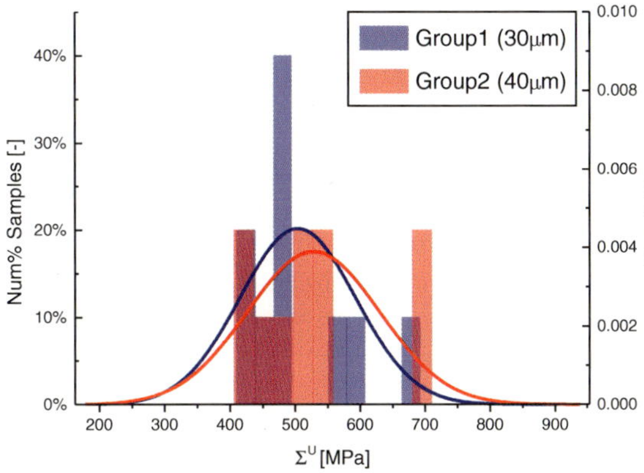

Fig. 6 Deoendence of Σ^U's distribution on size

to 0.193 (Fig. 6). More interestingly, the size effect also depends on properties being considered. For example, it happens simultaneously that $C^{\text{Group2}} < C^{\text{Group1}}$ for Σ^U, and in contract $C^{\text{Group2}} > C^{\text{Group1}}$ for the other properties including $\bar{E}$, $\bar{\upsilon}$, Σ^Y. A possible reason can be the effect of localization: when a global feature depends severely on some localized material behavior, the increase of RVE size can not fully expel the occurrence of such local effect and thus will not enhance the quality of a model.

Localization is the cause of another noteworthy phenomenon: for both model groups Σ^∞ are smaller than Σ^Y. In the first glance, this result may seems disturbing, nevertheless it can be well understood from the difference between Σ^Y and load leading to the onset of local plasticity. To a certain extent, the development of local plasticity does not contribute much to Σ^Y because of its limited share of volume. However, for Σ^∞, since it doesn't allow alternating plasticity to take place at any material point, its value is sensitive to stress concentration. In summary, $\Sigma^\infty < \Sigma^Y$ reflects serious stress concentration, and it also implies that alternating plasticity is the major cause of the plastic failure of current composite.

4 Conclusions

In this paper, it is presented how lower-bound DM, homogenization technique and statistical analysis can be used to study the strength of non-periodic PRMMC materials. The material investigated, WC-30 Wt.% Co, is a typical random PRMMC with complex microstructure. Main findings:

- By adopting an efficient algorithm, DM can be used to study composites with complex real microstructure.
- Although 2.5D models lead to optimization problems of greater size, this model type is more advantageous: first, the out-of-plane strengths in 2.5D models are more reasonable. Second, in shakedown analysis these models are more mesh insensitive.
- Σ^U is relatively mesh insensitive, increasing with the increase of the out-of-plane stiffness.
- Size effect of RVEs is not absolute; its increase does not necessarily leads to decrease of disparity among models. If a macroscopic material behavior to be studied depends strongly on localized behavior, then enlarging RVE size will not continuously make the model more objective.
- Highly localized alternating plasticity is the major cause of the failure and it leads to $\Sigma^Y > \Sigma^\infty$.

Finally, it should be emphasized that the material model and failure scenario assumed in this paper is over-restrictive. Future work should take into account kinematic hardening and material damage. Also, for reliable results, the statistical investigations have to be performed on an much larger number of numerical tests.

References

1. ABAQUS (2012) ABAQUS/CAE user's manual: version 6.12. Simulia, Dassault Systèmes
2. Akoa F, Hachemi A, An L, Said M, Tao P (2007) Application of lower bound direct method to engineering structures. J Glob Optim 37(4):609–630
3. Barrera O, Cocks A, Ponter A (2011) The linear matching method applied to composite materials: a micromechanical approach. Compos Sci Technol 71(6):797–804
4. Bisbos C, Makrodimopoulos A, Pardalos P (2005) Second-order cone programming approaches to static shakedown analysis in steel plasticity. Optim Methods Softw 20(1):25–52
5. Böhm HJ, Eckschlager A, Han W (2002) Multi-inclusion unit cell models for metal matrix composites with randomly oriented discontinuous reinforcements. Comput Mater Sci 25(1): 42–53
6. Carvelli V (2004) Shakedown analysis of unidirectional fiber reinforced metal matrix composites. Comput Mater Sci 31(1–2):24–32
7. Chawla N, Habel U, Shen YL, Andres C, Jones J, Allison J (2000) The effect of matrix microstructure on the tensile and fatigue behavior of SiC particle-reinforced 2080 Al matrix composites. Metall Mater Trans A 31(2):531–540
8. Chen G, Ozden U, Bezold A, Broeckmann C (2013) A statistics based numerical investigation on the prediction of elasto-plastic behavior of WC-Co hard metal. Comput Mater Sci 80:96–103

9. Chen H, Ponter A (2005) On the behaviour of a particulate metal matrix composite subjected to cyclic temperature and constant stress. Comput Mater Sci 34(4):425–441
10. Chen M, Hachemi A (2014) Progress in plastic design of composites. In: Spiliopoulos K, Weichert D (eds) Direct methods for limit states in structures and materials. Springer, Netherlands, pp 119–138
11. Chen M, Hachemi A, Weichert D (2013) Shakedown and optimization analysis of periodic composites. In: Saxcé G, Oueslati A, Charkaluk E, Tritsch JB (eds) Limit state of materials and structures. Springer, Netherlands, pp 45–69
12. Van Dang K, Griveau B, Message O (1989) On a new multiaxial fatigue limit criterion: theory and application. In: Brown M, Miller K (eds) Biaxial and multiaxial fatigue, EGF 3. Mechanical Engineering Publication, London, pp 479–496
13. Drucker D (1963) On the macroscopic theory of inelastic stress-strain-time-temperature behavior. In: AGARD advances in material researchs in the NATO Nations, pp 193–221
14. Drugan W, Willis J (1996) A micromechanics-based nonlocal constitutive equation and estimates of representative volume element size for elastic composites. J Mech Phys Solids 44(4):497–524
15. El-Bakry A, Tapia R, Tsuchiya T, Zhang Y (1996) On the formulation and theory of the Newton interior-point method for nonlinear programming. J Optim Theory Appl 89(3):507–541
16. Fleming W, Temis J (2002) Numerical simulation of cyclic plasticity and damage of an aluminium metal matrix composite with particulate SiC inclusions. Int J Fatigue 24(10): 1079–1088
17. Friend C (1987) The effect of matrix properties on reinforcement in short alumina fibre-aluminium metal matrix composites. J Mater Sci 22(8):3005–3010
18. Garcea G, Leonetti L (2009) Numerical methods for the evaluation of the shakedown and limit loads. In: 7th EUROMECH solid mechanics conference, Lisbon, Portugal
19. Gurobi Optimization I (2014) Gurobi optimizer reference manual. http://www.gurobi.com
20. Ibrahim I, Mohamed F, Lavernia E (1991) Particulate reinforced metal matrix composites—a review. J Mater Sci 26(5):1137–1156
21. Kanit T, Forest S, Galliet I, Mounoury V, Jeulin D (2003) Determination of the size of the representative volume element for random composites: statistical and numerical approach. Int J Solids Struct 40(13):3647–3679
22. Karmarkar N (1984) A new polynomial-time algorithm for linear programming. In: Proceedings of the 16th annual ACM symposium on theory of computing. ACM, pp 302–311
23. Koiter W (1956) A new general theorem on shake-down of elastic-plastic structures. In: Proceeding of eighth international congress of applied mechanics, pp 220–230
24. König J (1987) Shakedown of elastic-plastic structures. Fundamental Studies in Engineering. Elsevier, Warsaw
25. Krabbenhoft K, Lyamin A, Sloan S, Wriggers P (2007) An interior-point algorithm for elastoplasticity. Int J Numer Methods Eng 69(3):592–626
26. Li H (2013) A microscopic nonlinear programming approach to shakedown analysis of cohesivefrictional composites. Compos Part B-Eng 50:32–43
27. Magoariec H, Bourgeois S, Débordes O (2004) Elastic plastic shakedown of 3D periodic heterogeneous media: a direct numerical approach. Int J Plast 20(8–9):1655–1675
28. Maier G (2001) On some issues in shakedown analysis. J Appl Mech 68(5):799–808
29. MATLAB (2010) Version 7.10.0 (R2010a). The MathWorks Inc, Natick
30. Melan E (1938) Zur Plastizität des räumlichen Kontinuums. Arch Appl Mech 9(2):116–126
31. Nguyen AD, Hachemi A, Weichert D (2008) Application of the interior-point method to shakedown analysis of pavements. Int J Numer Methods Eng 75(4):414–439
32. Ozden UA, Chen G, Bezold A, Broeckmann C (2014) Numerical investigation on the size effect of a WC/Co 3D representative volume element based on the homogenized elasto-plastic response and fracture energy dissipation. Key Eng Mater 592:153–156
33. Pastor F, Loute E (2005) Solving limit analysis problems: an interior-point method. Commun Numer Methods Eng 21(11):631–642

34. Sadowski T, Nowicki T (2008) Numerical investigation of local mechanical properties of WC/Co composite. Comput Mater Sci 43(1):235–241
35. Schwabe F (2000) Einspieluntersuchungen von Verbundwerkstoffen mit periodischer Mikrostruktur. PhD thesis, RWTH Aachen
36. Segurado J, Llorca J (2002) A numerical approximation to the elastic properties of sphere-reinforced composites. J Mech Phys Solids 50(10):2107–2121
37. Simon JW, Weichert D (2011) Numerical lower bound shakedown analysis of engineering structures. Comput Methods Appl Mech 200(41):2828–2839
38. Suquet P (1987) Homogenization techniques for composite media. In: Sanchez-Palencia E, Zaoui A (eds) Introduction, Lecture Notes in Physics, vol 272. Springer, Berlin, pp 193–198
39. Swaminathan S, Ghosh S, Pagano N (2006) Statistically equivalent representative volume elements for unidirectional composite microstructures: part I-without damage. J Compos Mater 40(7):583–604
40. Torquato S (2002) Random heterogeneous materials: microstructure and macroscopic properties. Springer, New York
41. Wächter A, Biegler L (2006) On the implementation of an interior-point filter line-search algorithm for large-scale nonlinear programming. Math Program 106(1):25–57
42. Weichert D, Hachemi A, Schwabe F (1999) Application of shakedown analysis to the plastic design of composites. Arch Appl Mech 69(9–10):623–633
43. Wright M (2005) The interior-point revolution in optimization: history, recent developments, and lasting consequences. Bull Am Math Soc 42(1):39–56
44. You JH, Kim B, Miskiewicz M (2009) Shakedown analysis of fibre-reinforced copper matrix composites by direct and incremental approaches. Mech Mater 41(7):857–867
45. Zhang H, Liu Y, Xu B (2009) Plastic limit analysis of ductile composite structures from micro-to macro-mechanical analysis. Acta Mech Solida Sin 22(1):73–84
46. Zohdi T, Wriggers P (2008) An introduction to computational micromechanics. Springer, New York
47. Zouain N, Borges L, Silveira JL (2002) An algorithm for shakedown analysis with nonlinear yield functions. Comput Methods Appl Mech 191(23–24):2463–2481

A New Starting Point Strategy for Shakedown Analysis

Konstantinos Nikolaou, Christos D. Bisbos, Dieter Weichert and Jaan-W. Simon

Abstract Shakedown analysis is currently implemented by the coupling of finite element methods with techniques of computational optimization. Engineering structures problems contain a large number of variables and constraints, leading to large-scale nonlinear programming problems, since, usually, nonlinear yield criteria are preferred. The respective algorithms use iterative techniques to solve the problem at hand and the selection of a starting point is of crucial importance for their performance. To this goal the elastic limit solution could be applied, which yields a feasible point, since the zero residual stress identically satisfies the null space conditions. The present study proposes a mechanically motivated, simple technique to obtain an initial feasible point with nonzero residual stresses starting from the plastic shakedown analysis. The residual stresses obtained by this problem are generally infeasible and they are projected into the null space of the equilibrium conditions in order to yield a feasible set of self-equilibrating nonzero stresses. Next, this feasible point is completed by a safety factor, obtained from a one-dimensional optimization

K. Nikolaou (✉) · C.D. Bisbos
Department of Civil Engineering, Institute of Steel Structures,
Aristotle University of Thessaloniki, 54124 Thessaloniki, Greece
e-mail: konnikol@civil.auth.gr

C.D. Bisbos
e mail: cbisbos@civil.auth.gr

D. Weichert
Institute of General Mechanics, RWTH Aachen University,
Templergraben 64, 52062 Aachen, Germany
e-mail: weichert@iam.rwth-aachen.de

D. Weichert
Graduate School of Energy Science, Kyoto University, Kyoto, Japan

J.-W. Simon
Institute of Applied Mechanics, RWTH Aachen University,
Mies-van-der-Rohe-Straße1, 52074 Aachen, Germany
e-mail: jaan.simon@rwth-aachen.de

P. Fuschi et al. (eds.), *Direct Methods for Limit and Shakedown Analysis of Structures*,
DOI 10.1007/978-3-319-12928-0_7

problem of elastic limit type. The applicability and appropriateness of this approach is studied by numerical comparisons.

1 Introduction

Engineering structures have usually to be designed to withstand during their lifetime a large number of varying loads without knowing their exact history. These loads might exceed the elastic limit and several failure mechanisms are possible. In instantaneous collapse (plastic collapse) the static bearing capacity is exceeded and the system fails during the first load cycle. Ratcheting, also called incremental collapse happens if after every load cycle plastic strain is accumulated. Failure might occur even if total plastic strain is limited and small, but increments in an alternating manner occur. This failure mode is called alternating plasticity or plastic shakedown.

Direct methods [1–4] can be conveniently used to determine if a system is safe taking into account the aforementioned mechanisms. Shakedown analysis in particular can define the safety margins of a structure under loads varying arbitrarily but within a defined bounded region. Otherwise, a huge number of conventional step-by-step elastoplastic analyses with different loading and unloading scenarios is needed. Therefore, direct methods are efficient tools to calculate the bearing capacity avoiding cumbersome step-by-step calculations.

Quasi-lower bound techniques, based on the shakedown theorem by Melan [1], are currently implemented by the coupling of finite element methods with techniques of computational optimization. Depending on the yield criteria type, linear or nonlinear programming methods can be used. Convexity and differentiability are two features of major importance for nonlinear criteria, since their absence may provoke serious burdens to nonlinear programming algorithms. For engineering structures the arising problems are usually very large scale, constrained, nonlinear but convex. Such problems can be treated by a number of optimization software packages. Examples of widely available respective software are free codes as IPOPT [5], SeDuMi [6], SDPT3 [7] or commercial packages like MOSEK [8].

Due to the size of the problems encountered in engineering practice or the yield criterion type, algorithms applied might have low efficiency. Several techniques are presented in the literature to improve computational performance of the direct methods. For problems arising in J_2-plasticity, second order cone programming (SOCP) algorithms are very efficient, even for large scale problems. Christiansen and Andersen [9], Trillat and Pastor [10], Bisbos et al. [11] and Krabbenhøft et al. [12] have successfully used SOCP algorithms. However, not all problems fit within the SOCP scheme. Interestingly, the plastic interaction surfaces described in European structural steel design norm EC3, are generally nonlinear and for some cases non-convex. Limit and shakedown analysis under the EC3 criteria leads to large scale nonlinear problems very difficult to handle. In [13, 14] a systematic method to construct ellipsoidal approximations to these yield criteria is described. Using these ellipsoidal approximations, the arising problems become SOCP and upper and lower bounds to the safety factors can be derived.

Codes IPSA [15–17] and IPDCA [18, 19] belong to the family of Interior point algorithms specially tailored for limit (cf. e.g. [20–23]) or shakedown (cf. e.g. [24, 25]) problems. IPDCA and IPSA have proven to be efficient even for real scale engineering problems under the von Mises yield criterion. Moreover, a selective strategy was implemented in IPSA. The plastically most active zones of the structure are detected. Then, the entire system is reduced to a substructure containing only these zones. Subsystems are solved with a significantly lower cost.

Semidefinite programming (SDP) constitutes a straightforward generalization of SOCP framework. SDP can be used to mathematically model more complex yield criteria as the Tresca or the Mohr-Coulomb ones (see e.g. [26]). Also, SDP has been used to solve shakedown problems under ellipsoidal variable load domains [27]. More general load domains in parameterized form have been treated in [28] for polyhedral yield criteria.

Selection of appropriate starting points is crucial for the performance of all optimization algorithms used in the direct methods of plasticity. Appropriate starting points can significantly reduce computational time, while unfavorable ones may even lead to failure of convergence. Yildirim and Wright [29] and Gertz et al. [30] proposed warm start procedures for linear and nonlinear interior point algorithms respectively. Mehrotra [31] proposed a method based on the solution of quadratic optimization problems, which was later simplified by Stojkovic and Stanimirovic [32]. Recently starting point strategies are gaining importance in direct method problems [17, 18]. Akoa et al. [18] proposed an initial solution vector and some values for the Lagrange multipliers. Later Simon, Höwer and Weichert [17] took into account the underlying mechanical problem and used the elastic limit values as an initial guess. Both approaches lead to significantly improved computational times compared to using the zero point. In the present paper a new simple technique to produce a feasible starting point for shakedown problems is presented. The new starting point contains non-zero residual stresses and leads to improved solution times.

The plan of the paper is as follows. Shakedown analysis under the nonlinear von Mises yield criterion is discussed in next section. The third section constitutes the heart of the paper. Therein, the starting point strategies already used in the literature are shortly presented and the new strategy is introduced and discussed in detail. The fourth section contains numerical examples to illustrate the effectiveness of the proposed strategy. Finally some concluding remarks close the paper.

2 Shakedown Analysis Under von Mises Yield Criterion

Let us consider a structure discretized using the finite element method with NE elements, with NU free degrees of freedom and with NG numerical integration points (Gauss points) for the whole model. The quasi-lower bound elastic shakedown analysis problem in mathematical notation can be written:

$$P_{ESD} \quad \max \alpha \tag{1a}$$

$$s.t.: \quad \sum \mathbf{C}_j \boldsymbol{\rho}_j = 0 \tag{1b}$$

$$\mathbf{s}_j^{(i)} = \alpha \mathbf{v}_j^{(i)} + \boldsymbol{\rho}_j \tag{1c}$$

$$\mathbf{s}_j^{(i)} \in F_j \qquad \text{for } i = 1, \ldots, NV, \; j = 1, \ldots, NG \tag{1d}$$

α is the required safety factor. Subscript j runs over the NG Gauss points and superscript (i) runs over the NV vertices of the variable load domain. $\mathbf{v}_j^{(i)}$ represents the known elastic stress vector at the jth GP due to the ith loading and $\mathbf{s}_j^{(i)}$ is the respective local elastoplastic stress vector, which must be contained within the local yield criterion set F_j. The local vector $\boldsymbol{\rho}_j$ contains the respective residual stresses, and Eq. (1b) represents the condition, that the residual stresses must be self-equilibrating. In this equation, $\mathbf{C}_j$ is a known equilibrium matrix, depending on the discretization and the boundary conditions. The primary problem unknowns are: (α,$\boldsymbol{\rho}_j$) and the secondary ones $\mathbf{s}_j^{(i)}$. For NV$=1$, the load domain shrinks to a single point and the shakedown turns into a limit analysis problem.

Depending on the yield criteria sets F_j, P_{ESD} can be a linear or nonlinear programming problem. If the yield criteria sets are polyhedral, it becomes a linear programming problem. Otherwise, under nonlinear yield criteria, a nonlinear programming (NLP) problem must be solved. The algorithms deployed for the last depend on the type of nonlinearity of the criteria. In the present work, the von Mises criterion will be used. This yield criterion can be formulated as an Euclidean length constraint (cf. e.g. [11]), if the stresses are properly transformed. For the simple plane stress problem:

$$\hat{\mathbf{C}}_\mathbf{j} = \mathbf{C}_\mathbf{j}\, \mathbf{Q}_j^{-1}, \quad \hat{\mathbf{v}}_j^{(i)} = \mathbf{Q}_j\, \mathbf{v}_j^{(i)}, \quad \hat{\boldsymbol{\rho}}_j = \mathbf{Q}_j\, \boldsymbol{\rho}_j$$

where

$$\mathbf{Q}_j = \frac{1}{\sigma_Y} \begin{bmatrix} 1 & -0.5 & 0 \\ 0 & 0.5\sqrt{3} & 0 \\ 0 & 0 & \sqrt{3} \end{bmatrix}$$

After these transformations the resulting problem becomes:

$$P_{v.Mises,SOCP} \quad \alpha_{ESD} = \max \alpha \tag{2a}$$

$$s.t.: \quad \sum \hat{\mathbf{C}}_j \hat{\boldsymbol{\rho}}_j = 0 \tag{2b}$$

$$\hat{\mathbf{s}}_j^{(i)} = \alpha\, \hat{\mathbf{v}}_j^{(i)} + \hat{\boldsymbol{\rho}}_j \tag{2c}$$

$$\left\| \hat{\mathbf{s}}_j^{(i)} \right\| \leq 1.0 \qquad \text{for } i = 1, \ldots, NV, \; j = 1, \ldots, NG \tag{2d}$$

As already mentioned, the problem unknowns comprise: the safety factor α, the NG transformed local residual stress vectors $\hat{\boldsymbol{\rho}}_j$ and the $NGxNV$ local vertex transformed elastoplastic stress vectors $\hat{\mathbf{s}}_j^{(i)}$. By merging all the problem variables we get the following solution vector:

$$\mathbf{x} = [\hat{\boldsymbol{\rho}}_1, \ldots, \hat{\boldsymbol{\rho}}_j, \ldots, \hat{\boldsymbol{\rho}}_{NG}, \hat{\mathbf{s}}_1^{(1)}, \ldots, \hat{\mathbf{s}}_j^{(i)}, \ldots, \hat{\mathbf{s}}_{NG}^{(NV)}, \alpha]$$

Evidently, the resulting optimization problem can take the rather abstract form ($\mathbf{e}$ is a vector with zero entries, except the last one, which equals one):

$$P_{v.Mises,SOCP} \quad \max \mathbf{e}^T \mathbf{x} \tag{3a}$$

$$s.t.: \quad \mathbf{A}\mathbf{x} = \mathbf{0} \tag{3b}$$

$$c_I(\mathbf{x}) \geq \mathbf{0} \tag{3c}$$

The equalities Eq. (3b) comprise the null space condition and the linear dependencies defining the elastoplastic stresses $\hat{\mathbf{s}}_j^{(i)}$.

In the following, the software IPOPT [5] will be used to solve this optimization problem. IPOPT is a public domain package for general nonlinear optimization problems using the Interior Point Method (IPM). IPOPT requires smoothness of the involved functions and parameters have to be adjusted for each problem type. Particularly important for the present work is that IPOPT allows the user to define the starting point.

Selecting a user-defined starting point for the optimization problem depends on the algorithm used for the solution, i.e. some solvers require the starting point to be feasible, or require values for the Lagrange multipliers. IPOPT allows for infeasible initial values. It checks only if the initial point stays "far enough" from the bounds. If, for example a variable is set to its lower bound, then IPOPT modifies the user defined starting point so that it is sufficiently away from the boundary.

3 Starting Point Strategy

Several heuristics to select suitable points are studied in the literature. Yildirim and Wright [29], Gertz et al. [30] proposed several procedures for linear and nonlinear interior point algorithms for linear and nonlinear programming respectively. Another method was proposed by Mehrotra [31] based on the solution of quadratic optimization problems and was later simplified by Stojkovic and Stanimirovic [32]. The trivial option to select a feasible starting point is to use zeros for all the values of the variables. Then both the equality and inequality constraints are fulfilled. This is a common option from mathematical point of view, but still is far from the optimal starting point. The elastic limit [17] can be another feasible initial point that leads to improved solution times.

3.1 The Three Problem Types

Let us now summarize the three problem types, associated with the shakedown analysis of structures, namely the elastic shakedown problem, the elastic limit (first yield) problem and the plastic shakedown (alternating plasticity) problem.

3.1.1 Elastic Shakedown Analysis Problem

$$P_{ESD} \quad \alpha_{ESD} = \max \alpha \tag{4a}$$

$$s.t.: \quad \sum \mathbf{C}_j \boldsymbol{\rho}_j = 0 \tag{4b}$$

$$\mathbf{s}_j^{(i)} = \alpha \mathbf{v}_j^{(i)} + \boldsymbol{\rho}_j \tag{4c}$$

$$\mathbf{s}_j^{(i)} \in F_j \qquad \text{for } i = 1, \ldots, NV, \; j = 1, \ldots, NG \tag{4d}$$

The null-space condition Eq. (4b) introduces a coupling between the stress states at the various stress check points (Gauss points) of the structure. Omitting this linear coupling (complicating) constraint reduces the problem to the much simpler plastic shakedown problem.

3.1.2 Plastic Shakedown Analysis Problem

$$P_{PSD} \quad \alpha_{PSD} = \min \alpha_j^* \qquad \text{for } j = 1, \ldots, NG \tag{5a}$$

$$\alpha_j^* = \max \alpha \tag{5b}$$

$$s.t.: \mathbf{s}_j^{(i)} = \alpha \mathbf{v}_j^{(i)} + \boldsymbol{\rho}_j \tag{5c}$$

$$\mathbf{s}_j^{(i)} \in F_j \qquad \text{for } i = 1, \ldots, NV \tag{5d}$$

The local maximization problem Eqs. (5b)–(5d) constitutes the lower level of the minmax problem, where the global minimization problem Eq. (5a) is the upper level one.

A similar minmax problem is obtained if we set the residual stresses zero, satisfying the null space condition. This problem is the first yield or elastic limit problem.

3.1.3 Elastic Limit Problem

$$P_{ELM} \quad \alpha_{ELM} = \min \alpha_j^* \qquad \text{for } j = 1, \ldots, NG \tag{6a}$$

$$\alpha_j^* = \max \alpha \tag{6b}$$

$$s.t.: \mathbf{s}_j^{(i)} = \alpha \mathbf{v}_j^{(i)} \tag{6c}$$

$$\mathbf{s}_j^{(i)} \in F_j \qquad \text{for } i = 1, \ldots, NV \tag{6d}$$

These simple minmax problems are computationally very cheap. Moreover, P_{ELM} is a one-dimensional optimization problem. For convex yield criteria sets F_j, P_{ESD} is a convex optimization problem, but it is not a strictly convex one, since the objective function is linear. Consequently, the optimal value α_{ESD} is unique, but there are several solution points, i.e. pairs of $(\alpha_{ESD}, \boldsymbol{\rho}_{ESD})$ with different residual stress fields $\boldsymbol{\rho}_{ESD}$. This fact holds also for the plastic shakedown problem. The following inequality holds:

$$\alpha_{ELM} \leq \alpha_{ESD} \leq \alpha_{PSD} \tag{7}$$

In many cases the elastic and plastic shakedown analysis safety factors coincide. Moreover, if the load domain is a centrally symmetric bounded hyper-polyhedron centered at the origin it can be shown that $\alpha_{ELM} = \alpha_{ESD} = \alpha_{PSD}$.

3.2 The New Starting Point Strategy

Now, a new starting point strategy starting from the plastic shakedown problem is presented. Let $\boldsymbol{\rho}_j^{PSD}$ be the stresses obtained by the local plastic shakedown analysis problems. In general they are not self-equilibrating, since they satisfy the yield criteria conditions but the null space condition is violated, i.e.:

$$\sum \mathbf{C}_j \boldsymbol{\rho}_j^{PSD} \neq 0$$

We compute the self-equilibrating stresses $\boldsymbol{\rho}_j^*$, which are nearest to the stresses $\boldsymbol{\rho}_j^{PSD}$, by solving the simple linear projection problem.

3.2.1 Euclidean Projection into the Null Space of the Equilibrium Matrix

$$\textit{Find } \boldsymbol{\rho}_j^* \quad \min 1/2 \sum (\boldsymbol{\rho}_j^* - \boldsymbol{\rho}_j^{PSD})^T (\boldsymbol{\rho}_j^* - \boldsymbol{\rho}_j^{PSD}) \tag{8a}$$

$$s.t.: \quad \sum \mathbf{C}_j \boldsymbol{\rho}_j^* = 0 \tag{8b}$$

The pair $(\alpha_{PSD}, \boldsymbol{\rho}_j^*)$ generally does not satisfy the yield criterion, i.e. it represents an infeasible point for the elastic shakedown problem. However, we can combine $\boldsymbol{\rho}_j^*$ with an appropriate value of α to obtain a feasible point. This task is accomplished by solving the following minmax problem, which is a simple generalization of P_{ELM}. This one-dimensional problem has the residual stresses as data.

3.2.2 Generalized Elastic Limit Problem

$$P_{GELM} \quad \alpha_{PSD} = \min \alpha_j^* \qquad \text{for } j = 1, \ldots, NG \tag{9a}$$

$$\alpha_j^* = \max \alpha \tag{9b}$$

$$s.t.: \; \mathbf{s}_j^{(i)} = \alpha \mathbf{v}_j^{(i)} + \boldsymbol{\rho}_j^* \tag{9c}$$

$$\mathbf{s}_j^{(i)} \in F_j \qquad \text{for } i = 1, \ldots, NV \tag{9d}$$

The pair $(\alpha_{GELM}, \boldsymbol{\rho}_j^*)$ constitutes the new starting point we propose. The main reasons for our motivation are:

- The stresses $\boldsymbol{\rho}_j^{PSD}$ are well-centered, at least for yield criteria having the form of norm constraints, which is true for von Mises and Tresca criteria (cf. [11, 24]). This fact greatly facilitates the application of IPMs.
- Generally, we can complete some residual stress field $\boldsymbol{\rho}^{feas}$, satisfying the null space condition, with a value of α^{feas}obtained via P_{GELM} in order to obtain a feasible point of P_{ESD}. For feasible pairs $(\alpha^{feas}, \boldsymbol{\rho}^{feas})$, the following inequality holds:

$$a_{feas} \le a_{ESD} \le a_{PSD} \tag{10}$$

Obviously Eq. (10) is specialized to Eq. (7) if P_{ELM} is used as stating point.

- Very often, the case $\alpha_{ESD} = \alpha_{PSD}$ occurs. This is prone to appear especially in problems with free variables, as e.g. the spherical (volumetric) parts of the stress tensor in 3D shakedown analyses under deviatoric yield criteria.
- In several decomposition techniques for mathematical optimization [33], a common first solution step is to relax the complicating (coupling) constraints. Then, the original problem decomposes in a series of several, independent simpler problems. For the case of shakedown, the complicating constraints are the null space condition.

Figure 1 depicts a flow chart with the starting point strategy presented here. This methodology can be applied to general nonlinear criteria and frame or shell structures even for very large scale problems.

It is expected, that, using the plastic shakedown factor and the corrected residual stress field, the high stress zones of the structure can be determined in early phases of the whole iteration process. So, this initial point strategy can be combined with the selective algorithm [16] and further improve the solution efficiency. If the active zone is concentrated, then the selective algorithm would be very efficient.

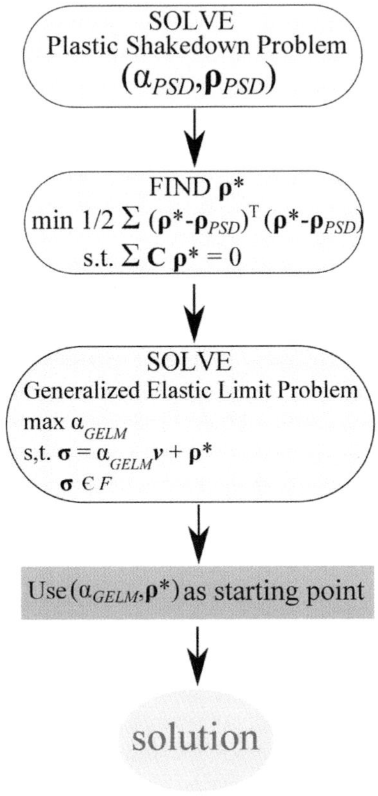

Fig. 1 Starting point strategy flow chart

3.3 Initial Values for the Lagrange Multipliers

To initialize the interior point algorithms efficiently initial values for the solution vector are not enough. Appropriate values for the Lagrange multipliers should be also assigned. This is not an easy task and the selection of incorrect Lagrange multipliers might have the opposite result and lead to increased solution times.

Akoa et al. [18] suggested useful estimations for the Lagrange multipliers separately corresponding to the equalities and the inequalities. Although the suggested values lead to better solution times compared to using simply zeros, all the equalities and inequalities are treated the same way and the Lagrange multipliers are not even close to their final values. According to [30] large initial values for the multipliers should be avoided, so regarding the equality multipliers smaller values should be more appropriate.

In [17] Simon, Höwer and Weichert the initial values for the Lagrange multipliers were selected to satisfy the complementary slackness conditions, hence:

$$\lambda_{I0} = \frac{\mu_0}{c_I(\mathbf{x}_0)}$$

μ_0 is the initial value for the barrier parameter. The multipliers connected to the equalities were set to a constant value since selecting other values would require solving a huge dense (not sparse) system of equations. So λ_{E0} are set equal to:

$$\lambda_{E0} = 10^{-2}$$

Using these values the starting point should be near the central path of the problem. For the currently presented starting point strategy, these values will be used.

IPOPT requires for the warm start the selection of some values for the Lagrange multipliers connected to the variable bounds. For our case no variable has finite bounds, so these values are fixed to zeros.

4 Numerical Examples

In this section some numerical examples are used to show the efficiency of the new starting point strategy. Materials are considered isotropic, homogeneous with elastic perfectly plastic behavior. Geometrical nonlinearities and material damage were ignored. All the elastic stresses calculations were carried out by a simple research FORTRAN FEM code using CST elements with one Gauss point per element. For the computations a Dell laptop with an Intel Core i5-3210M Processor @2.50 GHz and 4 GB RAM was used. The open source optimization package IPOPT [5] was used to solve the arising nonlinear optimization problems. Several IPOPT options, for example different barrier update strategies, initial values for the barrier or the Lagrange multipliers were tested and some comments on the optimal selection of these values are made. Since interior point algorithms are complex mathematical tools and warm-start strategies for interior point methods are still under development, from only a few examples we cannot derive general conclusions, but the presented results illustrate the method's potential.

4.1 Square Plate with Central Hole

The classical problem of a square plate with a circular hole (Fig. 2) subjected to biaxial uniform loads P_1 and P_2 varying proportionally in an arbitrary manner was considered. This example has been widely used in the literature (cf. e.g. [34–41]). 880 CST elements were used to model the structure; due to symmetry only one quarter of the plate was examined. The von Mises criterion is used without linearization.

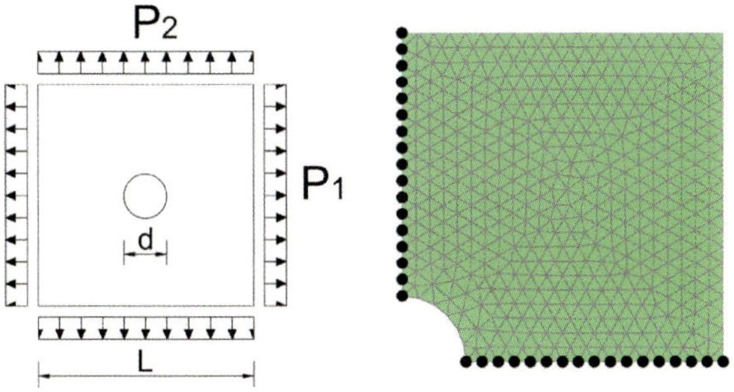

Fig. 2 *Square* plate with a central hole and FEM model

Table 1 Geometrical and material properties for the square plate

Geometry	L = 100 mm t = 10 mm D = 20 mm	Material	E = 210 GPa v = 0.30 σ_Y = 235 MPa

Table 1 presents the geometry of the problem and the properties of the material used. Two loading spaces will be considered. The first one, Ω_1 with only load P_1 varying arbitrarily between 0 and 100 MPa and $\Omega_2$with both load patterns P_1 and P_2 varying proportionally between 0 and 100.

$$\begin{aligned}\Omega_1 &= \{P : P = \mu P_1, \qquad \mu \in [0, 1] \;\; P_1 = 100MPa\}\\ \Omega_2 &= \{P : P = \mu(P_1 + P_2), \; \mu \in [0, 1] \;\; P_1 = P_2 = 100MPa\}\end{aligned}$$

Some results from the literature that were used to validate the model for the referenced load space are presented in Table 2. The values are normalized with respect to the yield stress: $\bar{\alpha} = \alpha P/\sigma_Y$. Table 3 contains the load factors for several limit states.

Figures 3 and 4 depict the residual and total von Mises stress fields of the structure for the two load cases respectively. All graphs are plotted on the same scale.

Four different starting point strategies were compared: the zero one, the one proposed by Akoa et al. [18] the elastic limit used by Simon et al. [17] and, finally, the present strategy. Moreover, a randomly-generated starting point has been included in order to get a better comparison of the results. For the stress variables random numbers between −1 and 1 were used since stresses in the optimization problem are normalized with respect to the yield stress. For the safety factor random numbers between 0 and 2 α_{ELM} were selected. All starting points yielded the same safety factor.

The results are presented in Table 4. Results concerning the random number starting points are the average values of a set of 20 problems that were solved. Although the mean time seems better than just using zeros, it has to be noted that there was a large deviation of the results. All times reported concern the solution of the warm-started optimization problem.

Table 2 Comparison of numerical results

Authors	Normalized shakedown factor $\bar{\alpha}$	
	$\boldsymbol{\Omega}_1$	$\boldsymbol{\Omega}_2$
Belytschko (1972) [34]	0.571	–
Hamilton et al. (1996) [35]	0.586	0.852
Gross-Weege (1997) [36]	0.611	0.882
Muscat et al. (2003) [37]	0.641	0.870
Zhang et al. (2004) [38]	0.634	0.893
Chen et al. (2008) [39]	0.649	–
Zhang et al. (2012) [40]	0.641	0.893
Present solution	0.633	0.857

Table 3 Load factors for different limit states

	Load factor α	Normalized load factor $\bar{\alpha}$
$\boldsymbol{\Omega}_1$		
Elastic limit	0.744	0.316
Gen. elastic limit	0.755	0.321
Elast. shakedown	1.487	0.633
Plastic shakedown	1.487	0.633
Limit analysis	1.928	0.820
$\boldsymbol{\Omega}_2$		
Elastic limit	1.137	0.484
Gen. elastic limit	0.830	0.353
Elast. shakedown	2.014	0.857
Plastic shakedown	2.273	0.967
Limit analysis	2.014	0.857

Several IPOPT option combinations were used; the results presented above refer to those leading to the biggest time reduction. This set of optimal options was used for all the problems solved. The adaptive barrier update strategy was preferred to the monotone one. Using the monotone barrier update strategy requires a bigger number of iterations, but every step is completed faster. $\mu_0 = 10$ (see par.3.3) was used to initiate the Lagrange multipliers. Not providing information for the Lagrange multipliers leads to a less significant time reduction and for some cases even to time increase. It was observed that the monotone update strategy is more sensitive to μ selection. Selecting inappropriate values for the Lagrange multipliers or for the barrier parameter can lead to computational times larger compared to those for using zeros. Mehrotra's predictor-corrector algorithm, which is incorporated in IPOPT, can improve the efficiency of linear and convex quadratic programming problems starting from zeros. But combined with non-zero starting points leads to

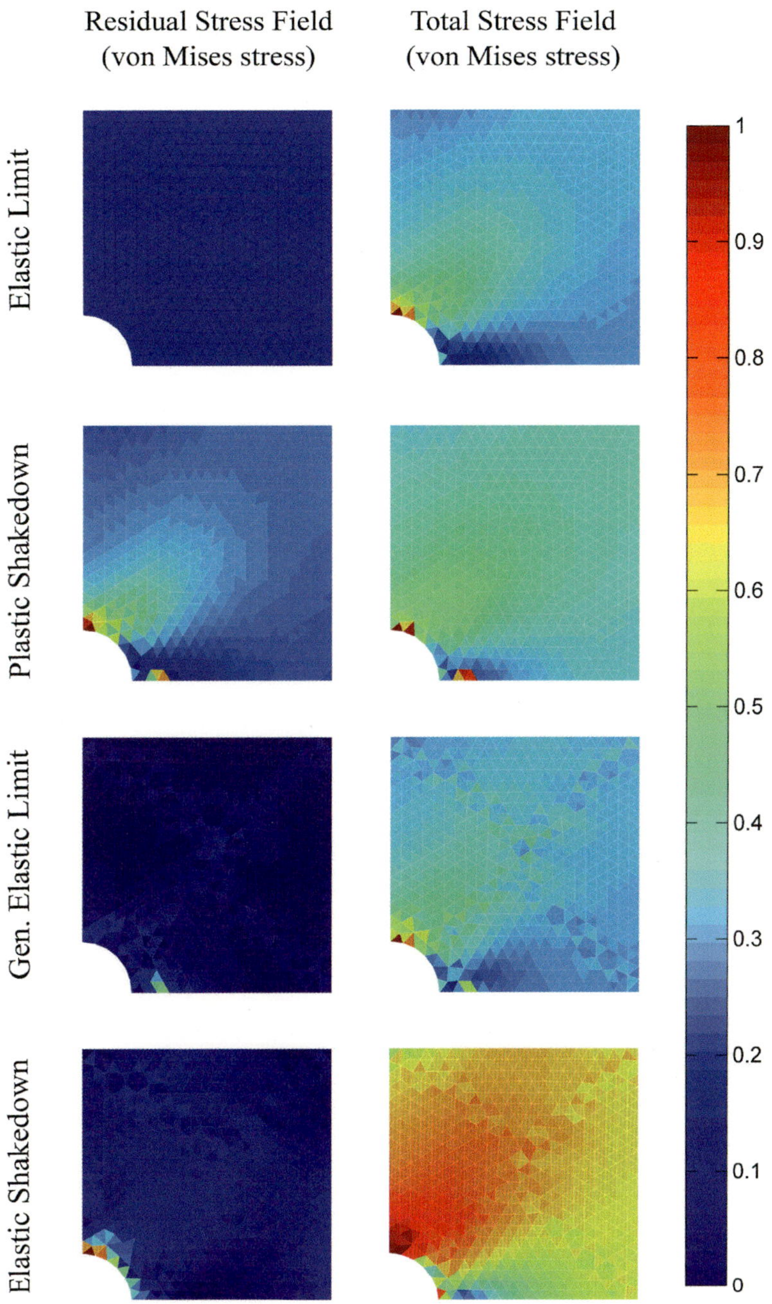

Fig. 3 Residual and total von Mises stress fields for $\boldsymbol{\Omega}_1$

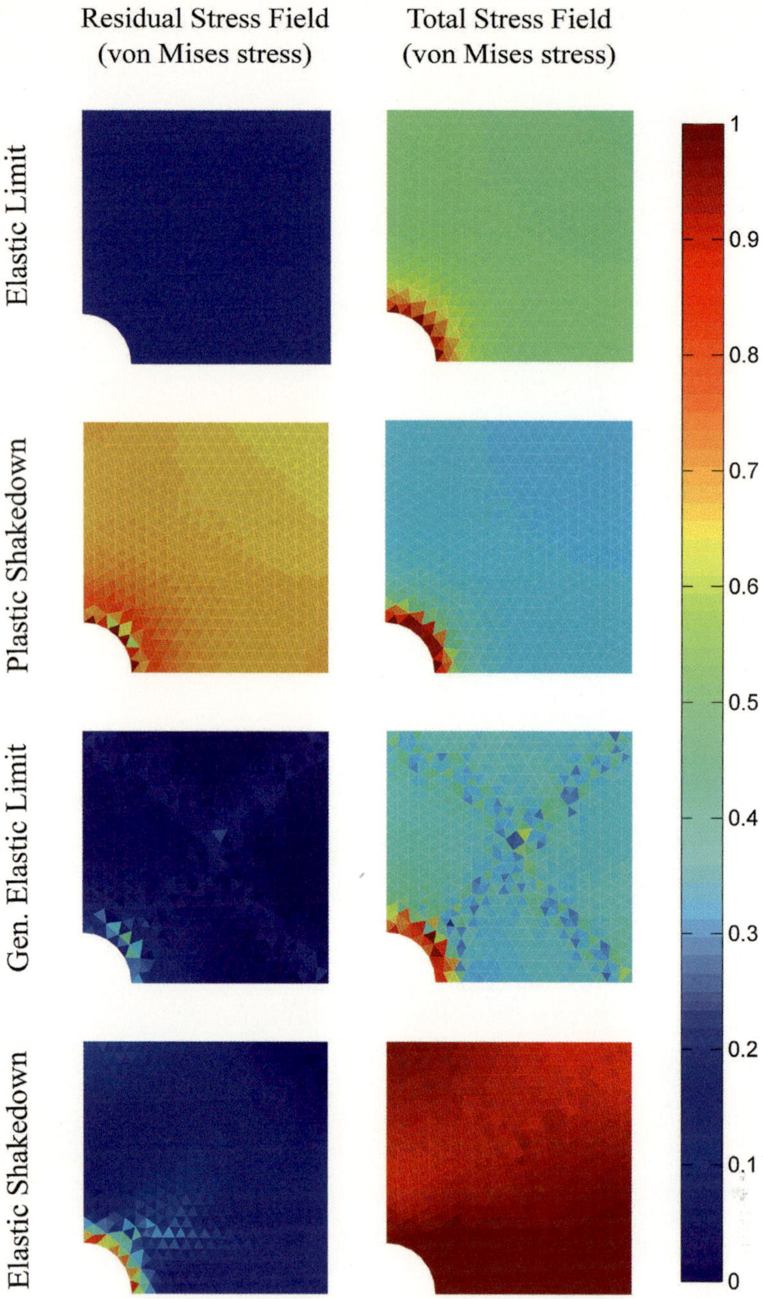

Fig. 4 Residual and total von Mises stress fields for $\boldsymbol{\Omega}_2$

Table 4 Shakedown results for different starting point strategies

	Shakedown factor	Iterations	CPU time (s)
$\boldsymbol{\Omega}_1$			
Zeros	1.487	30	165.251
Akoa	1.487	23	118.172
Random	1.487	31	30.997
Simon	1.487	26	23.218
Present	1.487	*17*	*15.678*
$\boldsymbol{\Omega}_2$			
Zeros	2.014	21	83.737
Akoa	2.014	18	71.278
Random	2.014	24	24.574
Simon	2.014	18	16.568
Present	2.014	*15*	*13.968*

relatively poor performance. Infeasible starting points can be used by IPOPT, but initial results show low efficiency. Using the elastic limit values leads to a significant reduction compared to the first strategy, but using the newly proposed strategy further reduces solution time, namely 16–32 % compared to the elastic limit and 83–90 % compared to zeros. By the results it can be seen that using random variables could be a better strategy than just selecting zeros. But as mentioned above times reported are mean times and the deviation is large. So random number points are considered inappropriate starting points (Figs. 5 and 6).

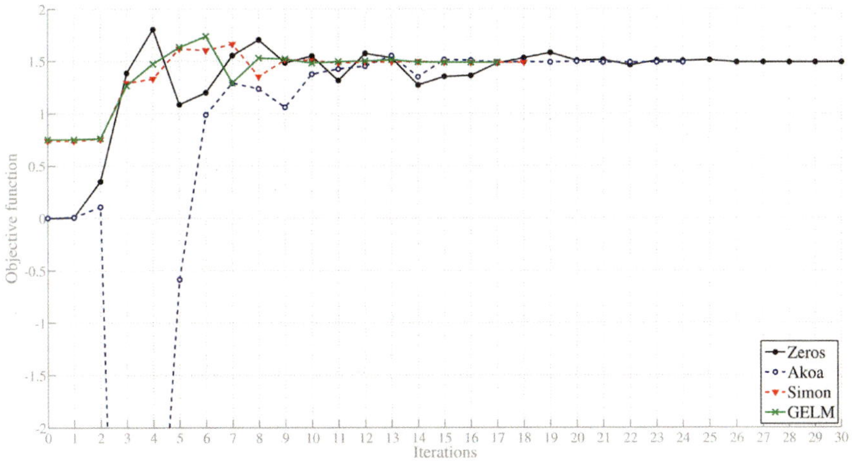

Fig. 5 Objective function versus number of iterations for $\boldsymbol{\Omega}_1$

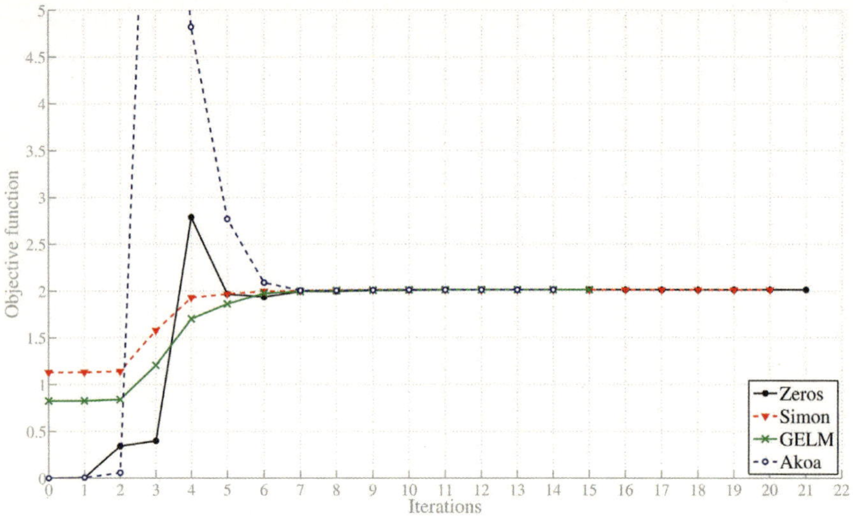

Fig. 6 Objective function versus number of iterations for $\mathbf{\Omega}_2$

5 Conclusion

Shakedown analysis of engineering structures leads to large scale nonlinear optimization problems. Computational cost is a major issue for the algorithms employed. The starting point has a strong influence on the computational time required. In the present paper a new starting point strategy was presented. Instead of using very small values for the variables or the elastic limit values, a feasible point with nonzero residual stresses was obtained by the plastic shakedown analysis. The stresses obtained this way are projected into the null space of the equilibrium conditions so that the equalities are fulfilled. Then a one-dimensional optimization problem of elastic limit type, called the generalized elastic limit, is used to define the shakedown factor that combined with the projected stresses satisfies the yield criterion at every Gauss point.

The main advantages of the presented method are that it is highly parallelizable, so even for large scale problems starting points can be calculated with insignificant computational cost. It is general enough to cover several linear and nonlinear criteria and different types of structures. Also it can be combined with the selective algorithm for further improved efficiency.

Appropriate algorithm options have been carefully selected in order to exploit the starting points potential. IPOPT requires a starting value for all the variables of the problem. Optionally the user can provide information for the Lagrange multipliers. Providing a different than zeros initial value for the variables can reduce the solution time. Moreover, providing appropriate Lagrange multipliers can further improve the algorithms performance. But, if the values are not selected carefully, the opposite result might occur. The effect of the starting points is also heavily influenced by

IPOPT options like the barrier update strategy, linear solver selection, use of special algorithms like Mehrotra's predictor corrector etc. Since nonlinear programming problems are very complex, we cannot jump to any conclusions for the optimal options easily. Further parametric studies and the use of different optimization algorithms could give a better insight into the still open subject of warm starting interior point methods.

As shown by the numerical examples studied, compared to other starting points the efficiency of the newly proposed strategy can be very high. 16–32 % faster than the elastic limit and more than 83 % compared to zeros. Having the ability to solve large scale problems fast is very important for practical applications and opens the way to shakedown analysis of improved/complicated structural models (i.e. taking into account hardening or use higher order elements). Further research is currently underway. Using different optimization solvers (for example IPSA or IPDCA) could give a very interesting insight in warm starting effects in interior point methods, a topic that is also interesting for the optimization community.

Acknowledgments The first author would like to thank DAAD for granting a short stay research scholarship at the Institute of General Mechanics (IAM) of RWTH-Aachen University, IAM for the hospitality and especially the head of IAM Prof. B. Markert for financially supporting the trip to Italy. Special thanks to Dr. Hachemi, Dr. Chen and MSc G.Chen from IAM and Dr Skordeli from Aristotle University Thessaloniki, for the fruitful discussions.

References

1. Melan E (1938) Zur Plastizität des räumlichen Kontinuums. Arch Appl Mech 9:116–126
2. König JA (1987) Shakedown of elastic-plastic structures. Elsevier, Amsterdam
3. Weichert D, Maier G (eds) (2000) Inelastic analysis of structures under variable repeated loads. Kluwer Academic, Dordrecht
4. Weichert D, Ponter ARS (eds) (2009) Limit states of materials and structures: direct methods. Springer, Wien
5. Wächter A, Biegler LT (2006) On the implementation of an interior-point filter line-search algorithm for large-scale nonlinear programming. Math Program 106(1):25–57
6. Sturm JF (1999) Using SeDuMi 1.02, a MATLAB toolbox for optimization over symmetric cones. Optim Methods Softw 11:625–653
7. Toh KC, Todd MJ, Tütüncü RH (1999) SDPT3—a MATLAB software package for semidefinite programming, version 1.3. Optim Methods Softw 11(1–4):545–581
8. Andersen ED, Roos C, Terlaky T (2003) On implementing a primal-dual interior-point method for conic quadratic optimization. Math Program 95(2):249–277
9. Christiansen E, Andersen KD (1999) Computation of collapse states with von Mises type yield condition. Int J Numer Methods Eng 46(8):1185–1202
10. Trillat M, Pastor J (2005) Limit analysis and Gurson's model. Eur J Mech-A/Solids 24(5):800–819
11. Bisbos CD, Makrodimopoulos A, Pardalos PM (2005) Second-order cone programming approaches to static shakedown analysis in steel plasticity. Optim Methods Softw 20(1):25–52
12. Krabbenhøft K, Lyamin AV, Sloan SW (2007) Formulation and solution of some plasticity problems as conic programs. Int J Solids Struct 44(5):1533–1549
13. Skordeli MA, Bisbos CD (2010) Limit and shakedown analysis of 3d steel frames via approximate ellipsoidal yield surfaces. Eng Struct 32(6):1556–1567

14. Nikolaou K, Skordeli MA-A, Bisbos CD (2013) Limit analysis of aluminium frames via approximate ellipsoidal yield surfaces. In: 10th HSTAM international congress on mechanics, Chania
15. Simon J-W, Weichert D (2011) Numerical lower bound shakedown analysis of engineering structures. Comput Methods App Mech Eng 200(41):2828–2839
16. Simon J-W, Kreimeier M, Weichert D (2013) A selective strategy for shakedown analysis of engineering structures. Int J Numer Methods Eng 94:985–1014. doi:10.1002/nme.4476
17. Simon J-W, Höwer D, Weichert D (2013) A starting point strategy for interior-point algorithms for shakedown analysis of engineering structures. Eng Optim ISSN:1029–0207. doi:10.1080/0305215X.2013.791816
18. Akoa F, Hachemi A, Said M, Tao PD (2007) Application of lower bound direct method to engineering structures. J Glob Optim 37(4):609–630
19. Hachemi A, Mouhtamid S, Tao P (2004) Large-scale nonlinear programming and lower bound direct method in engineering applications. In: Modelling, computation and optimization in information systems and management sciences. Hermes Sciences, London, pp 299–310
20. Zouain N, Herskovits J, Borges LA, Feijóo RA (1993) An iterative algorithm for limit analysis with nonlinear yield functions. Int J Solids Struct 30(10):1397–1417
21. Lyamin AV, Sloan SW (2002) Lower bound limit analysis using non-linear programming. Int J Numer Methods Eng 55(5):573–611
22. Krabbenhoft K, Lyamin AV, Sloan SW, Wriggers P (2007) An interior-point algorithm for elastoplasticity. Int J Numer Methods Eng 69(3):592–626
23. Pastor F, Loute E (2005) Solving limit analysis problems: an interior-point method. Commun Numer Methods Eng 21(11):631–642
24. Vu DK, Yan AM, Nguyen-Dang H (2004) A primal-dual algorithm for shakedown analysis of structures. Comput Methods Appl Mech Eng 193(42):4663–4674
25. Vu DK, Staat M, Tran IT (2007) Analysis of pressure equipment by application of the primal-dual theory of shakedown. Commun Numer Methods Eng 23(3):213–225
26. Bisbos CD, Pardalos PM (2007) Second-order cone and semidefinite representations of material failure criteria. J Optim Theory Appl 134(2):275–301
27. Skordeli MA-A (2010) Shakedown analysis of metal structures subjected to ellipsoidal variable repeated loading via robust optimization techniques. PhD Dissertation, Aristotle University of Thessaloniki
28. Bisbos CD, Ampatzis AT (2008) Shakedown analysis of spatial frames with parameterized load domain. Eng Struct 30(11):3119–3128
29. Yildirim EA, Wright SJ (2002) Warm-start strategies in interior-point methods for linear programming. SIAM J Optim 12(3):782–810
30. Gertz M, Nocedal J, Sartenar AA (2004) Starting point strategy for nonlinear interior methods. Appl Math Lett 17(8):945–952
31. Mehrotra S (1992) On the implementation of a primal-dual interior point method. SIAM J Optim 2(4):575–601
32. Stojković NV, Stanimirović PS (2001) Initial point in primal-dual interior point method. Facta Universitatis-Series: Mech Autom Control Robot 3(11):219–222
33. Castillo E, García-Bertrand R, Mínguez R (2006) Decomposition techniques in mathematical programming: engineering and science applications. Springer, Berlin
34. Belytschko T (1972) Plane stress shakedown analysis by finite elements. Int J Mech Sci 14(9):619–625
35. Hamilton R, Boyle JT, Shi J, Mackenzie D (1996) Shakedown load bounds by elastic finite element analysis. ASME Press Vessel Pip Div Publ PVP 343:205–211
36. Groß-Weege J (1997) On the numerical assessment of the safety factor of elastic-plastic structures under variable loading. Int J Mech Sci 39(4):417–433
37. Muscat M, Mackenzie D, Hamilton R (2003) Evaluating shakedown under proportional loading by nonlinear static analysis. Comput Struct 81(17):1727–1737
38. Zhang X, Liu Y, Cen Z (2004) Boundary element methods for lower bound limit and shakedown analysis. Eng Anal Bound Elem 28(8):905–917

39. Chen S, Liu Y, Cen Z (2008) Lower bound shakedown analysis by using the element free Galerkin method and nonlinear programming. Comput Methods Appl Mech Eng 197(45):3911–3921
40. Zhang W, Yang LF, Fu CX, Wang J (2012) Evaluating shakedown for structures based on the element bearing-ratio. Appl Mech Mater 137:16–23
41. Zouain N, Borges L (2002) An algorithm for shakedown analysis with nonlinear yield functions. Comput Methods Appl Mech Eng 191(23):2463–2481

Yield Design of Periodically Heterogeneous Plates

Jeremy Bleyer, Duc Toan Pham, Patrick de Buhan and Céline Florence

Abstract This work addresses the determination of the overall strength capacities of periodically heterogeneous plates within a yield design framework. Illustrative applications focus, notably, on reinforced concrete slabs in fire conditions. A homogenization procedure and related numerical tools are proposed to compute macroscopic strength criteria expressed in terms of generalized forces (membrane and bending solicitations). To this end, a yield design auxiliary problem is formulated on the representative three-dimensional unit cell and a numerical resolution by a static approach is presented, making use of simple 3D equilibrium finite elements. A particular emphasis is put on the link between the local strength criterion of steel and concrete and the resulting optimization problem, which can be formulated, either as a second-order cone programming (SOCP) problem or, more generally, as a semi-definite programming (SDP) problem. A first illustrative example of a concrete slab with a single array of steel bars will be used to validate the approach. Then, the influence of fire conditions on the strength capacities of reinforced concrete slabs will be investigated and numerical computations will be confronted to experimental results.

1 Introduction

Direct methods such as yield design or limit analysis of heterogeneous structures can be difficult to perform due to the presence of rapidly and strongly varying material properties on a large scale structure. Despite the efficiency of modern computational techniques, numerical computations on heterogeneous structures are, therefore, out of reach due to the high degree of local refinement needed to correctly capture such

J. Bleyer (✉) · D.T. Pham · P. de Buhan
Laboratoire Navier, Ecole des Ponts ParisTech-IFSTTAR-CNRS (UMR 8205),
Université Paris-Est, 6-8 Avenue Blaise Pascal, Cité Descartes,
77455 Champs-sur-Marne, France
e-mail: jeremy.bleyer@enpc.fr

D.T. Pham · C. Florence
Centre Scientifique et Technique du Bâtiment (CSTB),
84 Avenue Jean Jaurès, 77447 Marne-la-Vallée Cedex 2, France

P. Fuschi et al. (eds.), *Direct Methods for Limit and Shakedown Analysis of Structures*,
DOI 10.1007/978-3-319-12928-0_8

variations and provide sufficiently accurate solutions or bounds. To overcome these difficulties, it seems natural to replace the initially heterogeneous problem by an equivalent homogeneous one which is likely to be much easier to solve. Homogenization theory in yield design for periodically heterogeneous media, initially developed by Suquet [1] and de Buhan [2], is a proper framework which gives a consistent definition of the strength domain of the equivalent homogeneous medium. It relies on the resolution of an auxiliary yield design problem formulated on the unit cell with specific boundary conditions.

An analytical determination of such macroscopic strength criteria is, in general, limited to simple geometries or macroscopic loadings [2, 3]. Hence, numerical approaches have to be used to determine the macroscopic strength criterion. Even if elasto-plastic computations can be performed on the unit cell [4], direct limit analysis or yield design computations are more straightforward and easier to perform. This approach has been adopted for porous media [5, 6], periodic plates solicited in their own plane [3, 7], masonry walls [8] or stone column reinforced soils [9].

Such numerical approaches generally rely on a finite element discretization of the unit cell and the formulation of the auxiliary problem as an optimization problem. Since many yield criteria can be written using conic constraints [10–12] (either quadratic or semi-definite cones), the arising convex optimization problem belongs to the category of conic programming. Interior-point solvers, initially designed for linear programming only, have been developed to encompass this broader class of problems and implemented in commercial codes such as the MOSEK software package [13] for instance. Today, these solvers are considered by the scientific community as the most efficient numerical tools for such problems due to their excellent performance in practice.

This work will be devoted to the formulation and numerical resolution of auxiliary yield design problems for periodic plates in membrane and bending solicitations using a conic programming formulation. Section 2 is devoted to the formulation of the problem in a yield design homogenization framework. Section 3 treats the numerical aspects of finite element discretization and conic programming formulation. Finally, some illustrative applications are presented in Sect. 4 and a specific procedure designed to take into account the effect of fire conditions on the macroscopic strength domain, is also discussed.

2 Homogenization Theory in Yield Design for Periodic Plates

2.1 Initial Heterogeneous Yield Design Problem

We consider a heterogeneous plate of thickness h modeled as a three-dimensional continuum Ω, the middle plane of which, denoted by ω, is supposed to lie in the $(O, \underline{e}_1, \underline{e}_2)$ plane. The thickness direction is given by the vector $\underline{e}_3$ (Fig. 1).

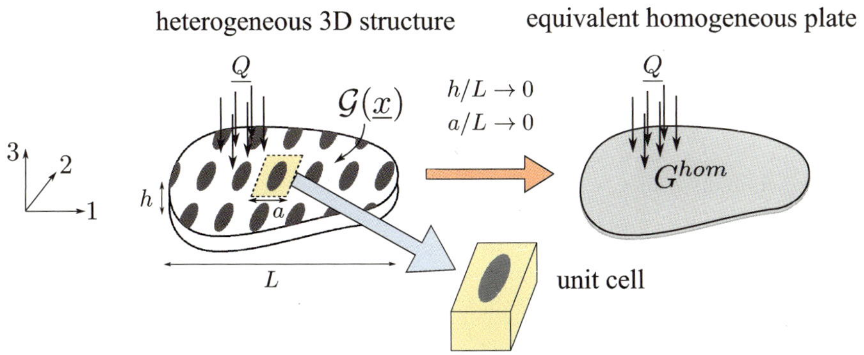

Fig. 1 Homogenization procedure for periodic plates

Assuming that the plate loading depends upon several loading parameters $\underline{Q}$, the domain K of potentially safe loads $\underline{Q}$ is defined, according to the yield design theory [14, 15], as the set of loads such that there exists at least one statically admissible (S.A.) stress field $\underline{\underline{\sigma}}(\underline{x})$ satisfying the strength criterion at each point of the plate:

$$K = \{\underline{Q} \mid \exists \underline{\underline{\sigma}}(\underline{x}) \text{ S.A. with } \underline{Q},\ \forall \underline{x} \in \Omega \quad \underline{\underline{\sigma}}(\underline{x}) \in \mathscr{G}(\underline{x})\} \tag{1}$$

where $\mathscr{G}(\underline{x})$ denotes the convex strength criterion at a point $\underline{x}$ in the plate expressed in terms of the local stress tensor $\underline{\underline{\sigma}}$.

2.2 Homogeneous Yield Design Plate Problem

The present work deals with plates for which the strength properties are periodic in their in-plane direction, that is, there exists two vectors $\underline{a}_1$ and $\underline{a}_2$ in ω such that $\mathscr{G}(\underline{x})$ can be reproduced by periodicity along $\underline{a}_1$ and $\underline{a}_2$:

$$\mathscr{G}(\underline{x} + n_1\underline{a}_1 + n_2\underline{a}_2) = \mathscr{G}(\underline{x}), \qquad \forall \underline{x} \in \Omega,\ \forall n_1, n_2, \in \mathbb{Z} \tag{2}$$

The two vectors $\underline{a}_1$ and $\underline{a}_2$ define the parallelepipedic unit cell of the periodic plate. The natural idea of homogenization theory is to substitute the local heterogeneous strength criterion $\mathscr{G}(\underline{x})$ formulated in terms of local stress tensors by a homogenized or macroscopic strength criterion G^{hom} with equivalent strength properties formulated in terms of generalized forces of the plate, as illustrated in Fig. 1. In the following, it will be supposed that the plate is infinitely resistant to shear forces so that the macroscopic strength criterion will be expressed in terms of a membrane force tensor $\underline{\underline{N}}$ and a bending moment tensor $\underline{\underline{M}}$. In this case, the set of potentially safe loads for the homogenized problem can be defined as:

$$K^{hom} = \{\underline{Q} \mid \exists \underline{\underline{N}}(\underline{x}), \underline{\underline{M}}(\underline{x}) \text{ S.A. with } \underline{Q}, \ \forall \underline{x} \in \omega \quad \underline{\underline{N}}(\underline{x}), \underline{\underline{M}}(\underline{x}) \in G^{hom}\} \tag{3}$$

The principal difference with respect to the definition of K is that the stress fields to be considered are now generalized forces consistent with a plate model and defined on the plate middle plane ω while the strength criterion G^{hom} is now homogeneous.

The main result of homogenization theory in yield design [1, 2] states that both sets K and K^{hom} becomes identical under an appropriate definition of G^{hom} and under the hypothesis that the different length scales are separated which reduces here to the following hypotheses:

- $h \ll L$, this hypothesis states that the thickness in the transverse direction of the continuum is sufficiently smaller than the characteristic length L in the plane ω so that Ω can be modeled as a thin plate;
- $a \ll L$, this hypothesis states that the typical length scale of the variation of the strength properties in the plane ω is much smaller than the characteristic length of the structure in this plane, so that the initially heterogeneous material can be replaced by an equivalent homogeneous one.

In the case when $h \ll a$, i.e. when the plate thickness is much smaller than the material properties variation length, it is possible to replace the initial heterogeneous 3D continuum by a heterogeneous plate (implicit up-scaling procedure) and perform the homogenization of the strength properties on a plate model directly [16]. In the present work, we will assume that $h \sim a$ so that both up-scaling procedures have to be performed in one step [8, 17, 18]. In particular, the unit cell will be modeled as a 3D continuum.

2.3 Definition of the Macroscopic Strength Criterion Form the Solution of a Yield Design Auxiliary Problem

The macroscopic strength criterion is defined and computed via the resolution of a specific auxiliary yield design problem formulated on the plate unit cell A. Its middle plane is denoted by ω_A. The appropriate conditions to formulate an appropriate auxiliary problem in the framework of periodic homogenization theory are the following:

- zero distributed body forces;
- periodic conditions on the lateral boundary of the unit cell and stress-free boundary conditions on the tip and bottom surfaces;
- averaging relation relating the microscopic stress fields to the macroscopic generalized forces.

Hence, we introduce the set $SA(\underline{\underline{N}}, \underline{\underline{M}})$ of tensor stress fields $\underline{\underline{\sigma}}$ defined on the unit cell which are statically admissible with a macroscopic membrane tensor $\underline{\underline{N}}$ and a macroscopic bending moment $\underline{\underline{M}}$ as follows:

$$\underline{\underline{\sigma}} \in SA(\underline{\underline{N}}, \underline{\underline{M}}) \Longleftrightarrow \left| \begin{array}{lrr} \operatorname{div} \underline{\underline{\sigma}} = \underline{0} & \forall \underline{\xi} \in A & \text{(4a)} \\ [\![\underline{\underline{\sigma}} \cdot \underline{n}]\!] = 0 & \text{through } \Gamma & \text{(4b)} \\ \underline{\underline{\sigma}} \cdot \underline{n} & \omega_A\text{−antiperiodic} & \text{(4c)} \\ \underline{\underline{\sigma}} \cdot \underline{e_3} = \underline{0} & \text{for } \xi_3 = \pm h/2 & \text{(4d)} \\ N_{ij} = \dfrac{1}{|\omega|} \displaystyle\int_A \sigma_{ij} d\xi & i,j = 1,2 & \text{(4e)} \\ M_{ij} = \dfrac{1}{|\omega|} \displaystyle\int_A (-\xi_3 \sigma_{ij}) d\xi & & \text{(4f)} \end{array} \right. \quad (4)$$

where $[\![\cdot]\!]$ denotes the jump of the quantity $\cdot$ through any possible discontinuity surface Γ. Note that the boundary conditions correspond to traction-free conditions on the top and bottom surfaces of the plate, whereas the stress vector $\underline{\underline{\sigma}} \cdot \underline{n}$ should be antiperiodic only on the lateral boundaries of the plate.

Finally, G^{hom} is defined as the set of all macroscopic membrane forces and bending moments for which there exists such a stress field $\underline{\underline{\sigma}}$, satisfying the local strength criterion every where. That is:

$$G^{hom} = \left\{ (\underline{\underline{N}}, \underline{\underline{M}}) \text{ s.t. } \exists \underline{\underline{\sigma}} \in SA(\underline{\underline{N}}, \underline{\underline{M}});\ \underline{\underline{\sigma}}(\underline{\xi}) \in \mathcal{G}(\underline{\xi}) \quad \forall \underline{\xi} \in A \right\} \tag{5}$$

Let us remark that a kinematic definition of G^{hom}, through the definition of its support function Π_{hom} where virtual velocity fields are considered, is also possible but will not be considered in this work.

3 Numerical Resolution of the Auxiliary Problem

In this section, the auxiliary yield design problem will be solved using a finite element lower bound static approach in association with conic programming.

3.1 Finite Element Discretization of the Unit Cell

In the following, the unit cell A is discretized using N_E tetrahedral equilibrium finite elements. Hence, the six components of the stress tensor are assumed to vary linearly. The values at the 4 vertices are then arranged in a 24 components vector $\boldsymbol{\sigma}^e = \{\sigma_{11}^{(1)}, \sigma_{22}^{(1)}, \ldots, \sigma_{23}^{(4)}\}^T$.

Due to the linear variation of $\underline{\underline{\sigma}}$, the equilibrium equation (4a) is exactly satisfied by enforcing it at only one point inside each element A_e through a 3×24 matrix $\mathbf{D}^e$:

$$\operatorname{div} \underline{\underline{\sigma}} = \underline{0} \quad \forall \underline{\xi} \in A_e \Longleftrightarrow \mathbf{D}^e \boldsymbol{\sigma}^e = \mathbf{0} \tag{6}$$

The continuity equation (4b) is enforced at each vertex of a face shared by two adjacent elements e and e' using a 9×48 matrix $\mathbf{C}^{e,e'}$:

$$[\![\underline{\underline{\sigma}} \cdot \underline{n}]\!] = \underline{0} \quad \forall \underline{\xi} \in A_e \cap A_{e'} \iff \mathbf{C}^{e,e'} \begin{Bmatrix} \boldsymbol{\sigma}^e \\ \boldsymbol{\sigma}^{e'} \end{Bmatrix} = \mathbf{0} \tag{7}$$

The enforcement of the periodic boundary condition (4c) on the lateral boundary A_{lat} can be expressed in a similar way:

$$\underline{\underline{\sigma}} \cdot \underline{n}^+ + \underline{\underline{\sigma}} \cdot \underline{n}^- = \underline{0} \quad \forall \underline{\xi} \in A_{lat} \iff \mathbf{P}^{e,e'} \begin{Bmatrix} \boldsymbol{\sigma}^e \\ \boldsymbol{\sigma}^{e'} \end{Bmatrix} = \mathbf{0} \tag{8}$$

whereas the traction-free condition on the top and bottom surfaces is written as:

$$\underline{\underline{\sigma}} \cdot \underline{n} = \underline{0} \quad \forall \xi_3 = \pm h/2 \iff \mathbf{S}^e \boldsymbol{\sigma}^e = \mathbf{0} \tag{9}$$

The averaging relations are computed as follows:

$$N_{ij} = \frac{1}{|\omega|} \int_A \sigma_{ij} d\xi \iff \mathbf{N} = \begin{Bmatrix} N_{11} \\ N_{22} \\ N_{12} \end{Bmatrix} = \underbrace{\left[\widetilde{\mathbf{A_N}}^1 \ \dots \ \widetilde{\mathbf{A_N}}^{N_E} \right]}_{\mathbf{A_N}} \begin{Bmatrix} \boldsymbol{\sigma}^1 \\ \vdots \\ \boldsymbol{\sigma}^{N_E} \end{Bmatrix} \tag{10}$$

$$M_{ij} = \frac{1}{|\omega|} \int_A \sigma_{ij} d\xi \iff \mathbf{M} = \begin{Bmatrix} M_{11} \\ M_{22} \\ M_{12} \end{Bmatrix} = \underbrace{\left[\widetilde{\mathbf{A_M}}^1 \ \dots \ \widetilde{\mathbf{A_M}}^{N_E} \right]}_{\mathbf{A_M}} \begin{Bmatrix} \boldsymbol{\sigma}^1 \\ \vdots \\ \boldsymbol{\sigma}^{N_E} \end{Bmatrix} \tag{11}$$

All the equilibrium, continuity and boundary condition constraints are finally assembled into a global constraint matrix $\mathbf{C}$ such that

$$\boldsymbol{\Sigma} \in SA(\mathbf{N}, \mathbf{M}) \iff \begin{array}{c} \mathbf{C}\boldsymbol{\Sigma} = \mathbf{0} \\ \mathbf{N} = \mathbf{A_N}\boldsymbol{\Sigma} \\ \mathbf{M} = \mathbf{A_M}\boldsymbol{\Sigma} \end{array} \tag{12}$$

where $\boldsymbol{\Sigma} = \begin{Bmatrix} \boldsymbol{\sigma}^1 \\ \vdots \\ \boldsymbol{\sigma}^{N_E} \end{Bmatrix}$ collects all stress components. It is to be noted that, since all equations are exactly satisfied, the previous relation defines a subset of the set of statically admissible stress fields.

3.2 Fulfillment of the Strength Criterion Using Conic Constraints

In the present work, illustrative examples are concerned with reinforced concrete slabs so that the local strength criterion will either be a von Mises criterion (for steel) or a Mohr-Coulomb criterion with tension cut-off (for concrete).

3.2.1 von Mises Strength Criterion

Introducing the following matrices:

$$\mathbf{P} = \frac{1}{3}\begin{bmatrix} 2 & -1 & -1 & 0 & 0 & 0 \\ -1 & 2 & -1 & 0 & 0 & 0 \\ -1 & -1 & 2 & 0 & 0 & 0 \\ 0 & 0 & 0 & 1 & 0 & 0 \\ 0 & 0 & 0 & 0 & 1 & 0 \\ 0 & 0 & 0 & 0 & 0 & 1 \end{bmatrix}, \quad \mathbf{D} = \frac{1}{2}\begin{bmatrix} 1 & 0 & 0 & 0 & 0 & 0 \\ 0 & 1 & 0 & 0 & 0 & 0 \\ 0 & 0 & 1 & 0 & 0 & 0 \\ 0 & 0 & 0 & 2 & 0 & 0 \\ 0 & 0 & 0 & 0 & 2 & 0 \\ 0 & 0 & 0 & 0 & 0 & 2 \end{bmatrix} \tag{13}$$

the von Mises strength criterion $\sqrt{\frac{1}{2}\underline{\underline{s}} : \underline{\underline{s}}} \leq k$, where $\underline{\underline{s}} = \underline{\underline{\sigma}} - \frac{1}{3}(\mathrm{tr}\,\underline{\underline{\sigma}})\underline{\underline{1}}$ is the deviatoric stress and k the ultimate shear stress, is written at each vertex i of all elements as follows:

$$\sqrt{\boldsymbol{\sigma}^{(i)\,T}\mathbf{PDP}\boldsymbol{\sigma}^{(i)}} \leq k_i \tag{14}$$

where k_i is the value of the ultimate shear stress at vertex i. It can also be written as:

$$\|\mathbf{D}^{1/2}\mathbf{P}\boldsymbol{\sigma}^{(i)}\| \leq k_i \tag{15}$$

which is a second-order cone (SOC) constraint. This type of constraint is particularly suited to interior point solvers such as the MOSEK software package.

3.2.2 Mohr-Coulomb Strength Criterion with Tension Cut-Off

A Mohr-Coulomb criterion with tension cut-off has been adopted to model the strength behavior of concrete:

$$f(\underline{\underline{\sigma}}) \leq 0 \Longleftrightarrow \begin{cases} K_p\sigma_I - \sigma_{III} \leq f_c \\ \sigma_I \leq f_t \end{cases} \quad \text{with}\, \sigma_I \geq \sigma_{II} \geq \sigma_{III}, \quad K_p = \frac{1+\sin\phi}{1-\sin\phi} \tag{16}$$

where f_c is the ultimate strength of concrete in compression, f_t its tensile strength, ϕ the friction angle and $\sigma_I, \sigma_{II}, \sigma_{III}$ the principal stresses.

Contrary to the von Mises criterion, this criterion is formulated in terms of principal stresses which are not known beforehand. To obtain another representation, let us introduce an auxiliary variable y such that $K_p\sigma_I - \sigma_{III} \leq K_p\sigma_I + y \leq f_c$. Recalling that $\sigma_I \geq \sigma_{II} \geq \sigma_{III}$, we then have $-\sigma_I \leq -\sigma_{II} \leq -\sigma_{III} \leq y$. Hence, this inequality can also be written in the form of a matrix inequality: $-\mathbf{S} \preceq y\mathbf{I}_3$ where $\mathbf{S}$ is the matrix of the components of $\underline{\underline{\sigma}}$ in a given basis, $\mathbf{I}_3$ the identity matrix of size 3×3 and $\mathbf{X} \preceq \mathbf{Y}$ means that the matrix $\mathbf{Y} - \mathbf{X}$ is semi-definite positive.

Similarly, we have also that $K_p\sigma_{III} + y \leq K_p\sigma_{II} + y \leq K_p\sigma_I + y \leq f_c$, which is equivalent to $K_P\mathbf{S} + y\mathbf{I}_3 \preceq f_c\mathbf{I}_3$. Finally, the tension cut-off condition gives: $\sigma_{III} \leq \sigma_{II} \leq \sigma_I \leq f_t$ or also $\mathbf{S} \preceq f_t\mathbf{I}_3$.

In the end, the Mohr-Coulomb criterion with tension cut-off can be written in the following form:

$$f(\underline{\underline{\sigma}}) \leq 0 \Longleftrightarrow \begin{cases} K_p\mathbf{S} + y\mathbf{I}_3 + \mathbf{X} = f_c\mathbf{I}_3 \\ -\mathbf{S} - y\mathbf{I}_3 + \mathbf{Y} = \mathbf{0} \\ \mathbf{S} \qquad + \mathbf{Z} = f_t\mathbf{I}_3 \end{cases} \quad \text{with } \mathbf{X}, \mathbf{Y}, \mathbf{Z} \succeq \mathbf{0} \tag{17}$$

This form, which has been previously obtained in [10, 11], involves linear matrix equalities and auxiliary positive semi-definite matrices $\mathbf{X}, \mathbf{Y}$ and $\mathbf{Z}$. These semi-definite constraints can also be accommodated using MOSEK in its version 7.0 [13].

3.3 Formulation of the Optimization Problem

We can now formulate the optimization problem corresponding to the resolution of the auxiliary yield design problem. In practice, G^{hom} will be obtained by finding the maximal amplification factor λ^+ along a prescribed direction $(\mathbf{N}_0, \mathbf{M}_0)$ such that $\lambda^+(\mathbf{N}_0, \mathbf{M}_0) \in G^{hom}$. Hence, the corresponding optimization problem reads as:

$$\begin{aligned} \lambda^+ = \max\ & \lambda \\ \text{s.t. } & \lambda\mathbf{N}_0 - \mathbf{A_N}\mathbf{\Sigma} = 0 \\ & \lambda\mathbf{M}_0 - \mathbf{A_M}\mathbf{\Sigma} = 0 \\ & \mathbf{C}\mathbf{\Sigma} = 0 \\ & \boldsymbol{\sigma}^{(i)} \in \mathscr{K}_i \quad i = 1, \ldots, 4N_E \end{aligned} \tag{18}$$

This problem consists in optimizing a linear function of the unknowns under linear equality constraints and conic constraints at each vertex where $\mathscr{K}_i$ is a cone representing the strength criterion. It can be either a second-order Lorentz cone (in the von Mises case (15)) or cones of semi-definite matrices in the case of a Mohr-Coulomb criterion (17). Hence, the problem reduces either to a SOCP problem or SDP problem depending on the strength criteria of the unit cell.

Finally, it is to be noted that since all equations are satisfied and since the criterion is satisfied inside each element (due to the linear variation of $\underline{\underline{\sigma}}$ and convexity), the maximum point $\lambda^+(\mathbf{N}_0, \mathbf{M}_0)$ belongs to G^{hom}, so that the macroscopic criterion obtained by varying the direction $(\mathbf{N}_0, \mathbf{M}_0)$ is a strict lower bound estimate for G^{hom}.

4 Illustrative Applications

4.1 Validating Example

The first example is concerned with a cubic unit cell of unitary side length made of concrete and a steel reinforcing bar oriented in direction 1, located at a distance $z_s = -0.3$ below the middle plane (Fig. 2). The concrete is modeled using a Mohr-Coulomb

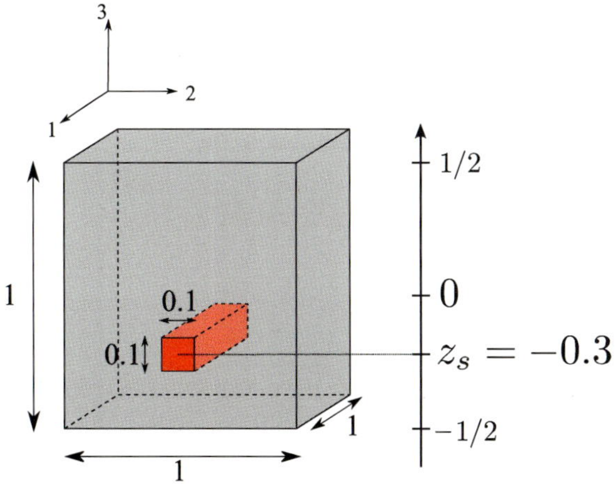

Fig. 2 Unit cell geometry of the first example

strength criterion with a friction angle of $\phi = 37°$, a compressive strength of $f_c = 30$ MPa and a negligible tensile strength $f_t \approx 0$. The steel rebar is modeled using a von Mises criterion with a uniaxial strength $f_s = 500$ MPa. The volume fraction of steel is 1 % (square cross section of 0.1×0.1).

In order to validate the computations, the strength criterion G_c of the unreinforced unit cell (concrete only) has been computed as well as the homogenized criterion G^{hom} of the reinforced unit cell. A simple analytical lower bound G_{lb} of the macroscopic strength criterion can also be obtained using the fact that the volume fraction of steel is small whereas $f_c \ll f_s$. In this case, it can be proved [19] that the state of stress is almost uniaxial in the reinforcement $\underline{\underline{\sigma}} = \sigma^s \underline{e}_1 \otimes \underline{e}_1$ and equal to $\underline{\underline{\sigma}}^c$ in the concrete. The membrane force and bending moment derived from the stress state in the concrete are denoted by $\underline{\underline{N}}^c$ and $\underline{\underline{M}}^c$. Then, the total membrane force and bending moment are given by:

$$\underline{\underline{N}} = \underline{\underline{N}}^c + A_s \sigma^s \underline{e}_1 \otimes \underline{e}_1 \tag{19}$$

$$\underline{\underline{M}} = \underline{\underline{M}}^c - A_s z_s \sigma^s \underline{e}_1 \otimes \underline{e}_1 \tag{20}$$

where A_s is the section of the steel bar. Now, if $\underline{\underline{\sigma}}^c \in \mathscr{G}^c$ satisfies the local concrete strength criterion and $|\sigma^s| \leq f_s$, then $(\underline{\underline{N}}, \underline{\underline{M}}) \in G^{hom}$. Therefore, denoting by G^c the concrete strength criterion expressed in terms of generalized forces, we have the following lower bound estimate for G^{hom}:

$$G_{lb} = G^c \oplus L^s \subseteq G^{hom} \tag{21}$$

where $L^s = \{(N_s \underline{e}_1 \otimes \underline{e}_1, -z_s N_s \underline{e}_1 \otimes \underline{e}_1);\ |N_s| \leq N_y = A_s f_s\}$. This construction means that the lower bound approach is obtained by translating the criterion of the concrete only along the generalized vector $\pm N_y(\underline{e}_1 \otimes \underline{e}_1, -z_s \underline{e}_1 \otimes \underline{e}_1)$. A kinematic approach would show that this simple construction coincides with G^{hom} in the limit of infinitely small volume fraction of steel and infinite contrast between steel and concrete resistances.

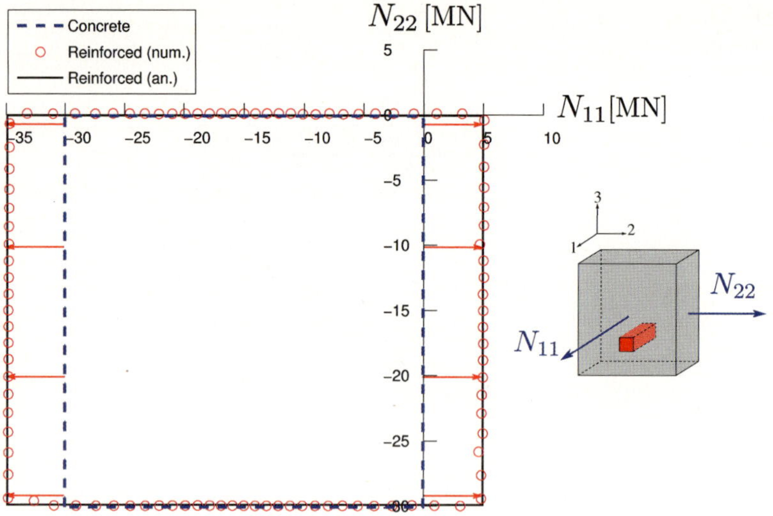

Fig. 3 Projections on the plane $N_{11} - N_{22}$

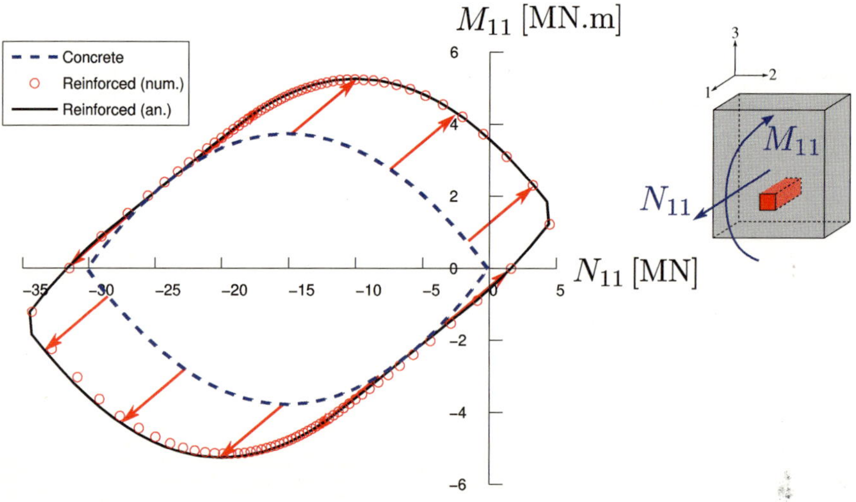

Fig. 4 $N_{11} - M_{11}$ interaction diagram

First, the projections of the different criteria onto the plane $N_{11} - N_{22}$ of membrane forces in directions 1 and 2 are represented in Fig. 3. Hence, all components of $\underline{\underline{M}}$ as well as the N_{12} component are left as free optimization variables. The criterion for concrete only G^c (blue dashed lines) corresponds to a square in the compression region only because of the fact that the tensile strength has been neglected. The inclusion of a reinforcement improves both traction and compression strengths in the N_{11} direction but not in direction 2, as expected. Furthermore, the numerically obtained macroscopic strength criterion (red circles) perfectly matches the analytical lower bound (black solid lines) which is obtained by translating the criterion of concrete by the previously mentioned vectors here drawn in red.

In Fig. 4, the axial-bending interaction diagram along the direction 1 has been represented (all $N_{ij} = M_{ij} = 0$ except for $i = j = 1$). Here, the classical result of the criterion for concrete only, consisting of two parabolas, is retrieved. The effect of the reinforcement expands the interaction diagram in a direction along vectors $\pm$(1 MN, 0.3 MN.m) as confirmed by the analytical lower bound.

4.2 Fire Resistance of Reinforced Concrete Slabs

The second illustrative application is concerned with the fire resistance of reinforced concrete (RC) slabs. Experiments have been conducted by the French Scientific and Technical Center for Building and were aimed at studying the evolution of strength properties (uniaxial bending capacity in a first step) of RC slabs in fire conditions. Four-points bending tests have been performed on RC slabs previously placed in an oven which heated the bottom surface of the RC slabs (Fig. 5). At a given stabilized temperature inside the oven and temperature gradient across the plate thickness, four-points bending tests have been performed up to failure. The aim of this example is to compare these experimental results to numerical computations using the previously described approach.

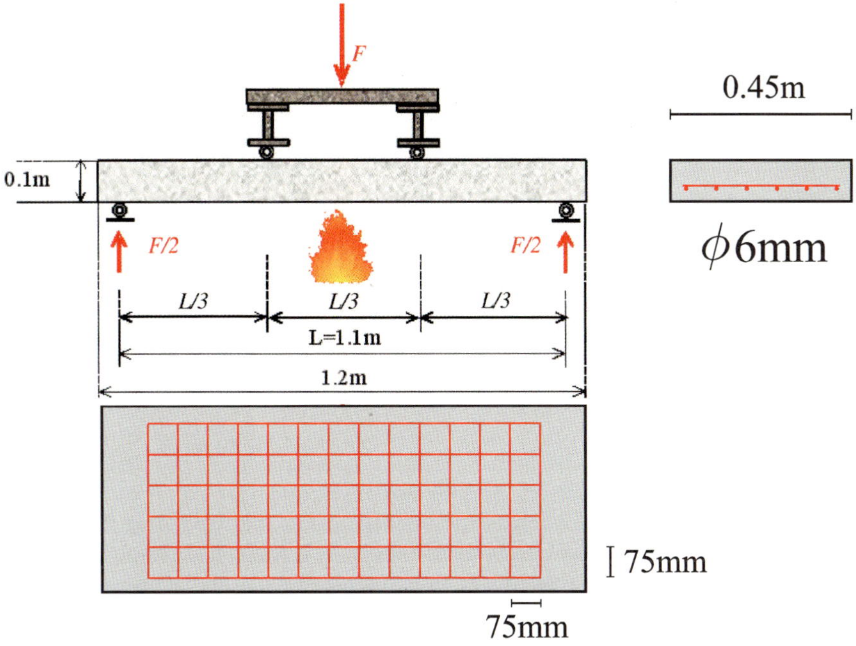

Fig. 5 Experimental setup of 4-points bending tests in fire conditions and RC slabs geometry

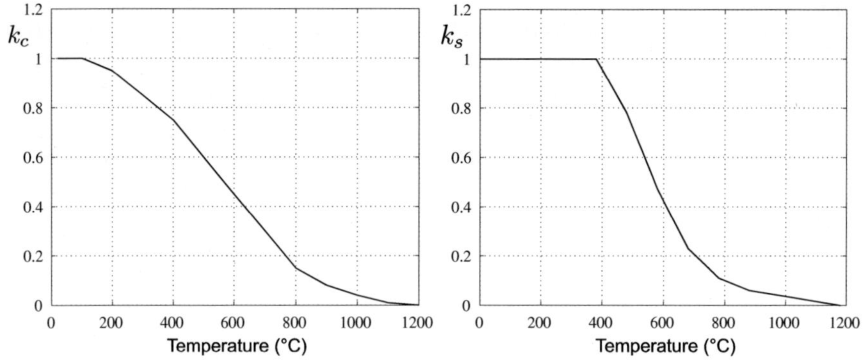

Fig. 6 Reduction factors for concrete (*left*) and steel (*right*) as a function of temperature

4.2.1 Computations in Fire Conditions

The general principle of the computations in fire conditions is rather straightforward. The strength characteristics of concrete and steel at ambient temperature are known from characterization experiments and thermocouples have been placed at various heights in the plate thickness during the tests so as to measure the temperature distribution through the thickness during the experiment.

The key point concerns the degradation of the strength properties of concrete and steel with the temperature. These strength properties introduced in the numerical simulations have been decreased according to Eurocodes prescriptions providing so-called reduction factors $k_c(\Theta)$ and $k_s(\Theta)$ which are decreasing functions of the temperature Θ, so that $f_c(\Theta) = k_c(\Theta)f_{c,amb}$ and $f_s(\Theta) = k_s(\Theta)f_{s,amb}$ where $f_{c,amb}$ (resp. $f_{s,amb}$) is the compressive strength of concrete (resp. uniaxial strength of steel) at ambient temperature. For the materials used in this experiment, such reduction factors are represented in Fig. 6. The concrete friction angle is assumed to remain unaffected by the temperature increase.

Therefore, to a given temperature profile across the slab thickness, is associated a non-uniform distribution of strength properties in the plate thickness. Yield design computations have then been performed on this unit cell with these non-uniform distributions so as to obtain a macroscopic strength criterion depending on the temperature field: $G^{hom}(\Theta(z))$.

4.2.2 Numerical Results

We performed computations on the unit cell represented in Fig. 7 with the four different temperature profiles represented in Fig. 8. The corresponding $N_{11} - M_{11}$ interaction diagrams have been represented in Fig. 9. It can clearly be observed that as the temperature increases, the interaction diagram reduces in size because of the reduction of all strength properties. Besides, it also "rotates" due to the fact that,

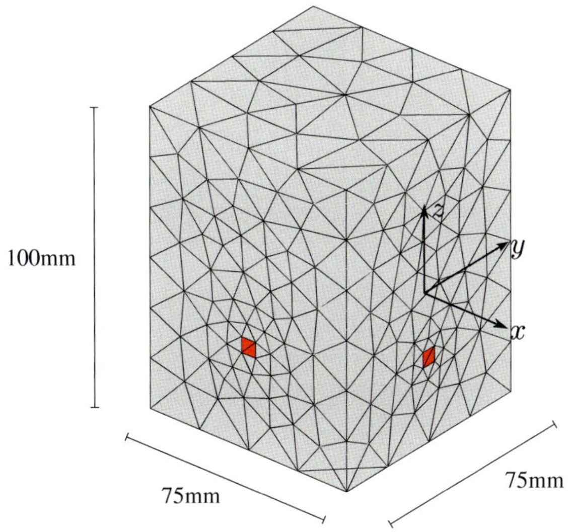

Fig. 7 Unit cell geometry and mesh for the second example

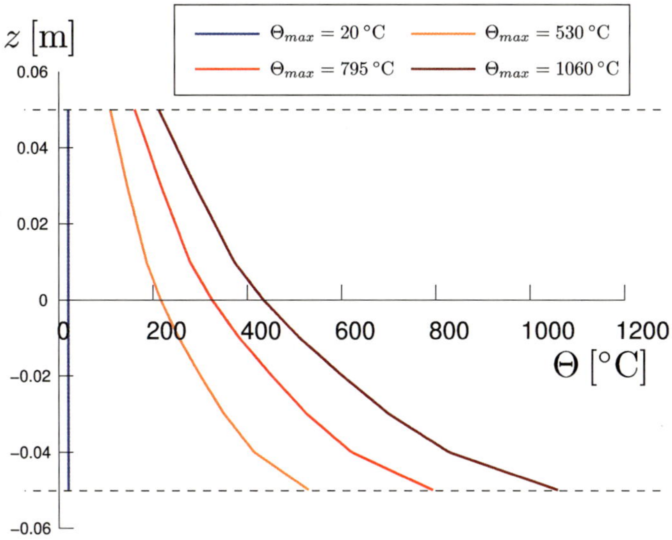

Fig. 8 Temperature fields used in the computations

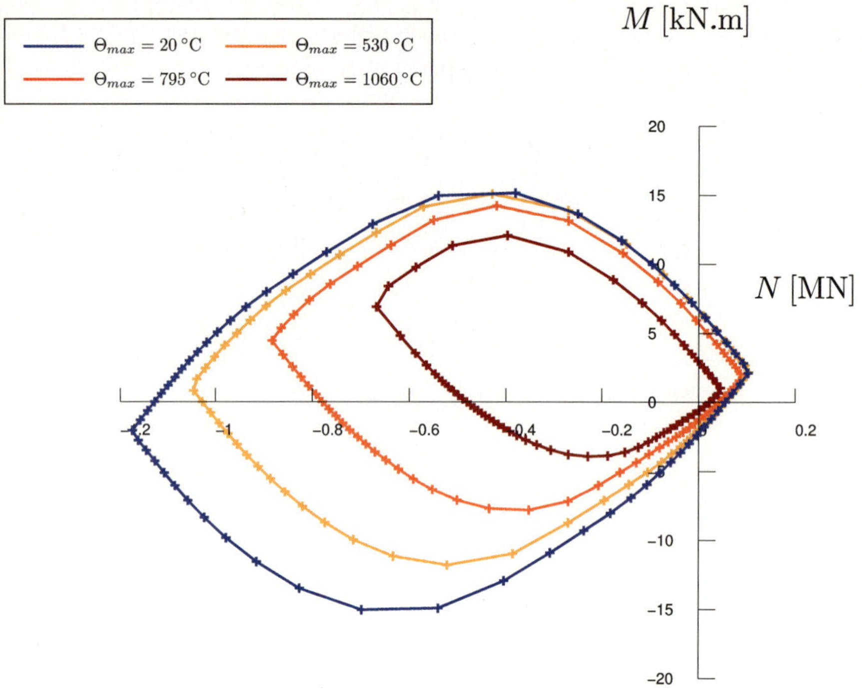

Fig. 9 Evolution of the $N_{11} - M_{11}$ interaction diagram with respect to the temperature

owing to the temperature gradient, the lower part of the slab becomes weaker than the upper part.

4.2.3 Confrontation to Experimental Results

In the experiments, two sets of materials for steel and concrete have been considered and 4-points bending tests have been performed until failure at ambient temperature as well as at an oven temperature approaching 1,000 °C. The temperature distribution in such fire conditions is close to the one represented in Fig. 8 for $\Theta_{max} = 1{,}060\,°C$. The choice of a 4-point bending test enables to relate directly the ultimate load at failure to the ultimate pure bending moment in the direction of the slab, since the bending moment is constant in the middle span of the structure. Hence, ultimate loads obtained from the experiments have been compared to those obtained from numerical computations of the ultimate bending moment. These results are reported in Fig. 10 for ambient and fire conditions and for the two different material sets. It can be seen that the numerical predictions, labeled in light blue and light red, compare very well to the ultimate loads measured in the experiments labeled in dark blue and dark red.

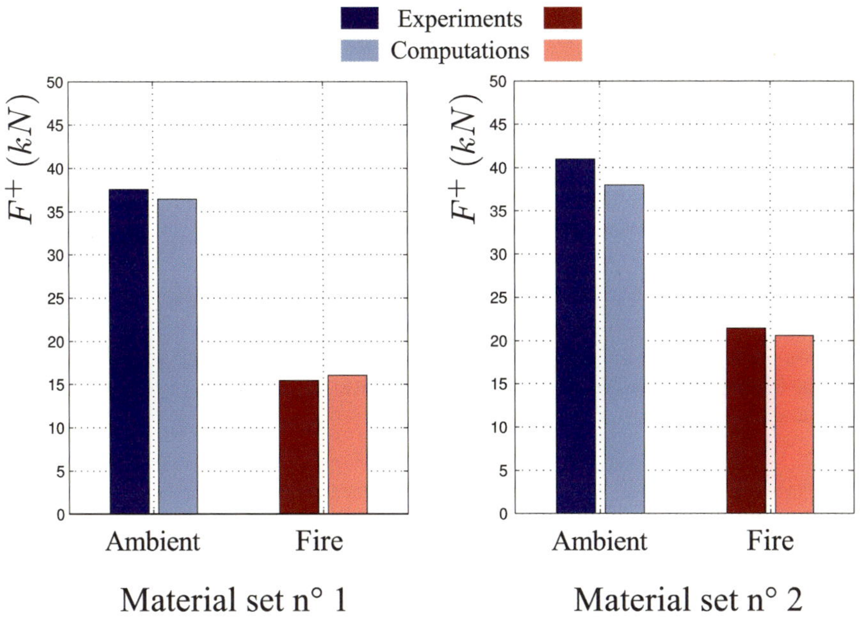

Fig. 10 Comparison of ultimate loads obtained from the experiments and numerical computations

5 Conclusions

A homogenization procedure for assessing the strength properties of periodically heterogeneous plates has been presented in the framework of yield design theory. The present work more specifically focuses on the case when the typical length of the plate heterogeneities and the plate thickness are of the same order. In this case, the unit cell is modeled as a 3D body on which an auxiliary yield design problem is formulated, the resolution of which enables to compute the macroscopic strength domain. A lower bound static approach using 3D equilibrium finite elements is presented and a conic optimization problem is formulated. Finally, the proposed method is applied to reinforced concrete slabs in fire conditions. It allows to predict, for instance, the evolution of interaction diagrams with respect to different temperature fields. These results are also confronted to ultimate loads obtained from experiments carried out in fire conditions.

It would be interesting to complete this work by an upper bound kinematic approach on the same problem, so as to bracket the exact macroscopic strength criterion, since the present method only yields lower bounds. Besides, further work could also investigate the use of such criteria in global computations on a complex structure.

References

1. Suquet P (1985) Homogenization techniques for composite media. 272:193
2. de Buhan P (1986) A fundamental approach to the yield design of reinforced soil structures, PhD thesis, Thèse d'Etat, Paris VI
3. Maghous S (1991) Détermination du critère de résistance macroscopique d'un matériau hétérogène à structure périodique. Approche numérique. These, Ecole Nationale des Ponts et Chaussées. http://tel.archives-ouvertes.fr/tel-00529369
4. Marigo JJ, Mialon P, Michel JC, Suquet P (1987) J de Mécanique Théorique et Appliquée 6(1):47
5. Pastor J, Turgeman S (1983) J de Mécanique Théorique et Appliquée 2:393
6. Turgeman S, Pastor J (1987) J de Mécanique Théorique et Appliquée 6(1):121
7. Francescato P, Pastor J (1998) Revue Européenne des Éléments Finis 7(4):421
8. Sab K (2003) Comptes Rendus Mecanique 331(9):641
9. Hassen G, Gueguin M, de Buhan P (2013) Eur J Mech—A/Solids 37:266. doi:10.1016/j.euromechsol.2012.07.003
10. Bisbos C, Pardalos P (2007) J Optim Theory Appl 134(2):275
11. Krabbenhøft K, Lyamin A, Sloan S (2007) Int J Solids Struct 44(5):1533
12. Makrodimopoulos A (2010) Int J Numer Methods Biomed Eng 26(11):1449
13. Mosek (2008) The Mosek optimization toolbox for MATLAB manual. http://www.mosek.com/
14. Salençon J (1983) Calcul à la rupture et analyse limite. Presses de l'Ecole Nationale des Ponts et Chaussées, Paris
15. Salençon J (2013) Yield design. Wiley, Hoboken
16. Bleyer J, de Buhan P (2014) Int J Solids Struct 28 (in press). http://dx.doi.org/10.1016/j.ijsolstr.2014.03.018
17. Bourgeois S (1997) Modélisation numérique des panneaux structuraux légers, PhD thesis, Université d'Aix-Marseille 2, France
18. Dallot J, Sab K (2008) J Mech Phys Solids 56(2):561
19. de Buhan P, Taliercio A (1991) Eur J Mech—A/Solids 10(2):129

RSDM-S: A Method for the Evaluation of the Shakedown Load of Elastoplastic Structures

Konstantinos V. Spiliopoulos and Konstantinos D. Panagiotou

Abstract To estimate the life of a structure, or a component, which are subjected to a cyclic loading history, the structural engineer must be able to provide safety margins. This is only possible by performing a shakedown analysis that belongs to the class of direct methods. Most of the existing numerical procedures addressing a shakedown analysis are based on the two theorems of plasticity and are formulated within the framework of mathematical programming. A different approach is presented herein. It is an iterative procedure and starts by converting the problem of loading margins to an equivalent loading of a prescribed time history. Inside an iteration, the recently published RSDM direct method is used, which assumes the decomposition of the residual stresses into Fourier series and evaluates its coefficients by iterations. It is proved that a descending sequence of loading factors is generated which converges, from above, to the shakedown load factor when only the constant term of the series remains. An elastic-perfectly plastic with a von Mises yield surface is currently assumed. The method may be implemented in any existing FE code and its efficiency is demonstrated by a couple of applications.

Keywords Direct methods · Shakedown analysis · RSDM

1 Introduction

High levels of cyclic loading are often applied on civil and mechanical engineering structures. The main source of such loadings on civil engineering structures, like bridges, pavements, buildings, offshore structures are heavy traffic, earthquakes or waves. Mechanical engineering structures, like nuclear reactors and aircraft gas turbine propulsion engines, also operate under high levels of cyclic mechanical and

K.V. Spiliopoulos (✉) · K.D. Panagiotou
Department of Civil Engineering, Institute of Structural Analysis and Antiseismic Research, National Technical University of Athens, Zografou Campus, Zografos, 157-80 Athens, Greece
e-mail: kvspilio@central.ntua.gr

K.D. Panagiotou
e-mail: kdpanag@central.ntua.gr

P. Fuschi et al. (eds.), *Direct Methods for Limit and Shakedown Analysis of Structures*,
DOI 10.1007/978-3-319-12928-0_9

temperature loads. Under all these kinds of loadings, these structures are forced to develop plastic strains.

The question to assess the life cycle of a structure, or a structural component, so that it can safely carry the applied loads, is answered, mostly, on the basis of cumbersome time stepping calculations. To this end, one has to know the exact time history. A better alternative, that requires much less computing time, is offered by the direct methods that may predict whether, under the given loading, the structure will become unserviceable due to collapse or excessive inelastic deformations. In addition, if the complete time history of loading in not known, but only its variation intervals, direct methods are the only way to establish safety margins. Typical examples of direct methods are the limit and shakedown analyses.

One may prove that for structures whose material is stable [1] an asymptotic state exists [2]. Direct methods are numerical approaches that attempt to estimate this state right from the start of the calculations. The search for the elastic shakedown state is based, for small displacements and elastic-perfectly plastic structures, on either the lower bound [3] or the upper bound theorems [4].

Various extensions of the above theorems to the large displacement regime have appeared (e.g. [5]) or to elastic-perfectly plastic cracked bodies (e.g. [6]). The problems of different elasto-plastic behavior like linear (e.g. [7]) or non-linear kinematic hardening (e.g. [8]) have also been considered. Non associated plasticity behaviors have also been addressed (e.g. [9]). Shakedown theorems have also been written on the basis of gradient plasticity concepts [10].

As also mentioned above, the vast majority of the direct methods for the solution of the shakedown problem make use of either the lower or the upper bound theorems of plasticity. They are formulated within the framework of mathematical programming (MP) and aim to minimize or maximize an objective function. Depending on whether either the objective function or the constraints or both are linear or non-linear one has to solve a linear (e.g. [11]) or a nonlinear (e.g. [12]) programming problem.

Various techniques have been used to solve the MP program, among which one could mention the reduced basis technique (e.g. [13]). Since the advent of the interior point algorithms, which proved valuable to solve large scale optimization problems, many researchers have formulated, used and applied them to the shakedown analysis (e.g. [14–16]). Both the fields of solid and soil mechanics are referred to. More recent applications have recently appeared in [17].

The Linear Matching Method (LMM) [18] is one of very few alternative procedures to the MP methods. The approach is a generalization of the elastic compensation method [19] and is based on matching a linear problem to a plasticity problem. A sequence of linear solutions, with spatially varying moduli, is generated that provide upper bounds that monotonically converge to the least upper bound. The method has been widely applied to several steel structural components and recently to the limit analysis of concrete beams [20]. The method was further extended beyond shakedown, for loadings that can be decomposed into constant and time varying components, so as to provide an upper bound estimation of the ratchet boundary [21, e.g.], for a recent version.

A direct method to predict any long-term steady cycle of an elastic-perfectly plastic structure under a given cyclic loading was suggested recently [22]. The physics of the steady cycle, which assumes the cyclic nature of the residual stresses, is the main ingredient of the method. It has been called the Residual Stress Decomposition Method (RSDM) and is based on decomposing the residual stresses in Fourier series inside a cycle of loading.

The numerical procedure, presented herein, is a direct method for shakedown analysis [23]; it makes use of the RSDM, and is thus called RSDM-S. Since only the variation intervals are now known, the problem is converted to a prescribed loading problem by drawing a curve between these intervals. Starting from a load factor calculated so as to be above shakedown an iterative procedure is generated which is proved to converge to the shakedown load factor from above. Two examples of application are given. The whole approach is shown to be stable and computationally efficient, with uniform convergence.

2 Cyclic Elastoplastic States

A structure having volume V and surface S is subjected to cyclic surface tractions on one part and on the other part of S to zero displacements.

If the set of loads **P**(t) that act on S is a cyclic loading we may write:

$$\mathbf{P}(\mathrm{t}) = \mathbf{P}(\mathrm{t} + \mathrm{nT}) \tag{1}$$

with t being a time point inside the cycle; T is the period of the cycle, $n = 1, 2, \ldots,$ denotes the number of full cycles. Bold letters are used, herein, to denote vectors and matrices.

Let us suppose that the structure is made of an elastic-perfectly plastic material. Let us further suppose that our structure has been discretized to finite elements and the stresses and strains refer to the Gauss points (GP).

The structure will develop, at a time point $\tau = t/T$ inside the cycle, a stress field $\boldsymbol{\sigma}(\tau)$, which may be decomposed into an elastic part $\boldsymbol{\sigma}^{\mathrm{el}}(\tau)$, that equilibrates the external loading $\mathbf{P}(\tau)$, assuming a completely elastic behavior, and a self-equilibrating residual stress part $\boldsymbol{\rho}(\tau)$ that is due to inelasticity. Therefore:

$$\boldsymbol{\sigma}(\tau) = \boldsymbol{\sigma}^{\mathrm{el}}(\tau) + \boldsymbol{\rho}(\tau) \tag{2}$$

The strain rates can also be decomposed analogously:

$$\dot{\boldsymbol{\varepsilon}}(\tau) = \dot{\mathbf{e}}^{\mathrm{el}}(\tau) + \dot{\boldsymbol{\varepsilon}}_{\mathrm{r}}(\tau) \tag{3}$$

The residual strain rate itself may be decomposed into an elastic and a plastic part [22]. Thus the final compatibility equation is expressed as:

$$\dot{\boldsymbol{\varepsilon}}(\tau) = \dot{\mathbf{e}}^{\mathrm{el}}(\tau) + \dot{\boldsymbol{\varepsilon}}_{\mathrm{r}}^{\mathrm{el}}(\tau) + \dot{\boldsymbol{\varepsilon}}^{\mathrm{pl}}(\tau) \tag{4}$$

The elastic strain rates are related to the stress rates through the elasticity matrix $\mathbf{D}$, whereas the plastic strain rate vector $\dot{\boldsymbol{\varepsilon}}^{\mathrm{pl}}(\tau)$ through the gradient to the flow rule:

$$\dot{\boldsymbol{\sigma}}^{\mathrm{el}}(\tau) = \mathbf{D} \cdot \dot{\mathbf{e}}$$
$$\dot{\boldsymbol{\rho}}(\tau) = \mathbf{D} \cdot \dot{\boldsymbol{\varepsilon}}_{\mathrm{r}}^{\mathrm{el}} \tag{5}$$
$$\dot{\boldsymbol{\varepsilon}}^{\mathrm{pl}} = \lambda \cdot \frac{\partial f}{\partial \boldsymbol{\sigma}}$$

where f is the yield surface and $\boldsymbol{\sigma}$ is a stress state on the yield surface, i.e. $f(\boldsymbol{\sigma}) = 0$. A state of stress that lies either inside or on the yield surface, i.e.$f(\boldsymbol{\sigma}_*) \leq 0$, is an allowable stress state, whereas a state of stress $\boldsymbol{\sigma}_*$that lies inside the yield surface, i.e.$f(\boldsymbol{\sigma}_*) < 0$, is a safe stress state.

As already mentioned in the introduction, for an elastoplastic material which is stable in the Drucker's sense [1], an asymptotic cyclic state always exists in which the stresses and strain rates stabilize and become periodic with the same period of the cyclic loading [2].

Depending on the amplitude of the load we may have unsafe conditions for the structure at hand, such as ratcheting or alternating plasticity, where we have asymptotically non vanishing plastic strain rates, or a safe long-term structural behavior, where, further plastic straining ceases to exist. This condition is the elastic shakedown.

The conditions for elastic shakedown to occur are given by Melan [3] which consists of two statements [24]:

(a) The structure will shake down under a cyclic loading, if there exists a time-independent distribution of residual stresses $\bar{\boldsymbol{\rho}}$ such that, under any combination of loads inside prescribed limits, its superposition with the elastic stresses $\boldsymbol{\sigma}^{\mathrm{el}}$, i.e. $\boldsymbol{\sigma}^{\mathrm{el}} + \bar{\boldsymbol{\rho}}$, results to a safe stress state at any point of the structure,
(b) Shakedown never takes place unless a time-independent distribution of residual stresses can be found such that under all the possible load combinations the sum of the residual and elastic stresses constitutes an allowable stress state.

These statements define the limit cycle for a structure subjected to a prescribed loading program. Parameters of this cycle are the shakedown load factor and the constant in time residual stress distribution which are unique and independent of the preceding deformation history [24].

It is these two parameters that the procedure RSDM-S presented underneath estimates. The method is based on the RSDM [22] which assumes the decomposition of the residual stresses of any cyclic elastoplastic state in Fourier series:

$$\boldsymbol{\rho}(\tau) = \frac{1}{2}\mathbf{a}_0 + \sum_{\mathrm{k}=1}^{\infty} \{\cos(2\mathrm{k}\pi\tau) \cdot \mathbf{a}_{\mathrm{k}} + \sin(2\mathrm{k}\pi\tau) \cdot \mathbf{b}_{\mathrm{k}}\} \tag{6}$$

with $\mathbf{a}_0$, $\mathbf{a}_{\mathrm{k}}$, and $\mathbf{b}_{\mathrm{k}}$ being the coefficients of the series found iteratively by the RSDM.

3 The Residual Stress Decomposition Method for Shakedown Analysis (RSDM-S)

The procedure is formulated for a maximum number of two loads P_1 and P_2, although it may be applied for more than two loads. The loads are assumed to vary between a minimum value which for simplicity is assumed zero and a maximum value, which are denoted by P_1^* and P_2^* respectively.

We may use any curve that passes through these two limits to express a cyclic loading variation. The two loads may vary proportionally (Fig. 1a) or independently (Fig. 2a).

(a) For proportional variation (Fig. 1)
The two loads in the time domain may be expressed by the following equation:

$$\mathbf{P}(\tau) = \begin{Bmatrix} P_1(\tau) \\ P_2(\tau) \end{Bmatrix} = \begin{Bmatrix} P_1^* \cdot \alpha(\tau) \\ P_2^* \cdot \alpha(\tau) \end{Bmatrix} \tag{7}$$

with $\alpha(\tau)$ being a time function common for both the loads. One possible smooth variation of a fourth order polynomial may be seen in Fig. 1c.

(b) For independent load variation (Fig. 2)

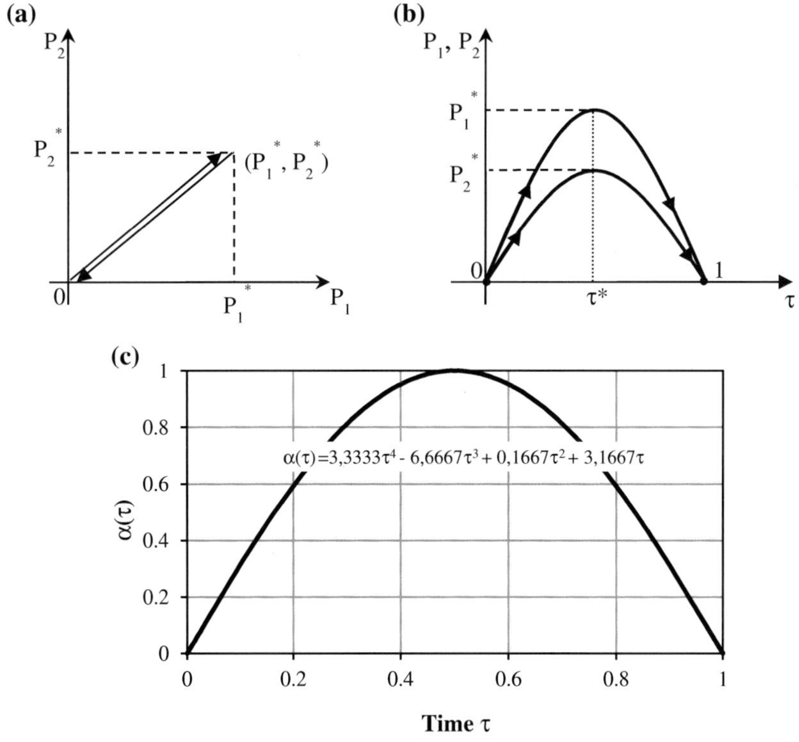

Fig. 1 Proportional loading variation **a** in load space, **b** in time domain **c** time function used

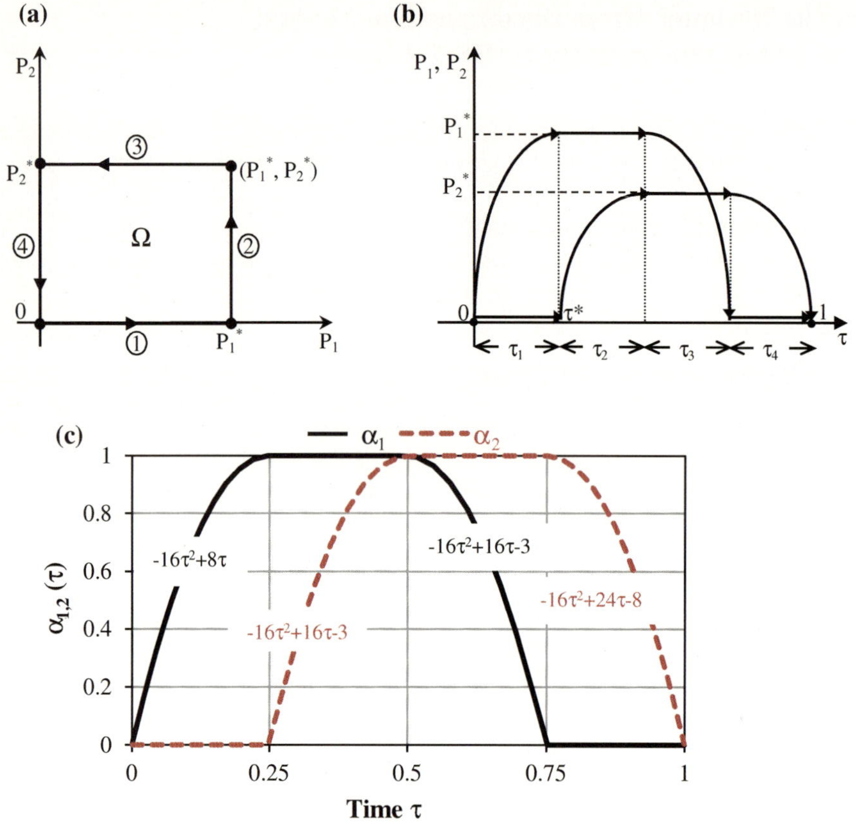

Fig. 2 Independent loading variation **a** in load space, **b** in time domain **c** time functions used

It has been proved in [25] that due to the convexity of the yield surface if a structure shakes down over the path that encloses the domain Ω, it certainly shakes down over any loading path contained inside this domain. This enclosing path may be described by the following equation:

$$\mathbf{P}(\tau) = \begin{Bmatrix} P_1(\tau) \\ P_2(\tau) \end{Bmatrix} = \begin{Bmatrix} P_1^* \cdot \alpha_1(\tau) \\ P_2^* \cdot \alpha_2(\tau) \end{Bmatrix} \tag{8}$$

Two time functions are used now (Fig. 2b) and a possible smooth variation is described in Fig. 2c, where a second degree polynomial is used for two quadrants of the cycle, whereas a constant non-zero and a zero value are employed inside the other two quadrants.

The load domains shown in Figs. 1 and 2 may be isotropically varied by multiplying them with a factor γ. Thus the numerical procedure is built, so that, starting from a load factor high above, this load factor is constantly lowered by shrinking the load domain in a continuous way up to the point that the shakedown load factor is reached.

The procedure is formulated herein for a von Mises yield surface. Although the procedure is general it will be formulated in the present work for two-dimensional structures of thickness d under plane stress conditions.

3.1 Initial Load Factor

From the way the loading time history is being constructed we may observe that when the loads vary proportionally (Fig. 1b, c), there always exists a cycle point τ^* that both the two loads attain their maximum values. On the other hand, for loads varying independently (Fig. 2b, c) the cycle point τ^* may be defined as the time point for which one of the two loads becomes maximum, with the other one being zero.

For either case, one may find the equivalent von Mises elastic stress $\bar{\sigma}^{el}$ at all the GPs of the structure at time τ^*. Denoting by $\min \bar{\sigma}^{el}$ the non-zero minimum of these stresses one may use an initial load factor equal to:

$$\gamma^{(0)} = \frac{\sigma_Y}{\min \bar{\sigma}^{el}} \tag{9}$$

where σ_Y is the yield stress of the material.

By choosing this load factor the starting conditions are definitely above the shakedown loading since the whole structure will have become plastic.

3.2 Development of the Procedure

Having found an initial solution, we use the RSDM [22] for the factored cyclic loading $\gamma^{(0)} \cdot \mathbf{P}(\tau)$, to produce an initial estimate for $\boldsymbol{\rho}_{(0)}, \mathbf{a}_0^{(0)}, \mathbf{a}_k^{(0)}, \mathbf{b}_k^{(0)}$. Then we enter the iterative phase, which consists of two iteration loops, one inside the other.

The following iterative steps are then followed:

1. Inside an iteration κ of the outer loop, starting with $\kappa = 1$, the following expression, which is the sum of the norms of the vectors of the coefficients of the trigonometric part of the Fourier series of the residual stresses, is found:

$$\varphi\left(\gamma^{(\kappa-1)}\right) = \sum_{k=1}^{\infty} \left\| \mathbf{a}_k^{(\kappa-1)} \right\| + \sum_{k=1}^{\infty} \left\| \mathbf{b}_k^{(\kappa-1)} \right\| \tag{10}$$

2. An update of the loading factor may be calculated:

$$\gamma^{(\kappa)} \cdot P_1^* = \gamma^{(\kappa-1)} \cdot P_1^* - \omega \cdot \left[\varphi\left(\gamma^{(\kappa-1)}\right) \cdot d\right] \tag{11}$$

The expression (11) actually monitors the level of the applied loads with the reminder that d is the thickness of the structure. The update of the loading factor

is chosen to be performed in conjunction with the maximum of the two peaks of the applied loads, although any other time point could be used.
It should also be noted that ω is a convergence parameter, which will be discussed below.

3. The following inequality is checked:

$$\frac{\left|\gamma^{(\kappa)} - \gamma^{(\kappa-1)}\right|}{\gamma^{(\kappa)}} \leq \mathrm{tol} \tag{12}$$

4. If Eq. (12) holds, the procedure stops and $\gamma^{(\kappa)} = \gamma^{(\kappa-1)} = \gamma_{sh}$, otherwise we set

$$\boldsymbol{\rho}^{(1)}_{(\kappa-1)}(\tau) = \boldsymbol{\rho}_{(\kappa-1)}(\tau) \tag{13}$$

Using the updated load factor of the outer loop, an inner loop of iterations controlled by μ starts with $\mu = 1$. The steps of this inner loop are virtually the same steps with the RSDM [22].

5. The following expression is computed for each cycle point τ and for each GP:

$$\boldsymbol{\sigma}^{(\mu)}(\tau) = \gamma^{(\kappa)}\boldsymbol{\sigma}^{el}(\tau) + \boldsymbol{\rho}^{(\mu)}_{(\kappa-1)}(\tau) \tag{14}$$

where:

$$\boldsymbol{\sigma}^{el}(\tau) = \alpha_1(\tau)\boldsymbol{\sigma}^{el}_{P_1^*} + \alpha_2(\tau)\boldsymbol{\sigma}^{el}_{P_2^*} \tag{15}$$

with $\alpha_1(\tau)$, $\alpha_2(\tau)$being the time functions for independent loading (Fig. 2b, c), whereas $\alpha_1(\tau) = \alpha_2(\tau) = \alpha(\tau)$is the time function for the proportional loading case (Fig. 1b, c).

6. It is checked whether the total effective stress $\bar{\sigma}^{(\mu)}(\tau) > \sigma_Y$; if this does not hold we set $\xi = 0$, otherwise:

$$\xi = \frac{\bar{\sigma}^{(\mu)}(\tau) - \sigma_Y}{\bar{\sigma}^{(\mu)}(\tau)} \Rightarrow \boldsymbol{\sigma}^{(\mu)}_{pl}(\tau) = \xi \cdot \boldsymbol{\sigma}^{(\mu)}(\tau) \tag{16}$$

This operation is a radial return [26] type rule and may be graphically seen in Fig. 3. For a more detailed discussion the reader is referred to [22].
Steps 5 and 6 are repeated for every GP.

7. Assemble for the whole structure the new rate vector of the nodal forces $\dot{\mathbf{R}}'(\tau)$:

$$\dot{\mathbf{R}}'(\tau) = \gamma^{(\kappa)}\dot{\mathbf{R}}(\tau) + \int_V \mathbf{B}^T \cdot \boldsymbol{\sigma}^{(\mu)}_{pl}(\tau)dV \tag{17}$$

with $\dot{\mathbf{R}}(\tau)$ being the equivalent nodal forces for the loading $\dot{\mathbf{P}}(\tau)$which may be evaluated by differentiating the Eqs. (7) or (8), depending on the case of loading. $\mathbf{B}$ is the compatibility strain-displacement matrix for the given FE mesh.

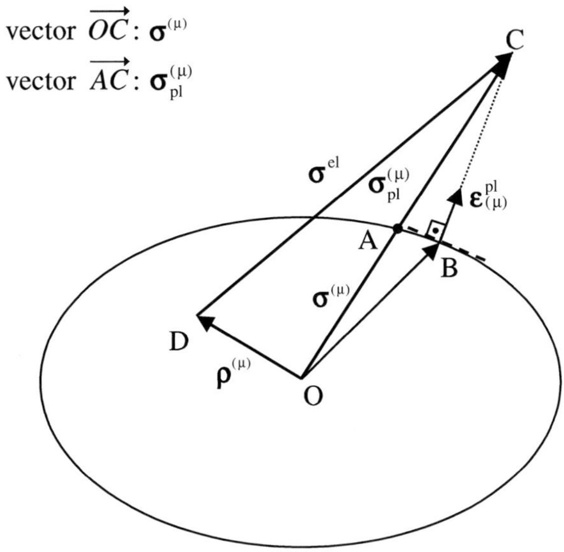

Fig. 3 Estimation of the plastic straining inside an iteration

8. We find an update for $\dot{\mathbf{r}}^{(\mu)}(\tau)$ using the relation

$$\mathbf{K}\dot{\mathbf{r}}^{(\mu)}(\tau) = \dot{\mathbf{R}}'(\tau) \tag{18}$$

with $\mathbf{K} = \int_V \mathbf{B}^T\mathbf{D}\mathbf{B}dV$ the structure's stiffness matrix.

9. A value for $\dot{\boldsymbol{\rho}}^{(\mu)}(\tau)$ is evaluated at each G.P.

$$\dot{\boldsymbol{\rho}}^{(\mu)}(\tau) = \mathbf{D}\mathbf{B}\dot{\mathbf{r}}^{(\mu)}(\tau) - \gamma^{(\kappa)}\dot{\boldsymbol{\sigma}}^{el}(\tau) - \boldsymbol{\sigma}_{pl}^{(\mu)}(\tau) \tag{19}$$

where $\dot{\boldsymbol{\sigma}}^{el}(\tau) = \dot{\alpha}_1(\tau)\boldsymbol{\sigma}_{P_1^*}^{el} + \dot{\alpha}_2(\tau)\boldsymbol{\sigma}_{P_2^*}^{el}$

The steps 5–9 are repeated for all the cycle time points.

10. By performing numerical time integration over the whole cycle we may obtain an update of the Fourier coefficients [22]:

$$\begin{aligned}
\mathbf{a}_k^{(\mu+1)} &= -\frac{1}{k\,\pi}\int_0^1 \left\{\left[\dot{\boldsymbol{\rho}}^{(\mu)}(\tau)\right](\sin 2k\pi\tau)\right\}d\tau \\
\mathbf{b}_k^{(\mu+1)} &= \frac{1}{k\,\pi}\int_0^1 \left\{\left[\dot{\boldsymbol{\rho}}^{(\mu)}(\tau)\right](\cos 2k\pi\tau)\right\}d\tau \\
\frac{\mathbf{a}_0^{(\mu+1)}}{2} &= -\sum_{k=1}^{\infty}\mathbf{a}_k^{(\mu+1)} + \frac{\mathbf{a}_0^{(\mu)}}{2} + \sum_{k=1}^{\infty}\mathbf{a}_k^{(\mu)} + \int_0^1\left[\dot{\boldsymbol{\rho}}^{(\mu)}(\tau)\right]d\tau
\end{aligned} \tag{20}$$

11. From these updates one may get an update for $\boldsymbol{\rho}^{(\mu+1)}(\tau)$using the iterative form of (6):

$$\boldsymbol{\rho}^{(\mu+1)}(\tau) = \frac{1}{2}\mathbf{a}_0^{(\mu+1)} + \sum_{k=1}^{\infty}\left\{\cos(2k\pi\tau)\cdot\mathbf{a}_k^{(\mu+1)} + \sin(2k\pi\tau)\cdot\mathbf{b}_k^{(\mu+1)}\right\} \quad (21)$$

12. Next we check whether the values of the residual stresses at the current and at the previous iteration differ within some tolerance at some cycle point [22], for example at the end of the cycle, where $\tau = t/T = 1$, i.e.:

$$\frac{\left\|\boldsymbol{\rho}^{(\mu+1)}(1)\right\| - \left\|\boldsymbol{\rho}^{(\mu)}(1)\right\|}{\left\|\boldsymbol{\rho}^{(\mu+1)}(1)\right\|} \leq \text{tolr} \quad (22)$$

In case this does not hold we set $\boldsymbol{\rho}_{(\kappa-1)}^{(\mu+1)}(\tau) = \boldsymbol{\rho}^{(\mu+1)}(\tau)$ and go back to step 5 and start a new iteration of the inner RSDM loop; otherwise we set $\boldsymbol{\rho}_{(\kappa)}(\tau) = \boldsymbol{\rho}^{(\mu+1)}(\tau)$ and go back to step 1 and start a new iteration of the outer loop with $\kappa = \kappa + 1$.

For accurate results, the values 10^{-4}–$10^{-5}$of *tolr* and 10^{-4} for *tol* are sufficient. The corresponding tolerance value for the function of φ (Eq. (10)) is 10^{-3}.

Due to the positive nature of φ, the algorithm generates a descending sequence of cyclic solutions which ends up with the parameters of the limiting cycle for elastic shakedown. Thus γ_{sh} is the elastic shakedown factor, and the constant term, which is the only remaining term in the Fourier series, is the constant in time distribution of the residual stresses, unique [24] for the adopted prescribed loading program.

3.3 Numerical Strategy

The numerical strategy adopted is to start the procedure with the convergence parameter $\omega = 1$. This normally leads to a monotonic convergence, from above, to the shakedown load. There could be cases, though, especially when we start from a high initial value that an overshooting of the shakedown factor occurs. In such a case, for the loading factor evaluated at the current iteration $\gamma^{(\kappa)}$, we would have $\varphi(\gamma^{(\kappa)}) < 10^{-3}$; this loading factor is then not accepted and the convergence factor is halved till we get a loading factor $\varphi(\gamma^{(\kappa)}) > 10^{-3}$ [23].

4 Application Examples

The method is applied to two examples of plated structures, a holed square plate under biaxial load and a grooved plate under tension and bending. Plane stress conditions are assumed.

4.1 Plate with a Central Hole

A square plate having a central hole is treated as a first example of application. The plate is subjected to two uniformly distributed loads acting at the two edges of the plate (Fig. 4). Due to the symmetry of the structure and the loading only a quarter of the plate is analyzed.

As far as the geometric characteristics of the plate, the ratio of the diameter D to the length L of the hole was considered equal to 0.2; the ratio of the thickness d of the plate to its length is taken 0.05. The case of a length $L = 20\,\text{cm}$ was used.

The finite element mesh is also shown in Fig. 4. Ninety-eight, eight-noded, isoparametric elements with 3 × 3 Gauss integration points were used.

The material data considered are Young's modulus $E = 210\,\text{Gpa}$, Poisson's ratio $\nu = 0.3$ and yield stress $\sigma_Y = 360\,\text{Mpa}$.

Both cases of proportional and independent variations of loads are presented. The shakedown loads for different ratios of the maximum values of the applied loads are given in Figs. 5 and 6.

For the proportional loading case the time function of Fig. 1c is used. One may see a good agreement with corresponding results (Fig. 5), reported in the literature [27].

For the independent load variation (Fig. 2), the time functions, whose equations for each part of the cycle, shown in Fig. 2c, are used. One may see good agreement (Fig. 6) with reported in the literature results that are based either on the LMM [27] or a mathematical programming algorithm [28]. The better agreement with [28] is due to the fact that a similar finite element mesh is constructed with the one used here.

One may realize the smoothness of the convergence of the method in Fig. 7 where the case of $P_1^* = P_2^*$ is shown. Analogous curves are obtained for all the other ratios.

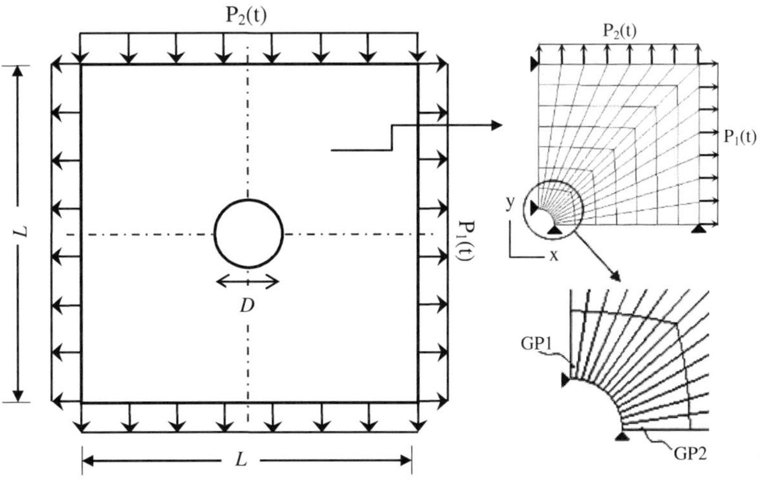

Fig. 4 Geometry, loading and FE discretization of the plate

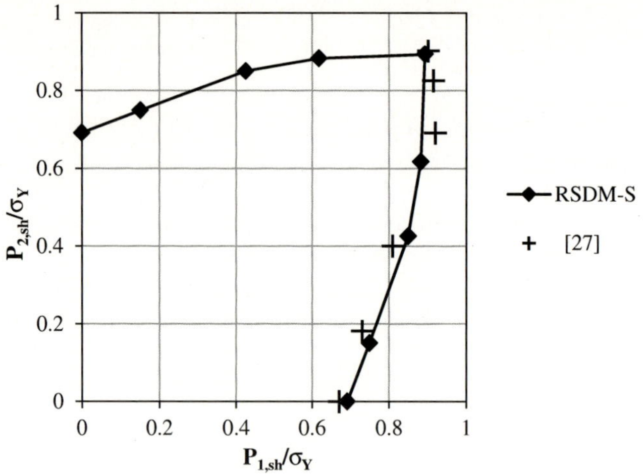

Fig. 5 Shakedown domain for the case of proportional loading

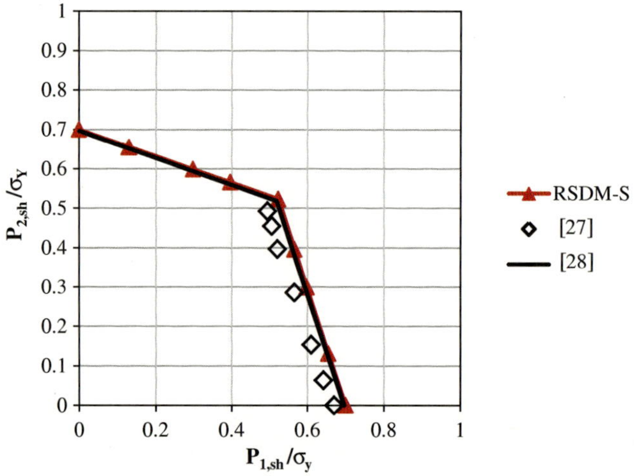

Fig. 6 Shakedown domain for the case of independent loading

Only the initial value of $\omega = 1$ was needed for convergence. In Fig. 8 one may also see a typical evolution of the plastic stress to zero at one of most stressed GPs, that are located at the two corners of the hole of the plate (Fig. 4). The CPU time reported to reach a solution for an Intel Core i7 at 2.93 GHz with 4,096 MB RAM was of the order of 40 s.

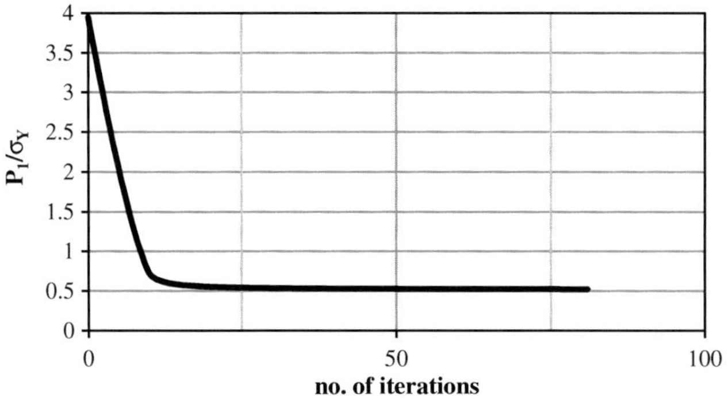

Fig. 7 Convergence of the RSDM-S for the holed plate, for equal maximum loads

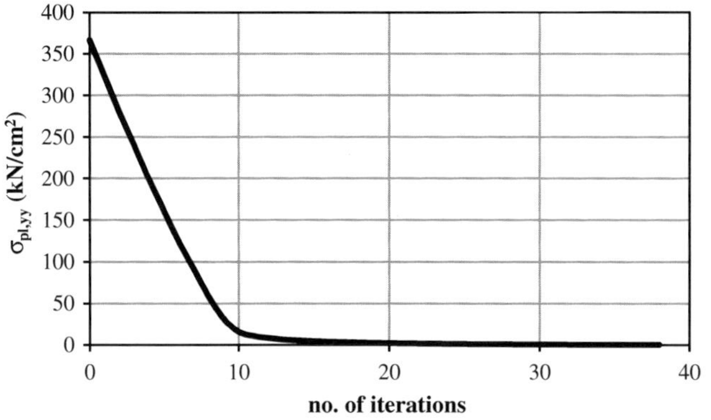

Fig. 8 Evolution of the yy component of the plastic stress vector at GP2 at the 3T/4

4.2 Grooved Rectangular Plate Under Varying Tension and Bending

The second example of application deals with a grooved rectangular plate subjected to an in plane constant tension $P_N(t)$ and a linearly varying bending moment $P_M(t)$ along the boundaries (Fig. 9). A rectangular loading domain is considered (Fig. 2) with the two loads varying independently, having maximum values $P_N^* = 1$ and $P_M^* = 1$, respectively. This example has been presented in [29].

The following geometrical data are used: $R = 25\,\text{cm}$, $L = 2\text{R}$ and $H = 4\text{R}$.

The plate is assumed homogeneous, isotropic, elastic-perfectly plastic with the following material data: Young's modulus $E = 210\,\text{GPa}$, Poisson's ratio $\nu = 0.3$ and yield stress $\sigma_Y = 111.62\,\text{MPa}$.

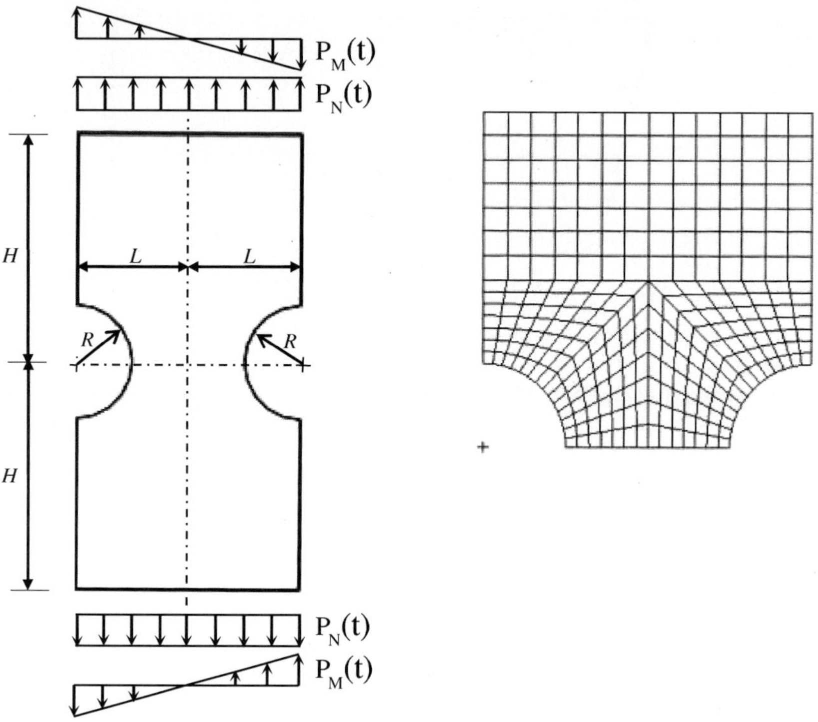

Fig. 9 Geometry, loading of the grooved rectangular plate and its finite element mesh

Due to the symmetry of the geometry and the applied loads, only a half of the structure has been modeled. The finite element mesh discretization of the plate is also shown in Fig. 9. Two hundred and ninety-four, eight-noded, iso-parametric elements with 3×3 Gauss integration points were used.

The shakedown domain obtained by the RSDM-S and its comparison with the results from [29] is shown in Fig. 10.

Specifically in the case we have both in-plane tension and bending with $P_N^* = P_M^*$ the proposed RSDM-S gives a shakedown factor equal to 0.227 which quite close to the value 0.236 of [29], where a different mesh and algorithm was used.

For this example, the initial convergence parameter $\omega = 1$, in the process of the iterations, had to be halved twice, due to an overshooting of the calculated shakedown load to negative values.

Although the starting point was quite high as compared to the final result, the descent was rapid as shown in the Fig. 11. In the insert of the figure, one may see, after the initial descent, the smooth convergence towards the shakedown value.

In this example the CPU time reported to reach the solution for a CPU with the same characteristics as above was of the order of 270 s.

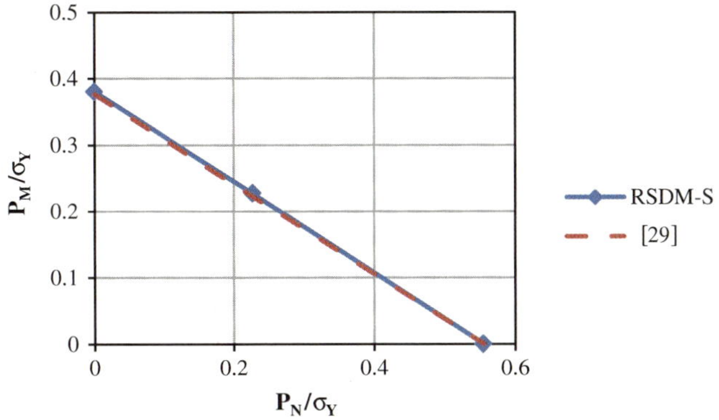

Fig. 10 Shakedown domain of the grooved rectangular plate in tension and bending

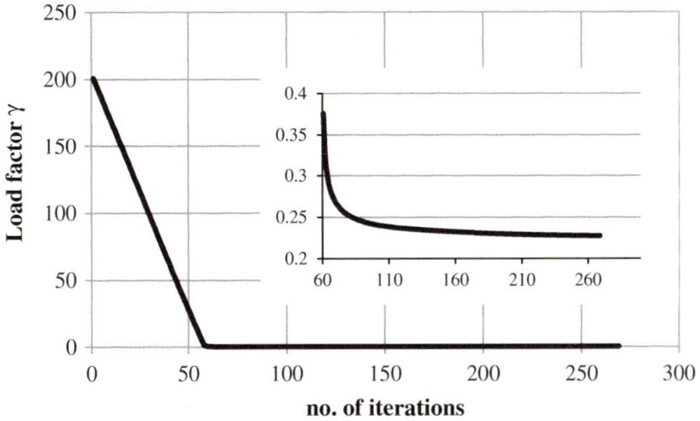

Fig. 11 Convergence of the RSDM-S towards the shakedown factor for the grooved rectangular plate

5 Concluding Remarks

A novel direct method to calculate the shakedown load of structures exhibiting elasto-plastic material behavior when subjected to cyclic loadings has been introduced.

The procedure is iterative in its nature and starts by converting a loading of prescribed limits to a loading of a prescribed cyclic history. This history is then multiplied by a factor. A sequence of decreasing factors, whose initial value is calculated to be above the shakedown load, is generated by the method. For each factored cyclic loading history the distribution of the residual stresses of the corresponding cyclic elasto-plastic state is estimated. This is done by decomposing these stresses into Fourier series and finding their coefficients also iteratively. The procedure stops when only the constant term of the series remains. Thus the method converges to the shakedown load factor from above.

It is a relatively simple method, formulated within the FE method. The stiffness matrix of the structure needs to be formed and decomposed only once. Three terms of the Fourier series are enough for accurate results. These two factors guarantee the method to be numerically efficient.

The main advantage of the method is that the solution procedure provides a better understanding of the physics of the problem than any method based on MP algorithms. At the same time, the absence of such an algorithm makes it directly implementable into any existing FE software.

Although the method has been applied here to two-dimensional structures, three-dimensional structures can be also considered. It may also be extended to loading domains of more than two loads. Other yield surfaces, except for von Mises, may be considered. Also, besides perfect plasticity, other plasticity laws, like strain hardening, may be taken care of.

References

1. Drucker DC (1959) A definition of stable inelastic material. ASME J Appl Mech 26:101–106
2. Frederick CO, Armstrong PJ (1966) Convergent internal stresses and steady cyclic states of stress. J Strain Anal 1:154–169
3. Melan E (1938) Zur Plastizität des räumlichen Kontinuums. Ing Arch 9:116–126
4. Koiter W (1960) General theorems for elastic-plastic solids. In: Sneddon IN, Hill R (eds) North-Holland, Amsterdam
5. Weichert D (1986) On the influence of geometrical nonlinearities on the shakedown of elastic-plastic structures. Int J Plast 2:135–148
6. Belouchrani MA, Weichert D (1999) An extension of the static shakedown theorem to inelastic cracked structures. Int J Mech Sci 41:163–177
7. Pham DC (2005) Shakedown static and kinematic theorems for elastic–plastic limited linear kinematic hardening solids. Eur J Mech A/Solids 24:35–45
8. Simon J-W (2013) Direct evaluation of the limit states of engineering structures exhibiting limited, nonlinear kinematical hardening. Int J Plast 42:141–167
9. Bousshine L, Chaaba A, de Saxcé G (2003) A new approach to shakedown analysis for non-standard elastoplastic material by the bipotential. Int J Plast 19:583–598
10. Polizzotto C (2008) Shakedown theorems for elastic–plastic solids in the framework of gradient plasticity. Int J Plast 24:218–241
11. Maier G (1969) Shakedown theory in perfect elastoplasticity with associated and nonassociated flow-laws: a finite element, linear programming approach. Meccanica 4:1–11
12. Stein E, Zhang G, König JA (1992) Shakedown with nonlinear strain-hardening including structural computation using finite element method. Int J Plast 8:1–31
13. Heitzer M, Staat M (2003) Basis reduction technique for limit and shakedown problems. In: Staat M, Heitzer M (eds) Numerical methods for limit and shakedown analysis. NIC Series, vol 15, pp 1–55
14. Andersen KD, Christiansen E, Overton ML (1998) Computing limit loads by minimizing a sum of norms. SIAM J Sci Comput 19:1046–1062
15. Vu DK, Yan AM, Nguyen-Dang H (2004) A primal-dual algorithm for shakedown analysis of structures. Comput Methods Appl Mech Eng 193:4663–4674
16. Simon J-W, Höer D, Weichert D (2014) A starting-point strategy for interior-point algorithms for shakedown analysis of engineering structures. Eng Optim 46(5):648–668
17. Spiliopoulos K, Weichert D (2014) Direct methods for limit states in structures and materials. Springer, Berlin

18. Ponter ARS, Carter KF (1997) Shakedown state simulation techniques based on linear elastic solutions. Comput Methods Appl Mech Eng 140:259–279
19. Mackenzie D, Boyle T (1993) A method of estimating limit loads by iterative elastic analysis. I—Simple examples. Int J Press Vessel Pip 53:77–95
20. Pisano AA, Fuschi P, de Domenico D (2013) A kinematic approach for peak load evaluation of concrete elements. Comput Struct 119:125–139
21. Chen H, Ponter ARS (2010) A direct method on the evaluation of ratchet limit. ASME J Press Vessel Technol 132(4):1–8
22. Spiliopoulos KV, Panagiotou KD (2012) A direct method to predict cyclic steady states of elastoplastic structures. Comput Methods Appl Mech Eng 223–224:186–198
23. Spiliopoulos KV, Panagiotou KD (2014) A residual stress decomposition based method for the shakedown analysis of structures. Comput Methods Appl Mech Eng 276:410–430
24. Gokhfeld DA, Cherniavsky OF (1980) Limit analysis of structures at thermal cycling. Sijthoff & Noordhoff, Alphen aan den Rijn
25. König JA, Kleiber M (1978) On a new method of shakedown analysis. Bull Acad Polon Sci Ser Sci Tech 26:165–171
26. Simo JC, Hughes TJR (1998) Computational inelasticity. Springer, New York
27. Ponter ARS, Engelhardt M (2000) Shakedown limits for a general yield condition: implementation and application for a von Mises yield condition. Eur J Mech A/Solids 19:423–445
28. Carvelli V, Cen ZZ, Liu Y, Maier G (1999) Shakedown analysis of defective pressure vessels by a kinematic approach. Arch Appl Mech 69:751–764
29. Tran TN, Liu GR, Nguyen XH, Nguyen TT (2010) An edge-based smoothed finite element method for primal-dual shakedown analysis of structures. Int J Numer Eng 82:917–938

An Efficient Algorithm for Shakedown Analysis Based on Equality Constrained Sequential Quadratic Programming

Giovanni Garcea, Antonio Bilotta and Leonardo Leonetti

Abstract A new iterative algorithm to evaluate the elastic shakedown multiplier is proposed. On the basis of a three field mixed finite element, a series of mathematical programming problems or steps, obtained from the application of the proximal point algorithm to the static shakedown theorem, are obtained. Each step is solved by an Equality Constrained Sequential Quadratic Programming (EC-SQP) technique that retain all the equations and variables of the problem at the same level so allowing a consistent linearization that improves the computational efficiency. The numerical tests performed for 2D problems show the good performance and the great robustness of the proposed algorithm.

1 Introduction

Directs methods are largely used for the shakedown analysis of elastic-plastic structure under variable [5, 12, 16, 19, 20, 22, 24, 25] or cyclic loading [23] because they an efficient alternative to time consuming incremental time-stepping calculations (see also [6]). They are essentially based on Interior Point Methods but alternative techniques as the linear matching method [8, 17] are efficiently employed too.

In [7, 11] an iterative algorithm has been proposed that makes possible to performs the finite element shakedown analysis using a formulation similar to that adopted for the evaluation of the equilibrium path of elastoplastic structure. In [9] has been shown has this algorithm can be obtained from a mathematical programming problem, consisting in the application of the proximal point algorithm to the static shakedown theorem so defining a convergent sequence of safe states or steps

G. Garcea (✉) · A. Bilotta · L. Leonetti
Dipartimento di Modellistica per L'Ingegneria, Università della Calabria, Cosenza, Calabria
e-mail: giovanni.garcea@unical.it

A. Bilotta
e-mail: antonio.bilotta@unical.it

L. Leonetti
e-mail: leonardo.leonetti@unical.it

P. Fuschi et al. (eds.), *Direct Methods for Limit and Shakedown Analysis of Structures*,
DOI 10.1007/978-3-319-12928-0_10

that are solved by means of dual decomposition methods. Each proximal point step coincides, in the elastoplastic case, with that defined using standard incremental iterative algorithms [4, 13] based on Riks arc-length method while the optimization subproblems, deriving from the dual decomposition technique, exactly correspond to the standard return mapping by closest point projection scheme (CPP). We denote from now on this method as SD-CPP (Strain Driven—Closest Point Projection) or *pseudo elastoplastic analysis*.

The major advantage of the dual decomposition approach is that the inequality constraints arising from the constitutive laws are eliminated from the step equations at the local level (Gauss point or finite element) using the CPP scheme, while the stresses and the plastic multipliers are implicitly defined in terms of the displacements. The finite step equations are so transformed into a nonlinear system of equations, without inequalities, easily solved by means of standard arc–length strategies. The global description of the algorithm is always performed in terms of displacement variables alone.

It is worth of noting that the more usual descriptions based on displacement variables alone can not be the best choice, while potentially more efficient and robust analysis algorithms can be obtained by directly solving the proximal point step equations maintaining all the variables of the problems at the same level. With this aim in the present work an approach similar to that presented in [4] for the evaluation of the equilibrium path of elastoplastic structures is proposed for shakedown analysis. The proximal point step equations are solved by using an Equality Constraints Sequential Quadratic Programming (EC-SQP) formulation instead of using dual decomposition with a great advantage in terms of both robustness and efficiency.

Each QP iteration of the algorithm is organized in two phases: (i) a suitable estimate of the active constraints at the current iteration is performed for a fixed value of the load multiplier and the displacements, solving a problem similar to that defined by the return mapping process; (ii) the new estimates of the step unknowns are obtained solving an equality constrained quadratic programming problem that retains only the active constraints. The second phase requires only the solution of a linear system of equations, and is far cheaper to solve than the complete QP problem arising from standard SQP strategies. The algorithm only require few modifications of existing codes that evaluate the equilibrium path of elastoplastic structures by means of path following incremental iterative algorithms (SD-CPP). From the computational point of view each iteration has almost the same computational cost as a standard SD-CPP step in the case of a single constraint, and a smaller cost in the multiple constraints case. Furthermore the QP problem is obtained on the basis of a consistent linearization of all the equations, so allowing the iterations to naturally evolve towards the solution. To improve the computational efficiency, the solution of each linear system required by the analysis is performed by means of a Gauss elimination of the locally defined quantities (stresses and plastic multipliers).

In order to validate the proposed method, the work presents a numerical experimentation by analyzing some 2D test problems under plane stress condition in both cases of fixed and variable loads. The numerical analyses are performed using the mixed finite elements proposed in [1–3]. These elements, which are based on a three field interpolation of displacements, stresses and plastic multipliers, have been adopted for their accuracy and performance properties both for elastic and elasto-plastic analysis as required by shakedown problems. This makes possible to avoid the use of two kinds of elements as done in [9, 11] one with a good elastic behaviour to correctly evaluate the plastic shakedown multiplier and one with a good plastic behaviour and free of volumetric locking phenomena to performs the nonlinear analysis. The obtained results highlight the improvement in terms of robustness and computational cost with respect to previous proposal [9, 11] based on dual decomposition but also with respect to the use of interior point methods, i.e. the most efficient methods for solving nonlinear convex programming problems, in all the cases considered.

2 A Mathematical Programming Formulation for the Shakedown Analysis

In the following the static shakedown theorem is rewritten in terms of the involved finite element quantities and of the total stress, making possible a unified treatment of shakedown and limit analysis. The chosen FEM format is based on the general three field interpolation presented in [3] but any other finite element, such as, for example, a standard compatible one based on Gauss integration points, can be cast in the framework proposed by giving the appropriate meaning to parameters and operators.

2.1 *The FEM Discrete Equation for the Static Shakedown Theorem*

We consider an elastoplastic body Ω subjected to bulk load $\mathbf{b}$ and tractions $\mathbf{t}$, that can varying with the time inside a given load domain.

Using the three fields interpolation proposed in [2, 3] we assume that displacement $\mathbf{u}[\mathbf{x}]$, stress $\boldsymbol{\sigma}[\mathbf{x}]$ and plastic multiplier $\gamma[\mathbf{x}]$ of a point $\mathbf{x} \in \Omega$ are interpolated as

$$\mathbf{u}[\mathbf{x}] := \mathbf{N}[\mathbf{x}]\mathbf{d}_e \quad \boldsymbol{\sigma}[\mathbf{x}] := \mathbf{S}[\mathbf{x}]\boldsymbol{\beta}_e \quad \gamma[\mathbf{x}] := \mathbf{G}[\mathbf{x}]\boldsymbol{\kappa}_e, \tag{1}$$

where $\mathbf{N}[\mathbf{x}]$, $\mathbf{S}[\mathbf{x}]$ and $\mathbf{G}[\mathbf{x}]$ are the matrices containing the interpolation functions and $\mathbf{d}_e$, $\boldsymbol{\beta}_e$ and $\boldsymbol{\kappa}_e$ are the vectors collecting the finite element parameters.

As usual for mixed finite elements we assume inter-element continuity only for the displacement field $\mathbf{u}$, while $\boldsymbol{\sigma}$ and γ will be defined locally inside the element, i.e. $\boldsymbol{\beta}_e$ and $\boldsymbol{\kappa}_e$ are local variables which are discontinuous across the elements. Furthermore the interpolation functions in $\mathbf{G}$ are assumed to be nonnegative allowing the condition $\gamma \geq 0$ to be easily expressed by making $\boldsymbol{\kappa}_e \geq \mathbf{0}$. From now on, vector inequality will be considered in a componentwise fashion while we omit the dependence of the quantities from $\mathbf{x}$ when clear from the context.

2.1.1 Equilibrium Equation

The interpolations introduced above allow to write the discrete form of the equilibrium equations as

$$\mathscr{A}_e \left\{ \mathbf{Q}_e^T \boldsymbol{\beta}_e - \lambda \mathbf{p}_e \right\} = 0 \tag{2}$$

$\mathscr{A}_e$ being the standard assembling operator which takes into account the inter–element continuity conditions on the displacement field and

$$\mathbf{Q}_e := \int_{\Omega_e} \mathbf{S}^T \mathbf{D} \mathbf{N}, \qquad \mathbf{p}_e := \int_{\Omega_e} \mathbf{N}^T \mathbf{b} + \int_{\partial\Omega_e} \mathbf{N}^T \mathbf{t} \tag{3}$$

are the element equilibrium operator and load vector while $\mathbf{D}$ is the continuum compatibility differential operator, Ω_e the element domain and $\partial\Omega_e$ its boundary. For the sake of the following discussion Eq. (2) can be rewritten as

$$\mathbf{Q}^T \boldsymbol{\beta} - \lambda \mathbf{p} = \mathbf{0} \tag{4}$$

where $\boldsymbol{\beta}$ and $\mathbf{p}$ denote the global vectors collecting all the stress parameters $\boldsymbol{\beta}_e$ and the applied loads $\mathbf{p}_e$, while $\mathbf{Q}^T$ the related global equilibrium matrix. From now on a subscript e denotes the finite element description of a quantity.

2.1.2 The Elastic Envelope of the Stresses

We assume that the external actions $\mathbf{p}[t]$, variable with the time t, are expressed as a combination of basic loads $\mathbf{p}_i$ belonging to the admissible closed and convex *load domain*

$$\mathbb{P} := \left\{ \mathbf{p}[t] \equiv \sum_{i=1}^{p} a_i[t] \mathbf{p}_i \; : \; a_i^{min} \leq a_i[t] \leq a_i^{max} \right\} \tag{5}$$

Denoting by $\hat{\boldsymbol{\beta}}_i$ the elastic stress solution corresponding to $\mathbf{p}_i$, the *elastic envelope* $\hat{\mathbb{S}}$

$$\hat{\mathbb{S}} := \left\{ \hat{\boldsymbol{\beta}}[t] \equiv \sum_{i=1}^{p} a_i[t] \hat{\boldsymbol{\beta}}_i \; : \; a_i^{min} \leq a_i[t] \leq a_i^{max} \right\} \tag{6}$$

defines the set of the elastic stresses $\hat{\boldsymbol{\beta}}[t]$ produced by each load path contained in $\mathbb{P}$.

By construction $\hat{\mathbb{S}}$ and $\mathbb{P}$ are convex polytopes and each $\hat{\boldsymbol{\beta}}[t] \in \hat{\mathbb{S}}$ can be expressed as a convex combination of the N_v elastic envelope vertexes $\hat{\boldsymbol{\beta}}^{\alpha}$ that can be usefully referred to the reference stress $\hat{\boldsymbol{\beta}}^{0}$ so obtaining

$$\hat{\boldsymbol{\beta}}[t] = \hat{\boldsymbol{\beta}}^{0} + \sum_{\alpha=1}^{N_v} s^{\alpha} \hat{\boldsymbol{\beta}}^{\alpha} \qquad s^{\alpha} \geq 0 \quad \sum_{\alpha=1}^{N_v} s^{\alpha} = 1 \tag{7}$$

If the external loads increase by a real number λ, called *load domain multiplier*, the elastic envelope becomes $\lambda\hat{\mathbb{S}} := \left\{\lambda\hat{\boldsymbol{\beta}} : \hat{\boldsymbol{\beta}} \in \hat{\mathbb{S}}\right\}$.

2.1.3 Plastic Admissibility for Shakedown Analysis

Following [2, 3] and due to the local nature of the stress interpolation, the plastic admissibility condition is rewritten on the element, in a weak form, as

$$\int_{\Omega_e} \delta\gamma\phi[\boldsymbol{\sigma}] \equiv \Delta\boldsymbol{\kappa}_e^T \boldsymbol{\Phi}_e[\boldsymbol{\beta}_e] = 0 \quad \forall\delta\gamma \geq 0, \tag{8}$$

where $\boldsymbol{\Phi}_e[\boldsymbol{\beta}_e] := \int_{\Omega_e} \mathbf{G}^T \phi[\boldsymbol{\beta}_e]$ and ϕ is the yield function. Equation (8) allows to control plastic admissibility in the N_e element so that $\boldsymbol{\beta}$ will be plastically admissible if

$$\boldsymbol{\Phi}[\boldsymbol{\beta}] \leq \mathbf{0} \iff \boldsymbol{\Phi}_e[\boldsymbol{\beta}_e] \leq \mathbf{0}, \qquad \forall e = 1 \ldots N_e\,. \tag{9}$$

Finally it is useful to express the plastically admissible condition for all the stresses contained in the amplified elastic envelope $\lambda\hat{\mathbb{S}}$ translated by $\bar{\boldsymbol{\beta}}$. Due to the convexity of $\boldsymbol{\Phi}$ and $\hat{\mathbb{S}}$ this can be easily expressed in terms of the plastic admissibility of all the α vertexes of the amplified elastic domain $\boldsymbol{\beta}^{\alpha} = \lambda(\hat{\boldsymbol{\beta}}^{\alpha} + \hat{\boldsymbol{\beta}}^{0}) + \bar{\boldsymbol{\beta}}$ as

$$\boldsymbol{\Phi}[\lambda\hat{\boldsymbol{\beta}} + \bar{\boldsymbol{\beta}}] \leq \mathbf{0}, \quad \forall\hat{\boldsymbol{\beta}} \in \hat{\mathbb{S}} \iff \boldsymbol{\Phi}[\boldsymbol{\beta}^{\alpha}] \leq \mathbf{0}, \quad \forall\alpha \tag{10}$$

where, from now on, a Greek superscript denotes vertex quantities.

2.1.4 The Static Theorem in Discrete Format and the Mathematical Programming Point of View

The Bleich–Melan static theorem states that a load domain multiplier λ_s will be safe if there exists a time-independent self-equilibrated stress field $\bar{\boldsymbol{\beta}}$ so that each stress in the amplified and translated domain $\lambda_s\hat{\mathbb{S}} + \{\bar{\boldsymbol{\beta}}\}$ is plastically admissible. The shakedown multiplier λ_a can be evaluated as the maximum of these safe multipliers. The static theorem can be reformulated in terms of total stress, instead of self–equilibrated ones, making possible a unified notation for shakedown and limit analysis, i.e.

$$\begin{aligned} &\text{maximize} && \lambda_s \\ &\text{subject to} && \mathbf{Q}^T\boldsymbol{\beta} = \lambda_s \mathbf{p}_0 \\ & && \boldsymbol{\Phi}[\boldsymbol{\beta} + \lambda_s \hat{\boldsymbol{\beta}}^{\alpha}] \leq \mathbf{0}, \quad \alpha = 1 \ldots N_v \end{aligned} \tag{11}$$

with $\mathbf{p}_0 \equiv \mathbf{Q}^T \hat{\boldsymbol{\beta}}^0$ and $\boldsymbol{\beta} \equiv \bar{\boldsymbol{\beta}} + \lambda_s \hat{\boldsymbol{\beta}}^0$. When $\hat{\boldsymbol{\beta}}^0 = \mathbf{0}$ we have the classic form in terms of the self–equilibrated stress. Furthermore, without any loss in generality, we can set $\hat{\boldsymbol{\beta}}^0$ as a generic vertex of $\hat{\mathbb{S}}$ so $\boldsymbol{\beta}$ becomes the total stress of this vertex. When the external load domain collapses in a single point ($a_i^{min} = a_i^{max}$) Eq. (11) directly transform into the standard form of the static theorem of limit analysis.

From now on we denote with $\boldsymbol{\Phi}^{\alpha}[\boldsymbol{\beta}, \lambda] \equiv \boldsymbol{\Phi}[\boldsymbol{\beta} + \lambda \hat{\boldsymbol{\beta}}^{\alpha}]$ the shakedown yield function.

3 Shakedown Analysis Using Dual Decomposition Methods

The mathematical programming problem in Eq. (11) can be solved using efficient interior point algorithms [5, 14, 19, 20] specialized to the shakedown case. We will now resume an alternative approach, already presented in [7, 9, 11] where further details can be found, that will be the basis for the new proposal. The approach uses the proximal point method to generate a convergent sequences of steps and a dual decomposition strategy to solve each step.

3.1 The Proximal Point Method and the Pseudo-elastoplastic Step

The proximal point method is applied to (11) by defining a sequences of subproblems or *steps* by adding a quadratic positive term to the objective function, i.e.

$$\begin{aligned} &\text{maximize} && \Delta\xi^{(n)}\lambda^{(n)} - \frac{1}{2}\Delta\boldsymbol{\beta}^T \mathbf{H} \Delta\boldsymbol{\beta} \\ &\text{subject to} && \mathbf{Q}^T\boldsymbol{\beta}^{(n)} - \lambda^{(n)}\mathbf{p}_0 = \mathbf{0} \\ & && \boldsymbol{\Phi}[\boldsymbol{\beta}^{(n)}, \lambda^{(n)}] \leq \mathbf{0} \end{aligned} \tag{12}$$

where the superscript $(\cdot)^{(n)}$ will denote quantities evaluated in the nth step, the symbol $\Delta(\cdot) = (\cdot)^{(n)} - (\cdot)^{(n-1)}$ is the increment of a quantity from the previous step and $\Delta\xi^{(n)} > 0$ is an assigned real positive number. To simplify the notation we collected all $\boldsymbol{\Phi}^{\alpha}[\boldsymbol{\beta}, \lambda]$ in the global vector $\boldsymbol{\Phi}[\boldsymbol{\beta}, \lambda] = \{\boldsymbol{\Phi}^1, \ldots, \boldsymbol{\Phi}^{N_v}\}$. Finally $\mathbf{H}$ is the compliance matrix and is defined by the following equivalence

$$\sum_e \boldsymbol{\beta}_e^T \mathbf{H}_e \boldsymbol{\beta}_e = \boldsymbol{\beta}^T \mathbf{H} \boldsymbol{\beta} \quad \text{with} \quad \mathbf{H}_e := \int_{\Omega_e} \mathbf{S}^T \mathbf{C}^{-1} \mathbf{S} \tag{13}$$

where $\mathbf{C}$ is the elastic matrix. Note how, due to the local nature of the stress interpolation, $\mathbf{H}$ has a block diagonal structure that couples only the finite element stress parameters. The finite element will be, from now on, the *local level* of the analysis.

3.1.1 First Order Conditions

Introducing the dual multipliers $\Delta\mathbf{d}$ and $\Delta\boldsymbol{\kappa}$ associated to the equalities and inequalities constraints of (12) respectively, the finite step equations are defined by the first order conditions of the following Lagrangian $\mathscr{L}^{(n)}$

$$\mathscr{L}^{(n)} = \Delta\xi^{(n)}\lambda^{(n)} - \frac{1}{2}\Delta\boldsymbol{\beta}^T \mathbf{H} \Delta\boldsymbol{\beta} + \Delta\mathbf{d}^T(\mathbf{Q}^T\boldsymbol{\beta}^{(n)} - \lambda^{(n)}\mathbf{p}_0) - \Delta\boldsymbol{\kappa}^T \boldsymbol{\Phi}[\boldsymbol{\beta}^{(n)}, \lambda^{(n)}]. \tag{14}$$

In order to simplify the notation the superscript (n) will be omitted from now on.

In particular from the stationary condition of (14) with respect to $\boldsymbol{\beta}$ and $\Delta\boldsymbol{\kappa}$ we obtain the finite step form of the constitutive law, i.e. the plastic admissibility and plastic consistence conditions for shakedown

$$\begin{cases} \mathbf{r}_\sigma \equiv -\mathbf{H}_e\Delta\boldsymbol{\beta}_e + \mathbf{Q}_e\Delta\mathbf{d}_e - \mathbf{A}_e[\boldsymbol{\beta}_e, \lambda]\Delta\boldsymbol{\kappa}_e = \mathbf{0} \\ \mathbf{r}_\mu \equiv \boldsymbol{\Phi}_e[\boldsymbol{\beta}_e, \lambda] \le 0, \quad \Delta\boldsymbol{\kappa}_e \ge 0, \quad \Delta\boldsymbol{\kappa}_e{}^T\boldsymbol{\Phi}_e[\boldsymbol{\beta}_e, \lambda] = 0, \end{cases} \tag{15a}$$

where $\mathbf{A}_e[\boldsymbol{\beta}_e, \lambda] := \left(\frac{\partial\boldsymbol{\Phi}_e[\boldsymbol{\beta}_e,\lambda]}{\partial\boldsymbol{\beta}_e}\right)^T$. In the fixed load cases Eq. (15a) coincide with the backward-Euler integration of the elasto-plastic constitutive equations.

Due to the discontinuity of $\boldsymbol{\beta}_e$ and $\boldsymbol{\kappa}_e$ across the elements Eq. (15a) are expressed with respect to element quantities alone when $\Delta\mathbf{d}_e$ and λ are assigned. For this reason they will be denoted, from now on, as *local equations* while $\boldsymbol{\beta}_e$ and $\boldsymbol{\kappa}_e$ will be denoted as *local variables*. Following [4] a task that uses only local variables and equations will be said to be at the *local level*.

In the same fashion the stationary condition with respect to $\Delta\mathbf{d}$ and λ furnishes the equilibrium equations and the normalization condition, coupling all the variables of the problem and defining the *global level* of the analysis,

$$\begin{cases} \mathbf{r}_u \equiv \mathbf{Q}^T\boldsymbol{\beta} - \lambda\mathbf{p}_0 = \mathbf{0} \\ r_\lambda \equiv \Delta\xi - \Delta\mathbf{d}^T\mathbf{p}_0 - \Delta\boldsymbol{\kappa}^T\boldsymbol{\Phi}_{,\lambda} = 0 \end{cases} \tag{15b}$$

where $\boldsymbol{\Phi}_{,\lambda} := \left(\frac{\partial\boldsymbol{\Phi}[\boldsymbol{\beta},\lambda]}{\partial\lambda}\right)$. Equation (15b) will be denoted, from now on, as *global equations* while $\mathbf{d}$ and λ will be denoted as *global variables*.

In the case of limit analysis Eq. (15a, 15b) exactly corresponds to a step of the arc-length algorithm used to solve the incremental elastoplastic problem [7, 11]. Due to its meaning in the case of fixed loads we call this kind of analysis *pseudo elastoplastic*. $\Delta\mathbf{d}$ and $\Delta\boldsymbol{\kappa}$ assume the meaning of displacements and plastic multipliers of the problem.

As for elastic perfectly plastic structures the limit load can be evaluated by recovering the complete equilibrium path by means of path-following algorithms, in the same fashion the skakedown multiplier can be obtained by evaluating a sequence of states, $\mathbf{z}^{(n)} := \{\lambda^{(n)}, \boldsymbol{\beta}^{(k)}, \mathbf{d}^{(n)}, \boldsymbol{\kappa}^{(n)}\}$, obtained by solving a series of problems (12), i.e. defining a pseudo-elastoplastic equilibrium curve [7]. In [9] it has been shown that starting from the known elastic limit $\mathbf{z}^{(0)}$, the sequence $\mathbf{z}^{(n)}$ generated in this way is safe in the sense of the static theorem and monotonously increasing in $\lambda^{(n)}$. In the case $\lambda^{(n)} = \lambda^{(n-1)}$ with $\Delta\mathbf{d} \neq \mathbf{0}$, it is simple to shown that $\Delta\boldsymbol{\beta} = \mathbf{0}$ and we have from (11) the convergence to the desired shakedown multiplier.

Finally note as the equation format reported in Eqs. (12) and (15a, 15b) is quite general and similar expressions could be obtained using other finite elements. In particular for a standard compatible finite element the local level coincides with the Gauss point, $\boldsymbol{\beta}_e$ becomes the Gauss point stress and $\boldsymbol{\kappa}_e$ the Gauss point plastic multiplier, plastic admissibility and consistency are imposed at each Gauss point and the operators consequently transform.

3.2 The Dual Decomposition Solution of the Pseudo Elastoplastic Step

The similarity of Eq. (15a, 15b) with standard strain driven path-following elastoplastic analysis suggests that also the same method of solution can be used. This is the approach followed in [7, 11] and it is based on an exact solution of the local conditions in (15a) for an assigned value of $\Delta\mathbf{d}_e$ and $\Delta\lambda$, so expressing $\boldsymbol{\beta}_e$ and $\boldsymbol{\kappa}_e$ as implicit functions of the displacements and of the load multiplier. This step is performed at the local level by using a return mapping by closest point projection process as in the case of the standard incremental elastoplastic analysis. This can be shown by noting that Eq. (15a) are the first order conditions of the following problem

$$\begin{cases} \max\limits_{\boldsymbol{\beta}_e} & \boldsymbol{\beta}_e^T \mathbf{Q}_e \Delta\mathbf{d}_e - \dfrac{1}{2}\Delta\boldsymbol{\beta}_e^T \mathbf{H}_e \Delta\boldsymbol{\beta}_e, \\ \text{subject to:} & \boldsymbol{\Phi}_e[\boldsymbol{\beta}_e, \lambda] \leq 0. \end{cases} \tag{16}$$

where the number of constraints depends on the parameters used to interpolate γ, that is on the dimension of $\boldsymbol{\kappa}_e$, and on the number of vertexes of the element elastic envelope $\hat{\mathbb{S}}_e$. As $\Delta\mathbf{d}_e$ is constant with respect to the maximization, it is possible to rewrite problem in (16) as

$$\begin{cases} \min\limits_{\boldsymbol{\beta}_e} & \frac{1}{2}\left(\boldsymbol{\beta}_e-\boldsymbol{\beta}_e^*\right)^T \mathbf{H}_e\left(\boldsymbol{\beta}_e-\boldsymbol{\beta}_e^*\right), \\ \text{subject to:} & \boldsymbol{\Phi}_e[\boldsymbol{\beta}_e,\lambda] \le 0. \end{cases} \tag{17}$$

that is the convex projection of the *trial stress* (or the elastic predictor) $\boldsymbol{\beta}_e^*$, defined by

$$\boldsymbol{\beta}_e^* = \boldsymbol{\beta}_e^{(n-1)} + \mathbf{H}_e^{-1}\mathbf{Q}_e\Delta\mathbf{d}_e, \tag{18}$$

onto the elastic shakedown domain bounded by the convex function $\boldsymbol{\Phi}_e[\boldsymbol{\beta}_e,\lambda]$. From the closest point projection (CPP) in Eq. (17) we have the stresses and the plastic multipliers as a function of $\Delta\mathbf{d}$ and λ

$$\boldsymbol{\beta} = \boldsymbol{\beta}[\mathbf{z}^{(n-1)},\Delta\mathbf{d},\lambda], \qquad \Delta\boldsymbol{\kappa} = \Delta\boldsymbol{\kappa}[\mathbf{z}^{(n-1)},\Delta\mathbf{d},\lambda]$$

and of the known initial step quantities collected in $\mathbf{z}^{(n-1)}$.

Omitting the dependence from $\mathbf{z}^{(n-1)}$ the global Eq. (15b) can be rewritten, in terms of $\Delta\mathbf{d}$ and λ, as

$$\begin{cases} \mathbf{r}_u \equiv \mathbf{Q}^T\boldsymbol{\beta}[\Delta\mathbf{d},\lambda] - \lambda\mathbf{p}_0 = \mathbf{0} \\ r_\lambda \equiv \Delta\xi - \Delta\mathbf{d}^T\mathbf{p}_0 - \boldsymbol{\Phi},_\lambda^T\Delta\boldsymbol{\kappa}[\Delta\mathbf{d},\lambda] = 0 \end{cases} \tag{19}$$

If the nonlinear system (19) is solved by means of a Newton iteration as in [7, 9], we obtain:

$$\begin{cases} \Delta\mathbf{d}^{j+1} = \Delta\mathbf{d}^j + \dot{\mathbf{d}}, \\ \Delta\lambda^{j+1} = \Delta\lambda^j + \dot{\lambda}, \end{cases} \quad \text{with} \quad \begin{cases} \mathbf{K}^j\dot{\mathbf{d}} - \dot{\lambda}\mathbf{y}^j = -\mathbf{r}_u^j, \\ -\mathbf{y}^{jT}\dot{\mathbf{d}} + \dot{\lambda}\,h_{\lambda\lambda}^j = -r_\lambda^j, \end{cases} \tag{20}$$

where $\mathbf{K}^j \equiv \partial\mathbf{r}_u/\partial\mathbf{d}$ is the algorithmic tangent matrix while r_λ^j and $\mathbf{r}_u^j$ are the residuals defined in Eq. (19) evaluated in $(\mathbf{d}^j,\lambda^j)$ after performing the return mapping process (17) to evaluate $\boldsymbol{\beta}[\Delta\mathbf{d}^j,\lambda^j]$ and $\Delta\boldsymbol{\kappa}[\Delta\mathbf{d}^j,\lambda^j]$, while

$$\mathbf{y}^j \equiv \left.\frac{\partial\mathbf{r}_u}{\partial\lambda}\right|_{(\lambda^j,\mathbf{d}^j)} = \left.\frac{\partial r_\lambda}{\partial\mathbf{d}}\right|_{(\lambda^j,\mathbf{d}^j)} \qquad h_{\lambda\lambda}^j \equiv \left.\frac{\partial r_\lambda}{\partial\lambda}\right|_{(\lambda^j,\mathbf{d}^j)}.$$

We have convergence to a new equilibrium point when the norm of $\mathbf{r}_u^j$ become sufficiently small, in this case we set $\mathbf{z}^{(n)} = \mathbf{z}^j$. We recall that the use of a modified Newton method that uses the initial elastic stiffness matrix assures global, even if simply linear, convergence [7].

3.3 The Optimization Point of View and the Motivation for a New Strategy

The dual decomposition solution strategy of the proximal point step exactly coincides in the fixed load case with the standard Strain Driven algorithms used in incremental elasto-plasticity with the stresses evaluated by means of a return mapping by Closest Point Projection scheme (SD-CPP). The reinterpretation previously described (see also [4, 9]), allows to extend the use of classical elastoplastic algorithms to the shakedown case.

The optimization point of view also makes simple to compare the previous described strategy with other direct methods in the evaluation of the shakedown multiplier. The dual decomposition, in fact, allows to split the optimization problem in the small CPP subproblems (17) defined at the local level that allows to obtain the dual function and in the solution of the global equations (19), that correspond to the first order condition of the so evaluated dual function, by a Newton method. In this way it is possible to simplify the solution of the optimization problem and a small cost for each iteration is required. On the other hand, due to the decomposition, the convergence when compared with interior point strategies, can be slow especially when the elastic stiffness matrix is used.

In following section we propose a new formulation to solve the proximal point step (12) that unlike the $SD - CPP$, attempts to use a consistent linearization of all the equations and variables at the same time.

4 A New Solution Scheme for the Pseudo Elastoplastic Step

This section presents the new solution algorithm based on the application of the SQP method to solve the mathematical programming problem in Eq. (12). By exploiting the decomposition point of view we devised a strategy which only solve, at the global level, a system of nonlinear equations similar to those presented in Eq. (20) and which is characterized by minimal algorithmic differences but greater robustness with respect to dual decomposition formulations.

4.1 The Linearized Equations for the Elastoplastic Step and the Sequential Quadratic Programming (SQP) Formulation

The starting point of the new algorithm is the linearization of the step equations (15a, 15b) with respect to all the involved variables, maintaining the full mixed format of the problem. In what follows we will denote with $\mathbf{z}^j = \{\boldsymbol{\kappa}_e^j, \boldsymbol{\beta}_e^j, \mathbf{d}^j, \lambda^j\}$ the current, known, estimate of the step solution $\mathbf{z}^{(n)}$ and with $\mathbf{z}^{j+1} = \mathbf{z}^j + \dot{\mathbf{z}}$ the new estimate we are searching for. From the linearization of the local equations (15a) we obtain:

$$\begin{cases} -\mathbf{H}_{et}\dot{\boldsymbol{\beta}}_e + \mathbf{Q}_e\dot{\mathbf{d}}_e - \mathbf{A}_e^j\dot{\boldsymbol{\kappa}}_e - \dot{\lambda}\mathbf{a}_\lambda^j = -\mathbf{r}_\sigma^j, \\ \boldsymbol{\Phi}_e{}^{j+1}[\boldsymbol{\beta},\lambda] \leq \mathbf{0}, \quad \Delta\boldsymbol{\kappa}_e{}^{j+1} \geq \mathbf{0}, \quad (\Delta\boldsymbol{\kappa}_e{}^{j+1})^T\boldsymbol{\Phi}_e{}^{j+1} = 0. \end{cases} \quad \forall e \tag{21a}$$

where $\mathbf{a}_\lambda^j = \mathbf{A}_e,_\lambda^j \, \Delta\boldsymbol{\kappa}_e{}^j$ a comma means derivatives with respect to the symbol that follows and

$$\mathbf{H}_{et} \equiv \mathbf{H}_e + \sum_k \Delta\kappa_{ek} \frac{\partial \Phi_{ek}}{\partial \boldsymbol{\beta}_e}\bigg|_{\boldsymbol{\beta}_e=\boldsymbol{\beta}_e^j}, \qquad \boldsymbol{\Phi}_e{}^{j+1} \equiv \boldsymbol{\Phi}_e{}^j + \mathbf{A}_e^{j\,T}\dot{\boldsymbol{\beta}}_e + \dot{\lambda}\boldsymbol{\Phi}_e{}^j,_\lambda$$

$\Delta\kappa_{ek}^j$ and Φ_{ek} are the kth components of $\Delta\boldsymbol{\kappa}_e^j$ and $\boldsymbol{\Phi}_e{}^j$ respectively. The linearization of the global finite step equation (4) gives:

$$\begin{cases} \mathbf{Q}^T\dot{\boldsymbol{\beta}} - \dot{\lambda}\mathbf{p}_0 = -\mathbf{r}_u^j \\ -\dot{\mathbf{d}}^T\mathbf{p}_0 - \dot{\boldsymbol{\kappa}}^T\boldsymbol{\Phi},_\lambda^j - \mathbf{a}_\lambda^{jT}\dot{\boldsymbol{\beta}} - a_{\lambda\lambda}^j\dot{\lambda} = -r_\lambda^j \end{cases} \tag{21b}$$

where $a_{\lambda\lambda}^j = (\Delta\boldsymbol{\kappa}^j)^T\boldsymbol{\Phi},_{\lambda\lambda}$.

Equation (21a, 21b) could also be obtained has the first order condition of the following jth QP subproblem obtained by applying the sequential quadratic programming (SQP) approach to (15a, 15b).

$$\begin{array}{ll} \max & \Delta\xi\dot{\lambda} - \dot{\boldsymbol{\beta}}^T\mathbf{H}\Delta\boldsymbol{\beta}_j - \frac{1}{2}\dot{\lambda}^2 a_{\lambda\lambda}^j - \dot{\lambda}\dot{\boldsymbol{\beta}}^T\mathbf{a}_\lambda^j - \frac{1}{2}\dot{\boldsymbol{\beta}}^T\mathbf{H}_{et}\dot{\boldsymbol{\beta}} \\ \text{subj.} & \mathbf{r}_u^j + \mathbf{Q}^T\dot{\boldsymbol{\beta}} - \dot{\lambda}\mathbf{p}_0 = \mathbf{0} \\ & \boldsymbol{\Phi}^{j+1}[\boldsymbol{\beta},\lambda] \leq \mathbf{0}, \end{array} \tag{22}$$

Equation (22) furnishes the new estimate $\mathbf{z}^{j+1}$ in the form

$$\mathbf{z}^{j+1} = \{\lambda^j + \dot{\lambda}, \boldsymbol{\beta}^j + \dot{\boldsymbol{\beta}}, \Delta\mathbf{d}^{j+1}, \Delta\boldsymbol{\kappa}^{j+1}\}. \tag{23}$$

4.2 The EC-SQP Formulation

A direct application of standard optimization algorithms to the solution of the QP sub-problems defined in (22) can hardly be competitive with respect to the dual decomposition approach because of the large number of degrees of freedom and constraints. For this reason we use the equality constraint sequential quadratic programming (EC-SQP) approach proposed in [4] for incremental elasto-plasticity to which we remind for further details.

Each iteration of the EC-SQP approach consists of two phases: (i) estimation of the active set of constraints; (ii) solution of an equality constrained quadratic program that imposes the apparently active constraints and ignores the apparently inactive ones.

The idea is to identify the active inequality constraints, i.e. the components of $\boldsymbol{\Phi}_e^{j+1}$ for which equality holds, using information available at a point $\bar{\mathbf{z}}^{j+1} = \mathbf{z}^j + \dot{\bar{\mathbf{z}}}$ near to $\mathbf{z}^{j+1}$ but which is less expensive to evaluate. Once the active set is known, the solution of an equality constrained QP, requiring only the solution of a linear system of equations, gives the new estimate $\mathbf{z}^{j+1}$. Moreover, by maintaining a pseudocompatible format in the solution of the linear system, a scheme similar to that presented in Eq. (20) is obtained at global level.

4.2.1 The Detection of the Active Set of Constraints

The estimation of the active constraints is performed by advocating the decomposition point of view, i.e. solving an optimization problem obtained by the original ones (21a) for a fixed, properly assumed, value of the global variables: $\bar{\mathbf{d}}^{j+1} = \mathbf{d}^j$ and $\bar{\lambda}^{j+1} = \lambda^j$, i.e. for $\dot{\mathbf{d}} = \mathbf{0}$ and $\dot{\lambda} = \mathbf{0}$ so obtaining from Eq. (21a)

$$\begin{cases} -\mathbf{H}_{et}\dot{\bar{\boldsymbol{\beta}}}_e - \mathbf{A}_e^j \dot{\bar{\boldsymbol{\kappa}}}_e = -\mathbf{r}_\sigma^j, \\ \bar{\boldsymbol{\Phi}}_e^{j+1} \leq \mathbf{0}, \quad \bar{\boldsymbol{\kappa}_e}^{j+1} \geq \mathbf{0}, \quad (\bar{\boldsymbol{\kappa}_e}^{j+1})^T \bar{\boldsymbol{\Phi}}_e^{j+1} = 0. \end{cases} \quad \forall e \tag{24}$$

where the symbols with a bar denote the estimates of the new quantities. In particular Eq. (24) is the first order conditions of the following QP problem:

$$\begin{cases} \min\limits_{(\dot{\bar{\boldsymbol{\beta}}}_e)} : \dfrac{1}{2}(\dot{\bar{\boldsymbol{\beta}}}_e)^T \mathbf{H}_{et}\dot{\bar{\boldsymbol{\beta}}}_e + (\dot{\bar{\boldsymbol{\beta}}}_e)^T \mathbf{g}^j, \\ \text{subj.:}\ \mathbf{A}_e^{jT}\dot{\bar{\boldsymbol{\beta}}}_e + \boldsymbol{\Phi}_e^j \leq \mathbf{0}, \end{cases} \quad \forall e \tag{25}$$

where $\mathbf{g}^j = \mathbf{H}_e(\boldsymbol{\beta}_e^j - \boldsymbol{\beta}_e^*)$ and $\boldsymbol{\beta}_e^* = \boldsymbol{\beta}_e^{(n-1)} + \mathbf{H}_e^{-1}\mathbf{Q}_e\Delta\mathbf{d}_e^j$. The decoupled QP problems (25), have the same form as a standard CPP scheme, and it can be easily solved at the local level by using the Goldfarb-Idnani active set method [11]. The evaluation of the set of active constraints is then continuously updated with the iterations and, if $\Delta\mathbf{d}^j$ converges to $\Delta\mathbf{d}^{(n+1)}$, the active set converges to that of the nonlinear problem.

Letting $\bar{\boldsymbol{\beta}}_e^{j+1} = \boldsymbol{\beta}_e^j + \dot{\bar{\boldsymbol{\beta}}}_e$ it is easy to show how problem (25) is also equivalent, apart from an inessential constant in the objective function, to the following CPP minimization

$$\begin{cases} \min\limits_{(\bar{\boldsymbol{\beta}}_e^{j+1})} : \dfrac{1}{2}(\bar{\boldsymbol{\beta}}_e^{j+1} - \boldsymbol{\beta}_e^{tr})^T \mathbf{H}_{et}(\bar{\boldsymbol{\beta}}_e^{j+1} - \boldsymbol{\beta}_e^{tr}), \\ \text{subject to:}\ \bar{\boldsymbol{\Phi}}_e^{j+1} \leq \mathbf{0}, \end{cases} \quad \forall e. \tag{26}$$

which allows a formulation of the problem in terms of a predictor/corrector strategy, i.e. the solution of problem (26) coincides with the trial stress defined as

$$\boldsymbol{\beta}_e^{tr} = \boldsymbol{\beta}_e^j + \mathbf{H}_{et}^{-1}(\mathbf{Q}_e\Delta\mathbf{d}^j - \mathbf{H}_e\Delta\boldsymbol{\beta}_e^j),$$

if $\boldsymbol{\Phi}_e[\boldsymbol{\beta}_e^{tr}] \equiv \boldsymbol{\Phi}_e^j + (\mathbf{A}_e^j)^T(\boldsymbol{\beta}_e^{tr} - \boldsymbol{\beta}_e^j) \leq \mathbf{0}$.

4.2.2 The Solution of the QP Equality Constraint Scheme

The second step of the algorithm consists in the solution of Eq. (21a, 21b) retaining, as equalities, only the active constraints evaluated in the previous step. When the active set is not void the solution is given by the following system of equations:

$$\begin{bmatrix} \cdot & \mathbf{A}_e^{jT} & \cdot & \boldsymbol{\Phi}_{e,\lambda}^j \\ -\mathbf{A}_e^j & -\mathbf{H}_{et} & \mathbf{Q}_e & -\mathbf{a}_\lambda^j \\ \cdot & \mathbf{Q}_e^T & \cdot & -\mathbf{p}_{0e} \\ -(\boldsymbol{\Phi}_{e,\lambda}^j)^T & -(\mathbf{a}_\lambda^j)^T & -\mathbf{p}_{0e}^T & -a_{\lambda\lambda} \end{bmatrix} \begin{bmatrix} \dot{\boldsymbol{\kappa}}_e \\ \dot{\boldsymbol{\beta}}_e \\ \dot{\mathbf{d}}_e \\ \dot{\lambda} \end{bmatrix} = - \begin{bmatrix} \mathbf{r}_\mu^j \\ \mathbf{r}_\sigma^j \\ \mathbf{r}_u^j \\ r_\lambda^j \end{bmatrix}, \qquad \mathbf{z}^{j+1} = \mathbf{z}^j + \dot{\mathbf{z}}, \tag{27}$$

where the further condition $\boldsymbol{\kappa}_{j+1} \geq 0$ needs to be imposed.

System (27) is easily solved by static condensation of the local defined quantities obtaining at the global level the same format as system (20) with different meaning of the operators. In particular, recalling that the QP scheme in (25) solves the first two equations of (27) zeroing the global variables we obtain

$$\begin{cases} \dot{\boldsymbol{\beta}}_e = \dot{\bar{\boldsymbol{\beta}}}_e + \mathbf{H}_{et}^{-1}(\mathbf{Q}_e\dot{\mathbf{d}}_e - \dot{\lambda}\mathbf{a}_\lambda^j) \\ \dot{\boldsymbol{\kappa}}_e = \dot{\bar{\boldsymbol{\kappa}}}_e + \mathbf{W}\left(\mathbf{A}_e^{jT}\mathbf{H}_{et}^{-1}(\mathbf{Q}_e\dot{\mathbf{d}}_e - \dot{\lambda}\mathbf{a}_\lambda^j) + \dot{\lambda}\boldsymbol{\Phi}_{e,\lambda}^j\right) \end{cases} \tag{28}$$

where $\mathbf{W} = \left[\mathbf{A}_j^T\mathbf{H}_{et}^{-1}\mathbf{A}_j\right]^{-1}$. At the global level then we have to assemble the condensed element contribution as

$$\mathscr{A}_e\left(\mathbf{Q}_e^T\mathbf{E}_t\mathbf{Q}_e\right)\dot{\mathbf{d}} - \dot{\lambda}\mathbf{y}^j = -\mathscr{A}_e\left(\mathbf{r}_{cu}^j\right), \qquad -(\mathbf{y}^j)^T\dot{\mathbf{d}}_e + h_{\lambda\lambda}\dot{\lambda} = -r_{c\lambda}^j, \tag{29}$$

where the quantities in Eq. (29) are so defined

$$\begin{cases} \mathbf{E}_t &= \mathbf{H}_{et}^{-1} - \mathbf{H}_{et}^{-1}\mathbf{A}_e^j\mathbf{W}\mathbf{A}_e^{jT}\mathbf{H}_{et}^{-1} \\ \mathbf{y}^j &= \mathbf{p}_0 + \mathscr{A}_e\left(\hat{\mathbf{y}}\right) \\ \mathbf{r}_{cu}^j &= \mathbf{r}_u^j + \mathbf{Q}_e^T\left(\hat{\mathbf{r}}_\sigma\right) \\ h_{\lambda\lambda} &= -(\boldsymbol{\Phi}_{e,\lambda}^j)^T(\mathbf{W}\boldsymbol{\Phi}_{e,\lambda}^j - \mathbf{Y}^T\mathbf{a}_\lambda) + \mathbf{a}_\lambda^T\hat{\mathbf{y}} - a_{\lambda\lambda} \\ r_{c\lambda}^j &= r_\lambda^j + (\boldsymbol{\Phi}_{e,\lambda}^j)^T(\mathbf{Y}^T\mathbf{r}_\sigma^j + \mathbf{W}\mathbf{r}_\mu^j) - (\mathbf{a}_\lambda^j)^T\hat{\mathbf{r}}_\sigma \end{cases}$$

being

$$\mathbf{Y} = \mathbf{H}_{et}^{-1}\mathbf{A}_e^j\mathbf{W}, \quad \hat{\mathbf{y}} = \mathbf{E}_t\mathbf{a}_\lambda^j + \mathbf{Y}\boldsymbol{\Phi}_{e,\lambda}^j, \quad \hat{\mathbf{r}}_\sigma = \mathbf{E}_t\mathbf{r}_\sigma^j - \mathbf{Y}\mathbf{r}_\mu^j$$

Note as $\mathbf{E}_t$ has the same expression as the algorithmic tangent matrix evaluated by dual decomposition methods. Also note as $\mathbf{E}_t$ and $\mathbf{Y}$ are evaluated at each step of the QP problem, by the optimization algorithm used. In the case of an element with zero active constraints the solution is simply obtained deleting the inequality constraints and so obtaining an elastic step:

$$\boldsymbol{\kappa}_e^{j+1} = \mathbf{0}, \quad \dot{\boldsymbol{\beta}}_e = \mathbf{H}_e^{-1}\mathbf{Q}_e\dot{\mathbf{d}}_e, \quad \mathbf{E}_t = \mathbf{H}_e^{-1}.$$

4.2.3 Comparison with SD-CPP Formulations

A comparison between the new EC-SQP and the SD-CPP methods can be useful at this stage. With this aim it is useful to recast the jth iteration of SD-CPP method in a format similar to that of the EC-SQP method as follows

Step 1 Obtain the values of $\boldsymbol{\beta}^j$ and $\Delta\boldsymbol{\kappa}_e^j$ by solving the closest point projection scheme in Eq. (17), that is by exactly solving Eq. (15a) with $\mathbf{r}_\mu^j = \mathbf{0}$ and $\mathbf{r}_\sigma^j = \mathbf{0}$ for the fixed, actual value, of the global variables $\mathbf{d}^j$ and λ^j.

Step 2 With the active set of constraints evaluted in the Step 1 solve the system (27) to obtain the increment in the global variables $\dot{\mathbf{d}}$ and $\dot{\lambda}$, as better described in the follows.

The Step 2 is obtained by solving system (27) with the condition $\mathbf{r}_\mu^j = \mathbf{0}$ and $\mathbf{r}_\sigma^j = \mathbf{0}$ being zeroed by the and the active set obtained from previous step. We presents the comparison in the more simple case of fixed loads (elasto-plasticity) but similar conclusions apply to variable loads case. We obtain

$$\begin{bmatrix} \cdot & \mathbf{A}_e^{jT} & \cdot & \cdot \\ -\mathbf{A}_e^j & -\mathbf{H}_{et} & \mathbf{Q}_e & \cdot \\ \cdot & \mathbf{Q}_e^T & \cdot & -\mathbf{p}_{0e} \\ \cdot & \cdot & -\mathbf{p}_{0e}^T & -a_{\lambda\lambda} \end{bmatrix} \begin{bmatrix} \dot{\boldsymbol{\kappa}}_e \\ \dot{\boldsymbol{\beta}}_e \\ \dot{\mathbf{d}}_e \\ \dot{\lambda} \end{bmatrix} = - \begin{bmatrix} \mathbf{0} \\ \mathbf{0} \\ \mathbf{r}_u^j \\ r_\lambda^j \end{bmatrix}, \quad \mathbf{z}^{j+1} = \mathbf{z}^j + \dot{\mathbf{z}}, \tag{30}$$

From the first two equations we obtain with same algebra

$$\dot{\boldsymbol{\kappa}}_e = \mathbf{W}\mathbf{A}_e^{jT}\mathbf{H}_{et}^{-1}\mathbf{Q}_e\dot{\mathbf{d}}_e, \quad \dot{\boldsymbol{\beta}}_e = \mathbf{E}_t\mathbf{Q}_e\dot{\mathbf{d}}_e \tag{31a}$$

where $\mathbf{E}_t$ and $\mathbf{W}$ and $a_{\lambda\lambda}$ have the same expression as for the EC-SQP algorithm previously described. From the global equation we obtain

$$\begin{cases} \mathbf{Q}_e^T \mathbf{E}_t \mathbf{Q}_e \dot{\mathbf{d}}_e - \dot{\lambda} \mathbf{p}_{0e} = -\mathbf{r}_u^j \\ -\mathbf{p}_{0e}^T \dot{\mathbf{d}}_e - a_{\lambda\lambda} \dot{\lambda} = -r_\lambda^j \end{cases} \tag{31b}$$

Eq. (31a, 31b) are the standard equation obtained with strain-driven strategies [18].

With this interpretation of the strain driven formulation it be clear that it evaluates the active set of constraints at the iteration $j+1$ by exactly solving the local equation for the value of the global unknown at the jth iteration. In our proposal we detect the active set by the linearized value of the local equations for a fixed value of the global equation at j then we solve the complete system in $j+1$.

4.3 A Final Remark

In the large series of numerical tests performed, only partially reported in the next section, the method has always shown an impressive robustness. In particular it was possible to verify, in all numerical tests, how the convergence to the correct new solution step $\mathbf{z}^{(n)}$ is achieved even when the algorithm is initialized with a point $\mathbf{z}^0$ very distant from the final solution $\mathbf{z}^{(n)}$, validating the choice of maintaining the cheap estimation of the active set as proposed without any supplementary improving strategy.

Finally, the adoption of a line search scheme can assure the global convergence of the algorithm with a little computational extra-cost, see for example [15, 21] and references therein.

5 Numerical Results

In order to evaluate the performance of the proposed algorithm, we propose some numerical tests regarding 2D problems in plane stress conditions, under the action of various kinds of loads and for von Mises materials.

To test the robustness of the new algorithm, for each test a series of equilibrium or pseudo-equilibrium paths at increasing values of the first arc–length parameters are evaluated. In particular, denoting by $\mathbf{z}_E$ the elastic limit solution, that for the shakedown is evaluated for the reference load $\mathbf{p}_0$, the analyses are performed for an initial extrapolation evaluated as $\Delta\mathbf{z}^{(1)} = \alpha^{(1)}\mathbf{z}_E$, selected so that the component of the displacements in a given point reaches a prescribed value. As the initial arc-length parameter is evaluated as a function of the extrapolated displacements using the second of Eq. (15b) we can force the analyses to perform large steps simply by increasing $\alpha^{(1)}$. The convergence to a new equilibrium point, in the sequel is considered as achieved when the norm of the residuals is less than a given tolerance, i.e. $\|\mathbf{r}_\mathbf{u}\| + \|\mathbf{r}_\sigma\| + \|\mathbf{r}_\mu\| \leq \text{toll}$, while the analysis is stopped when the displacement component of a specified point reaches a prescribed value. For each new step the variables are initialized as $\mathbf{z}^{(n)} = \alpha^{(n)}\mathbf{z}^{(n-1)}$ where α is selected as a function of the difference between the number of iterations required to achieve convergence in the previous step and the number of desired iterations, denoted with lps_d (see [10, 11]), that is the strategy is adaptive. If the residual norm increases by more than 10 times its first value the iterations are stopped and a failure in convergence is reported, that is a *false step*. In this case the analysis restarts from the last evaluated equilibrium point but using a half of the value of $\alpha^{(n)}$ used in the previous iteration. After 15 consecutive false steps the analysis is stopped. A false step can also occur if we don't reach convergence after more than a prescribed number of iterations denoted as lps_m.

The following indicators are compared in order to highlight the efficiency and the robustness of each nonlinear strategy: (I) the number of points in the evaluation of the pseuso-equilibrium path, denoted as "stps", the number of false steps due to both failure or slow convergence are reported in brackets; (II) the total number of the iterations required for each step to converge, denoted as "lps".

Note as in the case of the SD-CPP algorithm a backtracking line search is adopted in the solution of the global equations in order to also tackle with larger step sizes while the analysis with the new EC-SQP algorithm is always performed without any globalization technique. It is important to say that without the line search only in the case of the smallest step size initialization the SD-CPP analysis is capable to perform the analysis without failures for all the tests analyzed. With respect to the CPU time this penalizes the SD-CPP iteration that, as shown in [4], is already more expensive than the EC-SQP one.

The finite element used in the numerical tests was proposed in [3]. It is a four-node element with a bi-linear interpolation of the displacement field and a 5-parameter stress field interpolation which improves the in-plane bending behavior of the element. In [3] several elements with different interpolations of the plastic multiplier field were tested and among them we chose the more simple one, denoted as FC_1, and characterized by a constant interpolation of the plastic multiplier all over the element.

5.1 Description of the Test Problems

The first example regards a classical stress concentration test for a plate with a circular hole subject to biaxial uniform loads on the free edges. The geometry, the material, the applied loads and the mesh used in the analyses, are shown in Fig. 1. The test is analyzed with respect to both the fixed load cases and the variable one. In particular for the fixed load case the analysis is performed setting $\alpha_1 = 1$ and $\alpha_2 = 1$. For the shakedown case the load domain is evaluated by assuming $0.6 \le \alpha_1 \le 1$ and $0.6 \le \alpha_2 \le 1$.

The second test problem is the symmetric continuous beam depicted in Fig. 2 where all the data relative to the material and the applied loads are reported. The analyses are performed with respect to fixed and variable loads. In particular the limit analysis is performed assuming $\alpha_1 = 0.6$ and $\alpha_2 = 1.0$ while for the shakedown case the load domain is defined by by assuming $0.6 \le \alpha_1 \le 1$ and $0.6 \le \alpha_2 \le 1$.

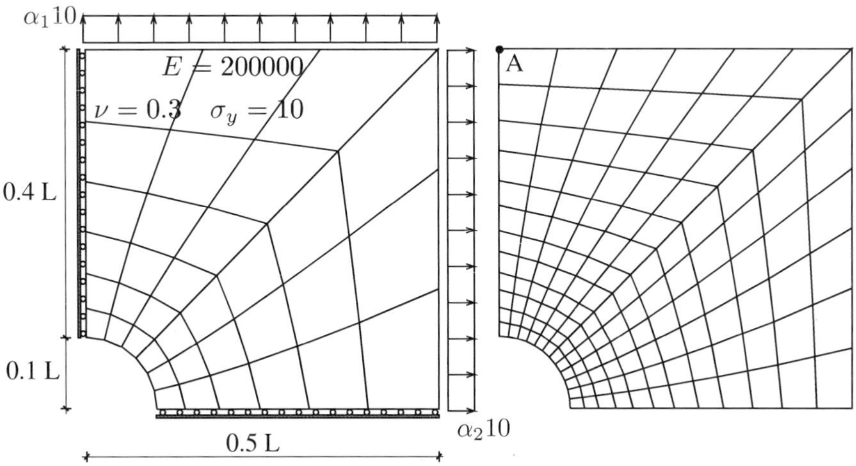

Fig. 1 Plate with a *circular hole*: problem data and meshes used in the analysis

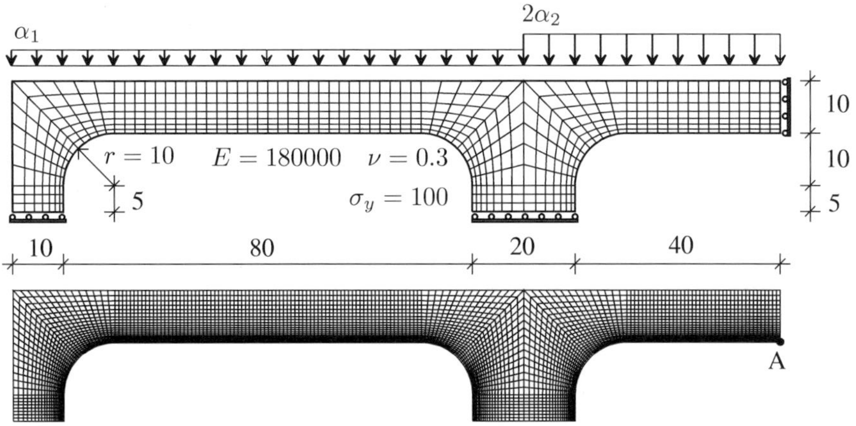

Fig. 2 Symmetric continuous beam: problem data and meshes used in the analysis

The collapse and shakedown multipliers are compared with those obtained by several authors as reported for example in [11, 25].

5.1.1 Limit Analysis

Table 1 reports the results obtained with the SD-CPP and EC-SQP algorithms for all the assigned initial increments of the observed displacement parameter, ranging from $2e-6$ to $1e-2$ and on the basis of two different meshes. The analyses are stopped when the max value $1e-2$ is reached or exceeded. As can be observed

Table 1 Plate with circular hole, mesh 12×12 (338 dofs)

	Mesh$_1$		Mesh$_2$		
	stps (fls)	lps	stps (fls)	lps	incr
SD-CPP	15(−)	51	17(1)	52	$2e-6$
	7(−)	29	14(4)	49	$1e-4$
	4(1)	17	14(7)	49	$1e-3$
	5(5)	18	12(9)	45	$1e-2$
EC-SQP	24(−)	69	26(−)	75	$2e-6$
	12(−)	42	13(−)	46	$1e-4$
	5(−)	23	5(−)	25	$1e-3$
	1(−)	14	1(−)	20	$1e-2$

Limit analysis report, $v_{A,max} = 5e-3$, $toll = 1e-4$, $desired = 12$, $max = 50$. Computed collapse multiplier: $\lambda_c = 0.9096$ FC$_1$(Mesh$_1$); $\lambda_c = 0.9011$ FC$_1$(Mesh$_2$)

Table 2 A symmetric continuous beam, mesh 12×12 (338 dofs)

	Mesh$_1$		Mesh$_2$		
	stps (fls)	lps	stps (fls)	lps	incr
SD-CPP	33(1)	96	33(−)	101	$2e-4$
	22(−)	74	22(−)	77	$5e-3$
	17(2)	61	16(1)	59	$5e-2$
	19(6)	66	18(6)	63	1
EC-SQP	33(−)	99	33(−)	100	$2e-4$
	23(−)	80	21(−)	77	$5e-3$
	14(−)	52	13(−)	53	$5e-2$
	1(−)	11	1(−)	13	1

Limit analysis report, $v_{A,max} = 1e-1$, $toll = 1e-4$, $desired = 12$, $max = 50$. Computed collapse multiplier: $\lambda_c = 5.585$ FC$_1$(Mesh$_1$); $\lambda_c = 5.535$ FC$_1$(Mesh$_2$)

the robustness of the algorithm proposed is good showing a smooth decrease in the number of required steps and iterations according to the assigned step size. For the greatest initial step size it performs the evaluation of the collapse state with a single step and without any loss in accuracy. On the contrary the standard SD-CPP algorithm is adversely affected by the increase in the step size registering occurrences of step failure already from the second size of the first step increment.

Table 2 reports the results obtained for the second test problem. Also in this case two different meshes have been used. The assigned initial increments of the observed displacement parameter range from $2e-4$ to 1. The analysis is stopped when the max value 1 is reached or exceeded. Also in this case for the greatest initial step size the evaluation of the collapse state is obtained in a single step.

5.1.2 Shakedown Analysis

Tables 3 and 4 report the results obtained for the shakedown analysis for the plate with a hole and the continuous symmetric beam, respectively. The analysis are performed with the same assigned initial increments of the observed displacement parameter of the corresponding limit analysis case.

Also for the shakedown case the SD-CPP algorithm it is unable to deals with large step sizes while the EC-SQP algorithm treats large step increment effectively. In particular for both the test problems and for the two finite element grids the analyses obtain the correct multipliers in a single step.

Table 3 Plate with circular hole, mesh 12 × 12 (338 dofs)

	$Mesh_1$		$Mesh_2$		
	stps (fls)	lps	stps (fls)	lps	incr
SD-CPP	17(1)	61	21(2)	74	$2e-6$
	14(4)	58	16(4)	61	$1e-4$
	14(8)	51	14(7)	55	$1e-3$
	15(11)	57	17(12)	64	$1e-2$
EC-SQP	25(−)	80	28(−)	90	$2e-6$
	13(−)	52	14(−)	58	$1e-4$
	6(−)	37	6(−)	38	$1e-3$
	1(−)	18	1(−)	21	$1e-2$

Shakedown analysis report, $v_{A,max} = 5e-3, toll = 1e-4, desired = 12, max = 50$. Computed collapse multiplier: $\lambda_c = 0.8946$ $FC_1(Mesh_1)$; $\lambda_c = 0.8838$ $FC_1(Mesh_2)$

Table 4 A symmetric continuous beam, mesh 12 × 12 (338 dofs)

	$Mesh_1$		$Mesh_2$		
	stps (fls)	lps	stps (fls)	lps	incr
SD-CPP	27(−)	93	26(−)	89	$2e-4$
	21(2)	76	20(2)	75	$5e-3$
	28(5)	72	19(5)	74	$5e-2$
	19(8)	68	18(8)	68	$5e-1$
EC-SQP	28(−)	102	26(−)	100	$2e-4$
	15(−)	60	14(−)	64	$5e-3$
	12(−)	62	12(−)	58	$5e-2$
	1(−)	27	1(−)	20	$5e-1$

Shakedown analysis report, $v_{A,max} = 1e-1, toll = 1e-4, desired = 12, max = 50$. Computed collapse multiplier: $\lambda_a = 3.294$ $FC_1(Mesh_1)$; $\lambda_a = 3.264$ $FC_1(Mesh_2)$

6 Conclusions

In this paper the method presented in [4], for the incremental elastoplastic analysis has been extended to shakedown. The method evaluate the shakedown multipliers by a sequences of pseudo elastoplastic steps obtained by applying a proximal point method to the Melan static theorem. Each step is solved by means of an EC-SQP that retains, at each iteration, all the variables of the problems. In the solution process the set of active constraints is obtained by solving a simple quadratic programming problem which has the same structure and variables of a standard return mapping by closest point projection scheme, i.e. it is decoupled and it can be solved at a local level (finite element, Gauss point). The solution of the equality constraint problem is performed by means of a static condensation of the locally defined variables, stress and plastic multiplier parameters, for which the inter element continuity is not required so obtaining, at the global level, a pseudo-compatible scheme of analysis that has the same structure as classic path following arc-length methods.

The numerical results are performed adopting the finite element interpolation proposed in [3]. This finite element uses a three field interpolation with a good accuracy with respect to both the elastic and elastoplastic response. This makes the proposed numerical framework particularly suitable for shakedown analysis. The numerical results show the improvement in robustness and efficiency with respect to previous proposals.

The presentation and the application are limited to the perfect plasticity case but its extension to other more complex cases is possible [20].

Acknowledgments The research leading to these results has received regional funding from the European Communitys Seventh Framework Programme FP7-FESR: "PIA Pacchetti Integrati di Agevolazione industria, artigianato e servizi" in collaboration with the Newsoft s.a.s. (www.newsoft-eng.it).

References

1. Bilotta A, Casciaro R (2002) Assumed stress formulation of high order quadrilateral elements with an improved in-plane bending behaviour. Comput Methods Appl Mech Eng 191(15–16):1523–1540
2. Bilotta A, Casciaro R (2007) A high-performance element for the analysis of 2d elastoplastic continua. Comput Methods Appl Mech Eng 196(4–6):818–828
3. Bilotta A, Leonetti L, Garcea G (2011) Three field finite elements for the elastoplastic analysis of 2D continua. Finite Elem Anal Des 47(10):1119–1130
4. Bilotta A, Leonetti L, Garcea G (2012) An algorithm for incremental elastoplastic analysis using equality constrained sequential quadratic programming. Comput Struct 102–103:97–107
5. Bisbos CD, Makrodimopoulos A, Pardalos PM (2005) Second-order cone programming approaches to static shakedown analysis in steel plasticity. Optim Methods Softw 20(1):25–52
6. Bouby C, De Saxcé G, Tritsch J-B (2006) A comparison between analytical calculations of the shakedown load by the bipotential approach and step-by-step computations for elastoplastic materials with nonlinear kinematic hardening. Int J Solids Struct 43(9):2670–2692

7. Casciaro R, Garcea G (2002) An iterative method for shakedown analysis. Comput Methods Appl Mech Eng 191(49–50):5761–5792
8. Chen HF, Ponter ARS (2001) Shakedown and limit analyses for 3-d structures using the linear matching method. Int J Press Vessel Pip 78(6):443–451
9. Garcea G, Leonetti L (2011) A unified mathematical programming formulation of strain driven and interior point algorithms for shakedown and limit analysis. Int J Numer Methods Eng 88:1085–1111
10. Garcea G, Trunfio GA, Casciaro R (1998) Mixed formulation and locking in path-following nonlinear analysis. Comput Methods Appl Mech Eng 165(1–4):247–272
11. Garcea G, Armentano G, Petrolo S, Casciaro R (2005) Finite element shakedown analysis of two-dimensional structures. Int J Numer Methods Eng 63(8):1174–1202
12. Krabbenhoft K, Damkilde L (2003) A general non-linear optimization algorithm for lower bound limit analysis. Comput Methods Appl Mech Eng 56(2):165–184
13. Krabbenhoft K, Lyamin AV, Sloan SW, Wriggers P (2007) An interior-point algorithm for elastoplasticity. Int J Numer Methods Eng 69(3):592–626
14. Ngo NS, Tin-Loi F (2007) Shakedown analysis using the p-adaptive finite element method and linear programming. Eng Struct 29(1):46–56
15. Nocedal J, Wright SJ (2000) Numerical optimization. Springer, New York
16. Pastor J, Thai TH, Francescato P (2003) Interior point optimization and limit analysis: an application. Commun Numer Methods Eng 19(10):779–785
17. Ponter ARS, Carter KF (1997) Shakedown state simulation techniques based on linear elastic solutions. Comput Methods Appl Mech Eng 140(3–4):259–279
18. Simó JC, Hughes TJR (1998) Computational inelasticity mechanics and materials. Interdisciplinary applied mathematics. Springer, Berlin
19. Simon J-W, Weichert D (2011) Numerical lower bound shakedown analysis of engineering structures. Comput Methods Appl Mech Eng 200(41–44):2828–2839
20. Simon J-W, Weichert D (2012) Shakedown analysis of engineering structures with limited kinematical hardening. Int J Solids Struct 49(15–16):2177–2186
21. Simon J-W, Weichert D (2011) Numerical lower bound shakedown analysis of engineering structures. Comput Methods Appl Mech Eng 200(41–44):2828–2839
22. Simon J-W, Kreimeier M, Weichert D (2013) A selective strategy for shakedown analysis of engineering structures. Int J Numer Methods Eng 94:985–1014
23. Spiliopoulos KV, Panagiotou KD (2012) A direct method to predict cyclic steady states of elastoplastic structures. Comput Methods Appl Mech Eng 223–224:186–198
24. Tran TN, Liu GR, Nguyen-Xuan H, Nguyen-Thoi T (2010) An edge-based smoothed finite element method for primal-dual shakedown analysis of structures. Int J Numer Methods Eng 82(7):917–938
25. Vu DK, Yan AM, Nguyen-Dang H (2004) A primal-dual algorithm for shakedown analysis of structures. Comput Methods Appl Mech Eng 193(42–44):4663–4674

Limit Analysis on RC-Structures by a Multi-yield-criteria Numerical Approach

Aurora Angela Pisano, Paolo Fuschi and Dario De Domenico

Abstract The present study proposes a multi-yield-criteria limit analysis numerical procedure for the prediction of peak loads and failure modes of reinforced concrete (RC) elements. The proposed procedure, which is a generalization of a previous one recently presented by the authors, is hereafter applied to structural elements reinforced either with traditional steel bars and stirrups or with fiber reinforced polymer (FRP) sheets used as strengthening system. The procedure allows to take into account the actual behaviour, at a state of incipient collapse, of steel, FRP and concrete by a finite element (FE) based plasticity approach where concrete is governed by a Menétrey-Willam-type yield criterion, FRP reinforcement obey to a Tsai-Wu-type yield criterion and steel reinforcement follow the von Mises yield criterion. To check the effectiveness and reliability of the numerically detected peak loads and failure modes a comparison with experimental laboratory findings, available in literature for large-scale specimens, is presented.

Keywords Limit analysis · Multi-yield-criteria · Reinforced concrete structures

1 Introduction

The design of concrete structures reinforced either with classical steel bars or with innovative FRP bars, as well as the strengthening or the rehabilitation of existing steel RC structures through externally bonded FRP sheets or strips, are subjects of great interest in the field of civil engineering. This interest is witnessed by a huge amount of analytical and experimental research studies proposed in the relevant literature in the last decades (see e.g. [2, 9]). Among the several methods already developed

A.A. Pisano (✉) · P. Fuschi · D. De Domenico
Department of PAU - via Melissari, University Mediterranea of Reggio Calabria, 89124 Reggio Calabria, Italy
e-mail: aurora.pisano@unirc.it

P. Fuschi
e-mail: paolo.fuschi@unirc.it

D.De Domenico
e-mail: dario.dedomenico@unirc.it

P. Fuschi et al. (eds.), *Direct Methods for Limit and Shakedown Analysis of Structures*,
DOI 10.1007/978-3-319-12928-0_11

and the different aspects analyzed, from an engineering point of view, it is of utmost importance the knowledge at ultimate (collapse) state of the load carrying capacity of such structures. This task can indeed be pursued, within a plasticity-based approach, using Limit Analysis. Although plain concrete is not a ductile material, experimental studies have shown that the presence of longitudinal web or stirrups reinforcement, as well as of FRP laminates, render the global behaviour of RC structural elements quite ductile so justifying the applicability of approaches based on plasticity theory [3, 5, 10]. It is the realm of the so called *direct methods* which often utilise finite element method (FEM) in conjunction with optimization algorithms such as linear [27] and nonlinear programming [11, 14, 28]. These approaches indeed do not allow the treatment of post elastic phenomena that may arise in concrete structures, such as localization, fracture, damage, creep etc., and that can be faced by coupling plasticity with fracture or damage mechanics theories within step-by-step analyses (see e.g. [12, 15, 30]); however, they can give information on the behaviour at limit (collapse) states of such structures, so resulting very useful for design purposes. In this context it has to be framed the present study.

The promoted approach belongs to a wider research program started by the authors in the context of laminates of FRP [18, 19, 21] and extended to RC structures with reference to a Menétrey-Willam (M-W)-type yield criterion with cap in compression [8, 20, 22]. In the latter studies, two limit analysis methods, namely the Linear Matching Method, LMM, [24], and the Elastic Compensation Method, ECM [16], have been applied under the hypothesis that reinforcement behave as indefinitely elastic. This assumption has inevitably produced some drawbacks. In particular only those structures whose behaviour at incipient collapse is dominated by crushing of concrete, that is the case of the so-called over-reinforced structures, can be appropriately analyzed, while for under-reinforced structures, where reinforcement often attain their limit capabilities, the methods result less accurate. To overcome this limitation and improve the overall analysis of the RC structural elements at collapse enhancements of the methods have been presented in [23]. The present study proposes a further advanced version of the methods that considers possible yielding of steel bars as well as collapse of FRP sheets. To this purpose the paper proposes a three yield criteria limit analysis formulation in which concrete is governed by a Menétrey-Willam-type yield criterion with cap in compression, steel reinforcement follow the von Mises yield criterion and FRP reinforcement obey to a Tsai-Wu-type yield criterion. This allows to predict the peak loads of those RC structures in which steel and/or FRP reinforcement play an important role in the behaviour exhibited at a state of incipient collapse. The effectiveness of the proposed approach is shown by comparison of the obtained numerical predictions with the experimental findings on steel reinforced beams strengthened in flexure using GFRP sheets.

It is worth noting that the promoted procedure is of general (wider) applicability, the only essential requisite being the strict convexity of the yield criteria assumed for the constituent materials.

The paper is organized as follows: after this introductory section, Sect. 2 gives the principles of the LMM and ECM employed in the limit analysis procedure and this with reference to a *generic, strictly convex, yield criterion*; Sect. 3 recalls few analytical expressions of the Menétrey-Willam-type yield criterion enriched with a cap in compression, of the Tsai-Wu-type yield criterion and of the von Mises yield criterion, just to set up the constitutive relations; Sects. 4 and 5 particularize the LMM and ECM to the three yield criteria here considered; Sect. 6 gives the results obtained by analyzing large scale RC beams strengthened with FRP sheet and compares these results with the available experimental tests; finally Sect. 7 draws some conclusions.

2 Limit Analysis: Basic Concepts and Numerical Issues

Limit analysis allows the direct evaluation of the load bearing capacity of a structure or of a structural element. In its classical formulation the theory of limit analysis refers to perfectly plastic structures, made of standard materials, and it is based on a lower and an upper bound theorem [6, 26]. The bound theorems allow the exact determination of the (unique) load value that will cause collapse.

The upper bound theorem states that an upper bound, say P_{UB}, to the collapse load multiplier for a given body of volume V is given by:

$$P_{UB} = \frac{\int_V \sigma_j^Y \, \dot{\varepsilon}_j^c \, \mathrm{d}V}{\int_{\partial V_t} \bar{p}_i \, \dot{u}_i^c \, \mathrm{d}(\partial V)}, \tag{1}$$

where: $\dot{\varepsilon}_j^c = \dot{\lambda}\, \partial f / \partial \sigma_j$ are the components of the strain rate at collapse having the direction of the outward normal to the yield surface $f(\sigma_j) = 0$ (with $\dot{\lambda} > 0$ a scalar multiplier); σ_j^Y are the stresses at yield associated to the given compatible strain rates $\dot{\varepsilon}_j^c$; $\dot{u}_i^c$ are the related displacement rates. Moreover, $\bar{p}_i$ are the surface force components of the reference load vector $\bar{\mathbf{p}}$ acting on the external portion ∂V_t of the body surface. For simplicity, only surface forces are considered. The set $(\dot{\varepsilon}_j^c, \dot{u}_i^c)$ defines a collapse mechanism.

On the other hand, the lower bound theorem states that if at every point within V exists a stress field $\widetilde{\sigma}_j$ which satisfies the condition $f(\widetilde{\sigma}_j) \leq 0$ and in equilibrium with the applied load $P\,\bar{\mathbf{p}}$ for a value of P, say P_{LB}, then P_{LB} is a lower bound to the collapse limit load multiplier.

The two assertions above, as known, lead to two classical approaches of limit analysis, namely: the kinematic and the static one. If the loads produced by their application are equal to each other, circumstance verifiable only for standard materials, then they equal the collapse load. As a matter of fact, the success of limit analysis approaches has been determined also by the possibility to apply the theory, with due attention, outside the realm of perfect plasticity so allowing the study of structures made of non standard materials. The non standard limit analysis theory is based on the two Radenkovic's fundamental theorems which can be summarized as: "every value of the limit load for a non standard body is located between two fixed boundaries defined by the values of the limit loads computed considering the body made by two standard materials whose yield surfaces are one outer, the other inner, to that of the nonstandard material." Obviously, within a nonstandard Radenkovic's approach only a range of collapse load multiplier values can be located between two computed bounds, the uniqueness of the limit load being missing.

The nonstandard approach will be followed hereafter, the main constituent materials of the addressed structural elements being indeed modelled by plasticity-based criteria of non associative type. The assumed yield surface playing the double role of outer and inner "standard" surfaces mentioned before. The promoted approach is based on sequences of FE elastic analyses as well as on the use, in concomitance, of two distinct limit analysis numerical methods, i.e.: the LMM which follows the kinematic approach of limit analysis theory [18, 20, 24] and the ECM which is grounded on the static approach [16, 21, 22]. For sake of brevity, the LMM and the ECM will be discussed with reference to a *generic, strictly convex, yield surface*, namely an ellipsoid in principal stress space. To this concern it is worth noting that the applicability of the whole approach is independent from the assumed stress field representation as will appear more clear in Sects. 4 and 5 where the adopted yield criteria will be explicited.

2.1 Upper Bound Evaluation via LMM

The LMM, firstly theorized by Ponter and Carter [24] for von Mises materials and then extended to more complex materials in [20–22], is an iterative procedure involving *one* sequence of *linear analyses*. The linear analyses are carried on the structure made, by hypothesis, of a *linear viscous fictitious material* with *spatially varying moduli*, D_I (I ranging over the elastic constants entering the considered material), and *imposed initial stresses*, $\bar{\sigma}_j$ (j ranging over the considered (needed) stress components). The adjective *fictitious* highlights the property of the material endowed with elastic parameters which may assume different values at different points; the latter being Gauss points (GPs) in a FE discrete model of the structure.

An easy understanding of the method can be achieved by looking at its geometrical interpretation sketched in Fig. 1. At the current iteration, say at the $(k-1)$th FE-analysis, the fictitious structure (i.e. the structure under study with its real geometry, boundary and loading conditions but made of fictitious material) is analyzed under loads $P^{(k-1)}\bar{p}_i$, with $P^{(k-1)}$ load multiplier and $\bar{p}_i$ assigned reference loads.

The fictitious linear solutions computed, at each Gauss Point of the FE mesh, can be represented by a point $\mathbb{P}_L^{(k-1)}$ lying on the complementary dissipation rate equipotential surface referred to the fictitious viscous material, say $W^{(k-1)}(\sigma_j^{\ell\,(k-1)}, D_I^{(k-1)}, \bar{\sigma}_j^{(k-1)}) = \bar{W}^{(k-1)}$, whose geometrical dimensions and center position depend on the fictitious values $D_I^{(k-1)}$ and $\bar{\sigma}_j^{(k-1)}$ fixed at the current GP. The point $\mathbb{P}_L^{(k-1)}$ with its coordinates, say $\sigma_j^{\ell\,(k-1)}$ in the chosen principal stress space, shown in the sketch of Fig. 1, represents the fictitious solution in terms of stresses while the outward normal at $\mathbb{P}_L^{(k-1)}$, say the normal of components $\dot{\varepsilon}_j^{\ell\,(k-1)}$, represents the fictitious solution in terms of linear viscous strain rates. At this stage the fictitious moduli and

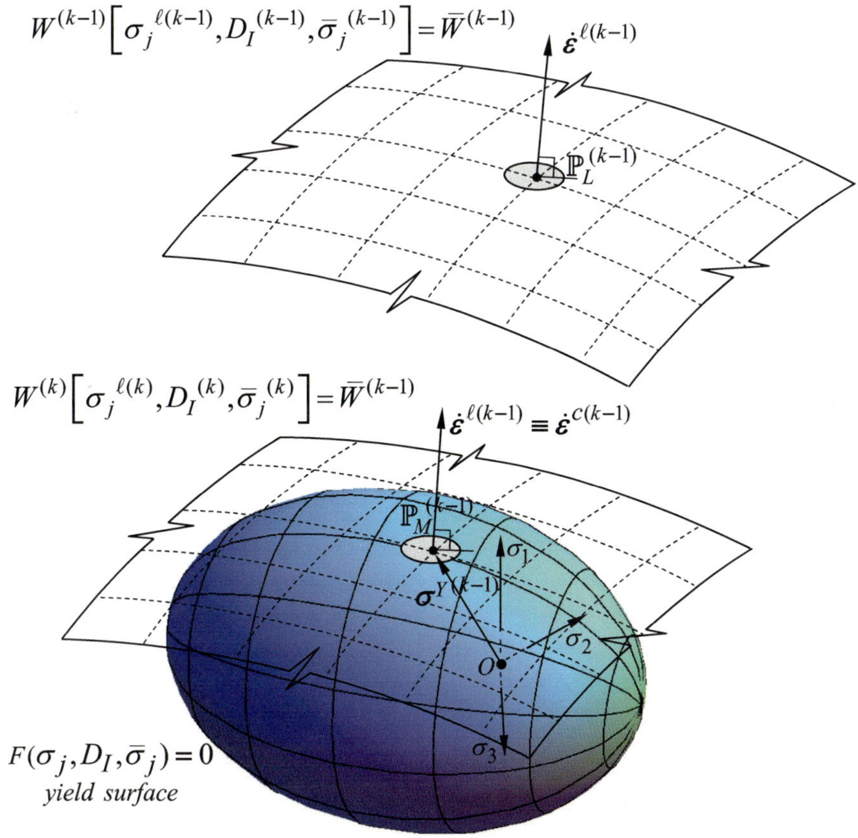

Fig. 1 Geometrical sketch, in the principal stress space, of the matching procedure, from iteration $(k-1)$ to (k), at the current GP within the current element

initial stresses are *modified* so that $\mathbb{P}_L^{(k-1)}$ is brought onto the yield surface of the real constitutive material the analyzed structure is made with. The latter surface is here presented by the ellipsoidal shaded surface of Fig. 1. Namely $\mathbb{P}_L^{(k-1)}$ is brought to identify with point $\mathbb{P}_M^{(k-1)}$, having the same outward normal of $\mathbb{P}_L^{(k-1)}$ but lying on the real material yield surface.

The fictitious solution in terms of strain rates, namely $\dot{\varepsilon}_j^{\ell\,(k-1)} \equiv \dot{\varepsilon}_j^{c\,(k-1)}$, where the apex c stands for "at collapse", as well as the stress coordinates of $\mathbb{P}_M^{(k-1)}$, say the stresses at yield $\sigma_j^{Y\,(k-1)}$, give all the information pertaining to a *state of incipient collapse built at the current GP.* In particular, the fictitious strain rates $\dot{\varepsilon}_j^{\ell\,(k-1)} \equiv \dot{\varepsilon}_j^{c\,(k-1)}$, with the associated displacement rates $\dot{u}_j^{\ell\,(k-1)} \equiv \dot{u}_j^{c\,(k-1)}$, define a *collapse mechanism*. The related stresses $\sigma_j^{Y\,(k-1)}$ are the pertinent *stresses at yield*. Indeed, for the formal analogy existing between the linear viscous problem and the linear elastic problem the strain rates $\dot{\varepsilon}_j^{\ell\,(k-1)}$ can be evaluated as linear *elastic* strain rates, viewing $W^{(k-1)} = const$ as the complementary energy equipotential surface of the fictitious material. While the described modification of $D_I^{(k-1)}$ and $\bar{\sigma}_j^{(k-1)}$ implies that the "modified" $W^{(k)}(\sigma_j^{\ell\,(k)}, D_I^{(k)}, \bar{\sigma}_j^{(k)}) = \bar{W}^{(k-1)}$ *matches* the yield surface at point $\mathbb{P}_M^{(k-1)}$, see again Fig. 1.

Operatively, the matching procedure performed at the current GP starts with the search of the stress point $\mathbb{P}_M^{(k-1)}$ on the yield surface having an assigned strain rate $\dot{\boldsymbol{\varepsilon}}^{\ell\,(k-1)} \equiv \dot{\boldsymbol{\varepsilon}}^{c\,(k-1)}$ as outward normal.

It is worth noting that if the yield surface is strictly convex such point $\mathbb{P}_M^{(k-1)}$ is uniquely determined by the given normal $\dot{\boldsymbol{\varepsilon}}^{\ell\,(k-1)}$. If this normal belongs to a cone of normals pertaining to a vertex (nonsmooth corner) if any, of the yield surface, point $\mathbb{P}_M^{(k-1)}$ simply identifies with this vertex.

If the expounded rationale is repeated at all GPs of the mesh, a collapse mechanism, $(\dot{\varepsilon}_j^{c\,(k-1)}, \dot{u}_i^{c\,(k-1)})$ with the related stresses at yield, $\sigma_j^{Y\,(k-1)}$ can be defined for the whole structure and, by Eq. (1), an upper bound value to the collapse load multiplier, say $P_{UB}^{(k)}$, can be evaluated at current $(k-1)$th FE elastic analysis. However, the above stress at yield, computed through the matching, do not meet the equilibrium conditions with the acting loads $P^{(k-1)}\bar{p}_i$ and the procedure is carried on iteratively until the difference between two subsequent P_{UB} values is less than a fixed tolerance. Convergence requires that the $W^{(k)} = \bar{W}^{(k-1)}$ matches the yield surface at $\mathbb{P}_M^{(k-1)}$ and otherwise lies outside the yield surface (see [25]). In the following the LMM is applied simultaneously to concrete, FRP and steel reinforcement.

2.2 Lower Bound Evaluation via ECM

The ECM conceived by Mackenzie and Boyle [16] with reference to steel and then modified to deal with more complex materials in Pisano et al. [18, 22] is aimed to construct an admissible stress field, suitable for the evaluation of a P_{LB}, in the spirit of the static approach of limit analysis. Also the ECM is an iterative procedure involving *many* sequences of linear elastic FE-based analyses, in which highly loaded regions of the structure are systematically weakened by reduction of the elastic moduli and this in order to simulate a *stress redistribution* arising within the structure before attaining its limit strength threshold. Also in this case the procedure can be more easily explained by means of a geometrical sketch as the one given in Fig. 2 with reference to a generic yield surface $F(\sigma_j, D_I, \bar{\sigma}_j) = 0$. The ECM starts with a first sequence, say $s = 1$, of FE elastic analyses, carried on the structure endowed with the proper (real) material elastic parameters and suffering applied initial loads $P_D^{(s)}\,\bar{p}_i = P_D^{(1)}\,\bar{p}_i$. At the current iteration, say at the $(k-1)$th FE analysis, the elastic stress solution is computed at the GPs of the mesh. Such values, averaged within the current element #*e*, allow to define a solution "at element level", which, as shown in the sketch of

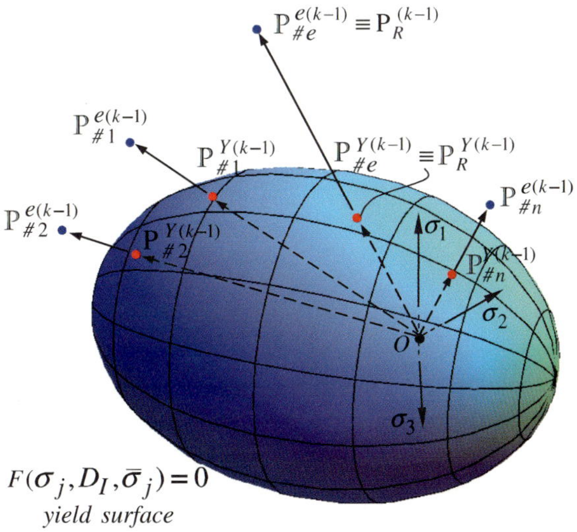

Fig. 2 Geometrical sketch, in the principal stress space, of the ECM at current iteration $(k-1)$ of the current sequence s. Stress points representing the elastic solution at elements #1, #2, ..., #e, ..., #n; with $\mathbb{P}_R^{(k-1)}$ denoting the "maximum stress" among all the elements

Fig. 2, locates in the principal stress space a stress point, say $\mathbb{P}_{\#e}^{e(k-1)}$. $\mathbb{P}_{\#e}^{Y(k-1)}$ denotes the corresponding stress point at yield (i.e. lying on the yield surface) measured on the direction $\overrightarrow{O\mathbb{P}}_{\#e}^{e}/|\overrightarrow{O\mathbb{P}}_{\#e}^{e}|$. In the figure are reported other stress points, representing the average stress elastic solution within the elements #1, #2, ..., #e, ..., #n. If the elastic solution at the #eth element is such that $|\overrightarrow{O\mathbb{P}}_{\#e}^{e}|^{(k-1)} > |\overrightarrow{O\mathbb{P}}_{\#e}^{Y}|^{(k-1)}$ then the element's Young modulus is reduced according to the formula:

$$E_{\#e}^{(k)} = E_{\#e}^{(k-1)} \left[\frac{|\overrightarrow{O\mathbb{P}}_{\#e}^{Y}|^{(k-1)}}{|\overrightarrow{O\mathbb{P}}_{\#e}^{e}|^{(k-1)}} \right]^2 \tag{2}$$

where the square of the updating ratio, within the square bracket, is used to increase the convergence rate.

After the above modulus variation, the *maximum stress value* has to be detected in the *whole FE mesh*, namely the value corresponding to the stress point farthest away from the yield surface, say $\mathbb{P}_R^{(k-1)}$ in the sketch of Fig. 2. If $|\overrightarrow{O\mathbb{P}}_R|^{(k-1)}$ is greater than $|\overrightarrow{O\mathbb{P}}_R^{Y}|^{(k-1)}$ (as drawn Fig. 2) a *new* FE analysis is performed within the current sequence trying to *re-distribute* the stresses within the structure; and this by keeping fixed the applied loads but with the updated $E_{\#e}^{(k)}$ values given by Eq. (2). The iterations are carried on, inside the given sequence, until all the stress points just reach or are below their corresponding yield values, which means that an admissible stress field has been built. Increased values of loads are then considered in the subsequent sequences of analyses, each one with an increased value of $P_D^{(s)}$, till further load increase does not allow the stress point $\mathbb{P}_R^{(k-1)}$ to be brought below yield by the re-distribution procedure. A P_{LB} load multiplier can then be evaluated at *last admissible stress field* attained for a maximum acting load $P_D^{(s)}\bar{p}_i$, say at $s = S$, and at last FE analysis, say at $k = K$, as:

$$P_{LB} = |\overrightarrow{O\mathbb{P}}{}^{Y}_{R}|^{(K)} \frac{P_D^{(S)}}{|\overrightarrow{O\mathbb{P}}_R|^{(K)}}. \tag{3}$$

3 Yield-Criteria for RC Structures

As outlined in the introductory section, this study is focused on concrete elements reinforced with steel bars and externally strengthened with FRP sheets. The main goal is indeed to apply the limit analysis approach, described in Sect. 2, to the yield criteria assumed for the constituent materials of the analyzed RC elements, namely: concrete, FRP laminates and steel. The main constitutive assumptions are given next.

Concrete is assumed as an isotropic nonstandard material obeying a plasticity model derived from the Menétrey-Willam [17] (M-W) failure criterion; this criterion is defined by the following expression:

$$F(\xi,\rho,\theta) = \left[\sqrt{1.5}\frac{\rho}{f_c'}\right]^2 + m\left[\frac{\rho}{\sqrt{6}f_c'} r(\theta,e) + \frac{\xi}{\sqrt{3}f_c'}\right] - 1 = 0; \tag{4}$$

where:

$$r(\theta,e) = \frac{4\left(1-e^2\right)\cos^2\theta + (2e-1)^2}{2\left(1-e^2\right)\cos\theta + (2e-1)\left[4\left(1-e^2\right)\cos^2\theta + 5e^2 - 4e\right]^{1/2}};$$

$$m := 3\frac{f_c'^2 - f_t'^2}{f_c' f_t'}\frac{e}{e+1}. \tag{5}$$

Equation (4) is given in terms of three stress invariants ξ, ρ, θ known as the Haigh Westergaard (H-W) coordinates; m is the friction parameter of the material depending, as shown in Eq. (5), on the compressive strength f_c', the tensile strength f_t' as well as on the eccentricity parameter e, whose value governs the convexity and smoothness of the elliptic function $r(\theta, e)$. The eccentricity e describes the out-of-roundness of M-W deviatoric trace and it strongly influences the biaxial compressive strength of concrete. To limit the concrete strength in high hydrostatic compression regime, a cap, closing in compression the surface defined by Eq. (4), is adopted. This cap, formulated in the H-W coordinates, can be given the shape:

$$\rho^{CAP}(\xi,\theta) = -\frac{\rho^{MW}(\xi_a,\theta)}{(\xi_a-\xi_b)^2}\left[\xi^2 - 2\xi_a(\xi-\xi_b) - \xi_b^2\right], \text{ with } \begin{cases} \xi_b \le \xi \le \xi_a \\ 0 \le \theta \le \frac{\pi}{3} \end{cases}; \tag{6}$$

where $\rho^{MW}(\xi, \theta)$ is the explicit form of the parabolic meridian of the M-W surface that, looking at Eq. (7), can be given by:

$$\rho^{MW}(\xi, \theta) = \frac{1}{2a}\left\{-b(\theta) + \left[b^2(\theta) - 4a\,c(\xi)\right]^{1/2}\right\}, \quad \text{with } \left\{\begin{matrix} \xi_a \leq \xi \leq \xi_v \\ 0 \leq \theta \leq \frac{\pi}{3} \end{matrix}\right. ; \tag{7}$$

and where:

$$a = \frac{1.5}{\left(f_c'\right)^2}; \quad b(\theta) = \frac{m}{\sqrt{6}f_c'} r(\theta, e); \quad c(\xi) = \frac{m}{\sqrt{3}f_c'}\xi - 1. \tag{8}$$

The values ξ_a, ξ_b and ξ_v entering Eqs. (6) and (7) locate the cap position and can be detected experimentally.

It is worth noting that the Menétrey-Willam surface equipped with a cap in compression is strictly convex and smooth, except for the vertices on the hydrostatic axis, and it is hereafter assumed as yield criterion for concrete. A realistic representation of concrete requires to take into account the dilatancy so a non-associated flow rule is also postulated.

The FRP laminate, the strengthening sheets are made with, are assumed as a composite, orthotropic, nonstandard material obeying a plasticity model derived from the Tsai-Wu failure criterion [29]. By denoting with 1 and 2 the principal directions of orthotropy in plane stress case as well as indicating, as usual in this context, $\sigma_6 \equiv \tau_{12}$, the adopted Tsai-Wu-type yield surface is given by:

$$\begin{aligned} F(\sigma_1, \sigma_2, \sigma_6) = {} & F_{11}\,\sigma_1^2 + F_{22}\,\sigma_2^2 + F_{66}\,\sigma_6^2 + 2\,F_{12}\,\sigma_1\sigma_2 \\ & + F_1\,\sigma_1 + F_2\,\sigma_2 - 1 = 0, \end{aligned} \tag{9}$$

where:

$$\begin{aligned} & F_1 := \frac{1}{X_t} + \frac{1}{X_c}; \quad F_2 := \frac{1}{Y_t} + \frac{1}{Y_c}; \quad F_{11} := -\frac{1}{X_t\,X_c}; \\ & F_{22} := -\frac{1}{Y_t\,Y_c}; \quad F_{66} := \frac{1}{S^2}; \quad F_{12} := -\frac{1}{2}\sqrt{F_{11}F_{22}}; \end{aligned} \tag{10}$$

with: X_t, X_c the longitudinal tensile and compressive strengths respectively; Y_t, Y_c the transverse tensile and compressive strengths respectively and S the longitudinal shear strength. In Eq. (10) the compressive strengths X_c and Y_c have to be considered intrinsically negative. Equation (9) represents in the space σ_1, σ_2, σ_6 an ellipsoid whose major axis lies on the $\sigma_6 = 0$ plane and it is rotated by an anticlockwise angle of 45° with respect to the σ_1 axis. This surface is hereafter assumed as yield criterion for the FRP composite reinforcement. Also in this case a non-associated flow rule is postulated.

Finally, steel follows the von Mises yield criterion which is one of the most popular criteria applied to describe the behaviour of perfectly plastic ductile materials. For a generic, multi-axial, loading condition and considering that $\rho^2 = 2J_2$ (with J_2

second stress deviatoric invariant) the von Mises yield surface, being σ_y the yield strength, is expressed in the form:

$$F(\rho) = \frac{3}{2}\rho^2 - \sigma_y^2 = 0, \tag{11}$$

that, in the case of uniaxial stress condition, like the one recorded in the reinforcement bars, simply reduces to $\sigma_1 = \sigma_y$, being σ_1 the first principal stress, measured along the bar longitudinal direction. As known, in the principal stress space the von Mises yield surface of Eq. (11) is represented by a cylinder, indefinite along the hydrostatic axis, circumstance that does not satisfy the required condition of strict convexity. However the limit analysis procedure will be pursued on the deviatoric plane where the von Mises criterion is a circle of radius $\rho_y = \sqrt{\frac{2}{3}}\sigma_y$.

4 Three-Yield-Criteria LMM

The LMM procedure presented in Sect. 2.1, with reference to a generic strictly convex yield surface, can be now easily specified for concrete, FRP and steel. To this aim, taking into account the assumed yield criteria, it will be sufficient to specify the apposite expressions of the complementary energy equipotential surface $W^{(k-1)}(\sigma_j^{\ell\,(k-1)}, D_I^{(k-1)}, \bar{\sigma}_j^{(k-1)}) = \bar{W}^{(k-1)}$ that has to be consistent with the yield criterion in use. Looking at concrete, governed by the M-W-type yield surface by Eqs. (4)–(8), the pertinent complementary energy equipotential surface, in the Haigh-Westergaard coordinates, depends on the bulk modulus K, on the shear modulus G as well as on initial stresses, $\bar{\xi}$ and $\bar{\rho}$. With reference to the rationale of Sect. 2.1, then, referring to Fig. 1, the elastic moduli D_I identify with G and K, while $\bar{\sigma}_j$ correspond to $\bar{\xi}$ and $\bar{\rho}$. At the current iteration $(k-1)$ of the LMM procedure and at the current GP within the mesh, the right complementary energy functional can be given the shape:

$$W^{(k-1)}(\xi, \rho) = \frac{\left(\xi - \bar{\xi}^{(k-1)}\right)^2}{6K^{(k-1)}} + \frac{\left(\rho - \bar{\rho}^{(k-1)}\right)^2}{4G^{(k-1)}}. \tag{12}$$

Equation (12), written as $W^{(k-1)} = \bar{W}^{(k-1)}$ represents, in the principal stress space, a *prolate spheroid* having semi axes proportional to the elastic parameters $(G^{(k-1)}, K^{(k-1)})$ and coordinates of the center depending on the initial stresses $(\bar{\xi}^{(k-1)}, \bar{\rho}^{(k-1)})$.

The LMM acts updating the elastic moduli and the initial stresses of Eq. (12) in such a way that the spheroid is modified in shape and position till matching is realized on the M-W-type surface at the stress point of given outward normal. In this context the normal components are the *volumetric* and *deviatoric* strain rates of the fictitious linear solution, namely: $\dot{\varepsilon}_v^{\ell\,(k-1)} \equiv \dot{\varepsilon}_v^{c\,(k-1)}$ and $\dot{\varepsilon}_d^{\ell\,(k-1)} \equiv \dot{\varepsilon}_d^{c\,(k-1)}$. The above matching point $\mathbb{P}_M^{(k-1)}(\xi^{Y\,(k-1)}, \rho^{Y\,(k-1)})$ gives, with its H-W coordinates, the associated stresses at yield. A nonlinear system of 5 equations provides the searched matching point. The quite complex geometrical problem associated to the LMM applied to M-W-type yield surface has been deeply discussed in a recent paper by the authors [20].

For the FRP strengthening sheets, at a current iteration $(k-1)$ and at a current GP, the complementary energy functional consistent with the T-W-type yield criterion, given by Eqs. (9) and (10), can instead be written as:

$$
\begin{aligned}
& W\left(\sigma_j, E_j^{(k-1)}, \nu_{12}^{(k-1)}, \bar{\sigma}_j^{(k-1)}\right) \\
& = \frac{1}{2}\left[\frac{\sigma_1^2}{E_1^{(k-1)}} + \frac{\sigma_2^2}{E_2^{(k-1)}} + \frac{\sigma_6^2}{E_6^{(k-1)}} - 2\,\nu_{12}^{(k-1)}\,\frac{\sigma_1\,\sigma_2}{E_2^{(k-1)}}\right. \\
& \qquad - 2\left(\frac{\bar{\sigma}_1^{(k-1)}}{E_1^{(k-1)}} - \nu_{12}^{(k-1)}\,\frac{\bar{\sigma}_2^{(k-1)}}{E_2^{(k-1)}}\right)\sigma_1 - 2\left(\frac{\bar{\sigma}_2^{(k-1)}}{E_2^{(k-1)}} - \nu_{12}^{(k-1)}\,\frac{\bar{\sigma}_1^{(k-1)}}{E_2^{(k-1)}}\right)\sigma_2 \\
& \qquad \left. -2\,\frac{\bar{\sigma}_6^{(k-1)}}{E_6^{(k-1)}}\sigma_6 + \frac{\bar{\sigma}_1^{(k-1)2}}{E_1^{(k-1)}} + \frac{\bar{\sigma}_2^{(k-1)2}}{E_2^{(k-1)}} + \frac{\bar{\sigma}_6^{(k-1)2}}{E_6^{(k-1)}} - 2\,\nu_{12}^{(k-1)}\,\frac{\bar{\sigma}_1^{(k-1)}\,\bar{\sigma}_2^{(k-1)}}{E_2^{(k-1)}}\right],
\end{aligned}
\tag{13}
$$

where E_6 is the shear modulus G_{12}. The surface obtainable by Eq. (13) represents in the stress space $(\sigma_1,\ \sigma_2,\ \sigma_6)$ an ellipsoid whose semi axes amplitude depend on the elastic moduli, while its center position depends on the initial stresses. Referring again to the sketch of Fig. 1, that can be easily redrawn with reference to the stress components $\sigma_1,\ \sigma_2,\ \sigma_6$, the elastic parameter D_I identify with $E_1,\ E_2,\ E_6,\ \nu_{12}$, while $\bar{\sigma}_j$ corresponds to $\bar{\sigma}_1,\ \bar{\sigma}_2,\ \bar{\sigma}_6$. The ellipsoid $W^{(k-1)} = \bar{W}^{(k-1)}$ is in all similar to the Tsai-Wu-type yield surface and this simplifies the matching procedure. By appropriately choosing the initial values of the elastic parameters and initial stresses of the fictitious material, the complementary energy surface may indeed result *homothetic* to the T-W-type surface and the matching procedure can be realized just by a rescaling of the complementary energy surface. To this aim it is sufficient to rescale by *only one* scalar coefficient the elastic parameters E_j, which govern the amplitude of the axes of $W^{(k-1)} = \bar{W}^{(k-1)}$, through the updating formula:

$$
E_j^{(k)} = E_j^{(k-1)}\,(\Lambda^{(k-1)})^2, \quad (j = 1,\ 2,\ 6)
\tag{14}
$$

being the scalar coefficient $\Lambda^{(k-1)}$ the homothetic ratio between the two ellipsoids, [18, 19]. In this case homothety implies that $W = \bar{W}$, at matching, simply *coincides* with the yield surface. The points $\mathbb{P}_L^{(k-1)}$ and $\mathbb{P}_M^{(k-1)}$ not only have the same normal of components $\dot{\varepsilon}_j^{c\,(k-1)}$, with $j = 1, 2, 6$, (as required by matching) but they belong, since the beginning, to the same straight line passing through the stress space origin. Eventually, at matching, the coordinates of $\mathbb{P}_M^{(k-1)}$, namely $\sigma_j^{Y\,(k-1)}$ with $j = 1, 2, 6$ give the pertinent stresses at yield.

The application of the LMM results even more simple with reference to the von Mises yield criterion. Nevertheless, the sketch of Fig. 1 becomes meaningless, the von Mises yield surface being in the principal stress space a cylinder indefinite along its axis coincident with the hydrostatic axis. The procedure can indeed be carried on with reference to the deviatoric plane where the essential requisite of strict convexity is recovered, the von Mises surface being a circle. Looking at the geometrical sketch of Fig. 3a, under the hypothesis of incompressible fictitious material the pertinent complementary energy potential functional consistent with a von Mises material, at a current iteration $(k-1)$, can be written as:

$$W(\rho, E^{(k-1)}) = \frac{3\rho^2}{4E^{(k-1)}} \tag{15}$$

which, if written in the shape $W^{(k-1)} = \bar{W}^{(k-1)}$, represents a circle concentric to the von Mises one. With reference to Fig. 3a, the linear fictitious solution at a current iteration, $(k-1)$, and at the current GP can be represented by the deviatoric stress

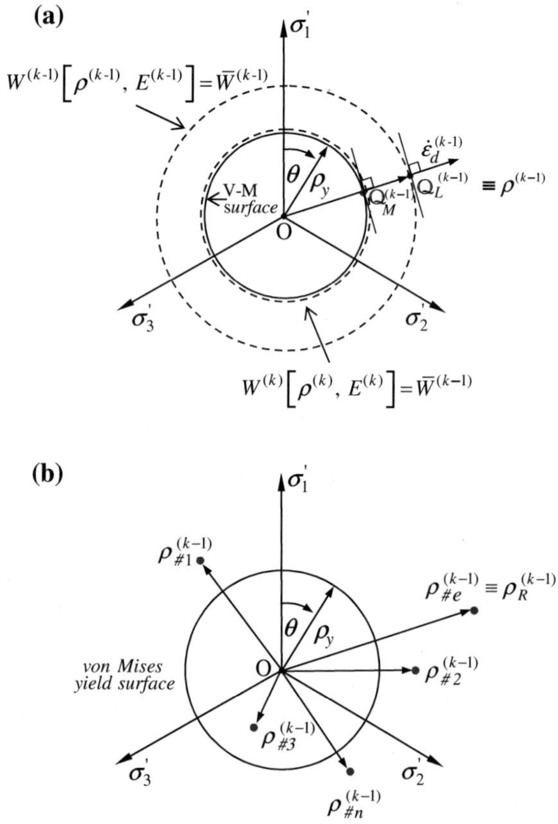

Fig. 3 Geometrical sketch of the limit analysis procedure in the deviatoric plane for steel bars: **a** matching procedure from iteration $(k-1)$ to (k) performed at current GP of current bar elements; **b** stress points measured within the steel bar elements $\#1, \#2, \#3, \ldots, \#e, \ldots, \#n$ to apply ECM, $\rho_R^{(k-1)}$ denoting the "maximum stress" among the bar elements

invariant, point $\mathbb{Q}_L^{(k-1)}$ ($|O\,\mathbb{Q}_L^{(k-1)}| \equiv \rho^{(k-1)}$), having outward normal the deviatoric strain rate $\dot{\varepsilon_d}^{(k-1)}$.

To find the corresponding matching point $\mathbb{Q}_M^{(k-1)}$ it is sufficient to move along the radius $\overrightarrow{O\,\mathbb{Q}_L}^{(k-1)}$ that is to re-scale the complementary energy surface by modifying the Young modulus $E^{(k-1)}$ with the updating formula:

$$E^{(k)} = E^{(k-1)} \frac{\rho_y}{\rho^{(k-1)}} \tag{16}$$

that brings $W^{(k-1)} = \bar{W}^{(k-1)}$ to coincide with the von Mises deviatoric circle of radius ρ_y. In the case of uniaxial stress condition, as the one detected in the steel bars, Eq. (16) is further simplified, i.e.:

$$E^{(k)} = E^{(k-1)} \frac{\sigma_y}{\sigma_1^{(k-1)}}. \tag{17}$$

The key idea of the present approach is now achievable: the LMM is applied simultaneously to *all* the FEs of the discrete model of the RC structure, i.e. to each GP of each element. The proper yield condition and matching procedure will be used at the GPs of the FEs describing concrete, FRP sheets and steel bars.

When, at current iteration and *for each* GP of the FE mesh the stresses at yield, the corresponding strain rates at collapse, together with the associated displacement rates at collapse, are known, it is possible to compute an upper bound multiplier, say $P_{UB}^{(k)}$, by applying Eq. (1) that particularizes as follows:

$$P_{UB}^{(k)} = \frac{\int_V (\xi^{Y\,(k-1)}\,\dot{\varepsilon}_v^{c\,(k-1)} + \rho^{Y\,(k-1)}\,\dot{\varepsilon}_d^{c\,(k-1)} + \sigma_j^{Y\,(k-1)}\dot{\varepsilon}_j^{c\,(k-1)} + \rho_y\,\dot{\varepsilon}_d^{c\,(k-1)})\mathrm{d}V}{\int_{\partial V_t} \bar{p}_i\,\dot{u}_i^{c\,(k-1)}\,\mathrm{d}\,(\partial V)}. \tag{18}$$

As said, the iterative procedure is carried on until the difference between two subsequent P_{UB} values is less than a fixed tolerance.

5 Three-Yield-Criteria ECM

Following the same rationale of the previous section, the ECM, presented in its general formulation in Sect. 2.2, is also applied to the whole structure. The admissible stress field for a given maximum applicable load has to be detected with respect to each of constituent materials yield surface.

With reference to concrete and FRP sheets, the point $\mathbb{P}_{\#e}^{e(k-1)}$ of Fig. 2 has to be simply interpreted with the pertinent coordinates, namely: ($\xi_{\#e}^{e(k-1)}$, $\rho_{\#e}^{e(k-1)}$, $\theta_{\#e}^{e(k-1)}$) in the Haigh-Westergaard coordinates when dealing with M-W-type yield surface or ($\sigma^{e(k-1)}_{\# e_1}$, $\sigma^{e(k-1)}_{\# e_2}$, $\sigma^{e(k-1)}_{\# e_6}$) when working on FRP sheets within the (σ_1, σ_2, σ_6) stress space. $\mathbb{P}^{Y(k-1)}_{\# e}$ denotes the corresponding stress point on the yield surface, measured on the direction $\overrightarrow{O\mathbb{P}}^e_{\# e}/|\overrightarrow{O\mathbb{P}}^e_{\# e}|$ and the reduction formula given by Eq. (2) holds true and it has to be applied if the elastic solution at the #eth element is such that $|\overrightarrow{O\mathbb{P}}^e_{\# e}|^{(k-1)} > |\overrightarrow{O\mathbb{P}}^Y_{\# e}|^{(k-1)}$.

In particular, assuming that the concrete Poisson ratio v remains constant, the updating of the Young modulus by Eq. (2) is equivalent to modify the bulk modulus $K^{(k-1)}_{\# e}$ and shear modulus $G^{(k-1)}_{\# e}$ by the same reducing factor. When dealing with FRP reinforcement the reduction is applied instead to the three element's Young moduli of #e, namely to $E^{(k)}_{\# e_j}, j = 1, 2, 6$.

With reference to steel, at the current iteration $(k-1)$, the deviatoric stress invariant $\rho^{(k-1)}_{\# e}$ evaluated within the steel bar elements #1, #2, ..., #e, ..., #n, can be represented as in Fig. 3b. As said, in the elements where $\rho^{(k-1)}_{\# e}$ is greater than ρ_y the elastic modulus must be modified (reduced). This goal can be achieved by varying the elastic modulus as in Eq. (16). It is worth to remark that the "modulus variation" realized by Eq. (16) possesses, in this case, a completely different meaning. When it is applied within the LMM, the above variation (reduction or increse), is driven by a fixed strain rate and, as said, it is oriented to build a collapse mechanism. On the other hand when such modulus variation is applied within the ECM it is always a reduction necessary to bring a not admissible stress onto the yield surface so realizing a stress redistribution oriented to build an admissible stress field. It has also to be noted that the LMM acts on all the GPs of the FE mesh. All the GPs are viewed as possible sites where a mechanism might arise. The confinement of the plasticized zone is obtained only at convergence. On the contrary, the ECM acts only on the elements characterized by stress quantities greater than the yielding ones in the attempt to mimic a stress redistribution within the structure. Also in this case, it has been numerically experienced that the convergence rate increases if the square of the updating ratio, $\rho_y/\rho^{(k-1)}_{\# e}$, is used so the updating formula becomes:

$$E^{(k)}_{\# e} = E^{(k-1)}_{\# e}\left[\frac{\rho_y}{\rho^{(k-1)}_{\# e}}\right]^2. \tag{19}$$

Within the current sequence, after the described moduli redistribution on concrete, FRP laminates and steel bars, the three *maximum stress values* have to be detected in the *whole FE mesh*. Precisely, the three stress points farthest away from the yield surfaces considered, say $\mathbb{P}^{(k-1)}_R$ for concrete or FRP and $\rho^{(k-1)}_R$ for steel bars.

Referring again to Figs. 2 and 3b if $|\overrightarrow{O\mathbb{P}}_R|^{(k-1)}$ is greater than $|\overrightarrow{O\mathbb{P}}^Y_R|^{(k-1)}$ or $\rho^{(k-1)}_R$ is greater than ρ_y a new FE analysis is performed within the current sequence keeping fixed the applied load but with the updated $E^{(k)}_{\# e}$ given by Eqs. (2) and (19). The iterations are carried on, inside the given sequence, until all the stress points just reach or are below their corresponding yield values, which means that an admissible stress field has been built. Increased values of loads are then considered in the subsequent sequences, each one with an increased value of $P_D^{(s)}$, till further load increase does not allow the maximum stresses to be brought below yield by redistribution. A P_{LB} load multiplier can then be evaluated at *last admissible stress field* attained for a *maximum acting load* $P_D^{(s)}\bar{p}_i$, say at $s = S$, and at last FE analysis, say at $k = K$, as the minimum between the three values:

$$P_{LB} = \min\left\{\left[|\overrightarrow{O\mathbb{P}}_R^Y|^{(K)}\frac{P_D^{(S)}}{|\overrightarrow{O\mathbb{P}}_R|^{(K)}}\right]_{CONCR};\left[|\overrightarrow{O\mathbb{P}}_R^Y|^{(K)}\frac{P_D^{(S)}}{|\overrightarrow{O\mathbb{P}}_R|^{(K)}}\right]_{FRP};\rho_y\frac{P_D^{(S)}}{\rho_R^{(K)}}\right\}. \tag{20}$$

6 Applications

The main goal of the following applications is to verify the reliability of the expounded three-yield-criteria limit analysis numerical procedure in predicting the limit state (peak load and failure mechanism) of structural RC elements strengthened by FRP sheets. Experimental findings on large scale specimens, taken from the relevant literature, have been numerically reproduced and the obtained results have been compared with those given by the laboratory tests.

In all the numerical analyses a perfect bonding between steel bars and concrete as well as between FRP sheets and concrete has been assumed. Reference is made to the experimental works of Almusallam and Al-Salloum [1] who tested, up to failure, steel reinforced concrete beams strengthened with Glass FRP (GFRP) sheets. Indeed, the above quoted paper faces a number of experimental tests, carried out to investigate the effects of the GFRP strengthening on the flexural capacity and central deflection of the beams, so that the experimental campaign considers beams with and without GFRP reinforcements. Only some of the available experimental results are then taken into consideration, namely those in which the sheets of GFRP are present. In particular, using the same label of Almusallam and Al-Salloum [1], the analyzed beams are: GB2, GB4 and GB6.

The beams are simply supported and subjected to four-point bending test through two line loads $P\bar{p}/2$ placed at the same distance with respect to the mid-span. For all the considered specimens, P denotes the load multiplier, while $\bar{p}$ denotes the line reference load whose resultant is assumed equal to 100 kN. Figure 4a shows the mechanical scheme used for beams together with all the geometrical details. Only half specimen is modeled due to symmetry with respect to the longitudinal direction. The beams are reinforced with different arrangement of internal steel bars and stirrups, and strengthened with different arrangement of GFRP sheets as reported in Fig. 4b. Furthermore, the relevant material properties for concrete and GFRP reinforcement are given in Tables 1 and 2 respectively, while, for what concerns steel a Young modulus of 200 GPa and a yield strength σ_y of 537 MPa have been assumed for all specimens.

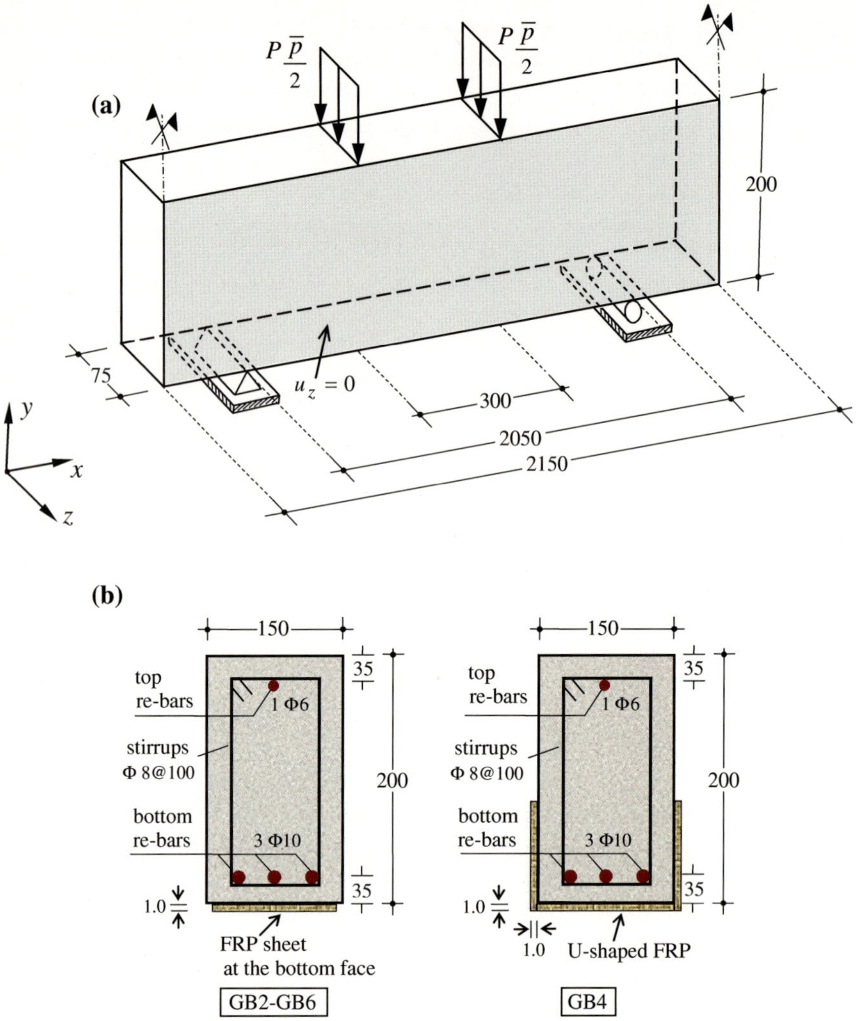

Fig. 4 Mechanical model of the analyzed simply-supported beams GB2, GB4, GB6: **a** geometry, loading and boundary conditions; **b** cross section geometry with reinforcement arrangement (after Almusallam and Al-Salloum [1]); all the dimensions are in mm

Table 1 Mechanical parameters of concrete for the analyzed beams

Specimen label	Concrete properties		
	f_c' (MPa)	f_t' (MPa)	E_c (GPa)
GB2	36.00	1.98	32.31
GB4	36.60	2.00	32.47
GB6	33.80	1.92	31.70

Table 2 Mechanical parameters of FRP laminae of the strengthening sheets

FRP lamina moduli				FRP lamina strengths				
E_1 (GPa)	E_2 (GPa)	G_{12} (GPa)	ν_{12}	X_t (MPa)	X_c (MPa)	Y_t (MPa)	Y_c (MPa)	S (MPa)
30.0	30.0	3.80	0.28	600	600	600	600	89

It has to be noted that in Table 1 the uniaxial tensile strength value has been computed as $f_t^{'} = 0.33\sqrt{f_c^{'}}$ as suggested by Bresler and Scordelis [4], and the elastic concrete modulus has been assumed $E_c = 22(f_c^{'}/10)^{0.3}$ according to Eurocode 2. Moreover, for what concerns the choices related to the adopted concrete constitutive model, the value of the eccentricity parameter e of the M-W-type yield surface has been evaluated by the expression $e = \left[2 + f_t^{'}/f_c^{'}\right] \Big/ \left[4 - f_t^{'}/f_c^{'}\right]$, the $f_t^{'}/f_c^{'}$ being assumed as a measure of the material brittleness. Finally, the value of ξ_v can be expressed as $\xi_v = \sqrt{3}f_c^{'}/m$, while ξ_a and ξ_b have been set equal to $\xi_a = 0.7923 f_c^{'}$ and $\xi_b = 1.8964 f_c^{'}$ as suggested by Li and Crouch [13].

Concerning the numerical model, it is worth noting that in both LMM and ECM the elastic analyses can be carried out by *any* commercial FE code. In the following applications the ADINA code has been used while a Fortran main program has been used to drive the FE analyses within the sequences. The elastic analyses performed within the LMM and ECM have been carried out using FE meshes of 3D-solid 8-nodes elements with 2 × 2 × 2 GPs per element for modeling concrete, 2-D-solid plane stress 4-nodes elements for modeling GFRP laminates and 2-nodes, 1-GP, truss elements for modeling steel bars and stirrups. To set up the FE model of each analyzed specimen a preliminary mesh sensitivity study, to assure an accurate FE elastic solution, has been performed. More precisely, the number of 3D-solid elements is 504, that of truss elements is 246, while the number of 2D-solid elements ranges from 80 to 160.

The obtained numerical results are reported in Table 3 in which, for the three analyzed specimens, are given: the peak load multiplier experimentally detected P_{EXP}; the predictions in terms of upper and lower bound values, P_{UB} and P_{LB}, and the relative errors (in %), $Err(P_{UB})$ and $Err(P_{LB})$, computed as the difference

Table 3 Peak load multipliers for the analyzed RC beams: values experimentally detected, P_{EXP}; values of the upper and lower bounds, P_{UB} and P_{LB}; relative errors, $Err(P_{UB})$ and $Err(P_{LB})$

Specimen label	Peak load multipliers			Relative error %	
	P_{EXP}	P_{UB}	P_{LB}	$Err(P_{UB})$	$Err(P_{LB})$
GB2	0.562	0.596	0.498	6.05	−11.39
GB4	0.587	0.649	0.550	10.56	−6.30
GB6	0.635	0.671	0.580	5.67	−8.66

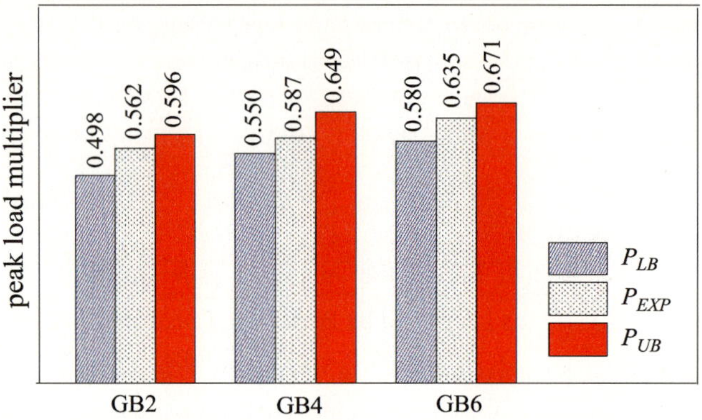

Fig. 5 Comparison between numerical and experimental peak load multipliers for the analyzed beams

between the numerical detected values and the experimental ones over the experimental one.

With the adopted definition of the relative errors, the upper bound values are expected to have a positive relative error, while the lower bound values are expected to have a negative relative error. The inspection of the numerical findings highlights the very good prediction obtained with the proposed three-yield-criteria limit analysis approach. In order to have a more immediate perception of the quality of the numerical predictions, the data of Table 3 are drawn in Fig. 5 as histograms. Other quite encouraging results have been obtained on FRP strengthened RC beams and slabs (see e.g. [7]), but are not reported here for lack of space.

The LMM gives also some hints on the type of failure mechanisms. Such a prediction in fact is given by the possibility to point out the plastic zone (portions of the FE-mesh where the collapse mechanism has been eventually located) at "last converged solution" of the LMM. Just to show one of the analysed cases, the plots of the principal (compressive) strain rates $\dot{\varepsilon}_3^c$ within RC concrete elements and of the principal strain rates $\dot{\varepsilon}_1^c$ within GFRP elements have been considered for beam GB4, at convergence. Plots are shown in Fig. 6a, b. By inspection of Fig. 6 it is possible to observe a plasticized zone, corresponding to a plastic hinge spread at beam center. This zone appears reasonably narrow and located where the damaged zones have been actually experimentally detected, as confirmed by the photograph given in Fig. 6c concerning beam GB4 at failure. It is worth to mention that the band plots of Fig. 6a, b provides only qualitative information of the failure mechanisms, but they can be anyway useful to localize critical zones or weaker members within larger structural systems.

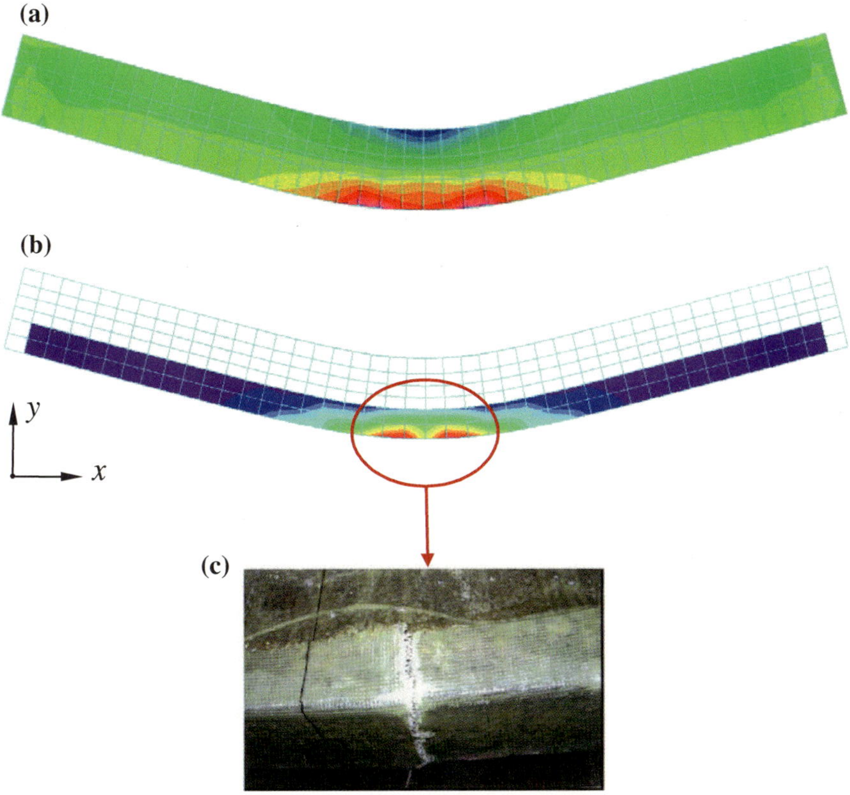

Fig. 6 Prediction of the failure mechanism for the beam GB4. Band plots of principal strain rates, in the deformed configuration, at last converged solution of the LMM. **a** RC elements; **b** FRP laminate elements; **c** photograph at midspan of beam GB6 at failure (after Almusallam and Al-Salloum [1])

7 Concluding Remarks

A multi-yield-criteria limit analysis numerical procedure for the prediction of peak loads and failure modes of steel-reinforced concrete elements strengthened with FRP laminates has been proposed. The behavior at ultimate state of the constituents, namely: concrete, FRP sheets and steel bars has been taken into account by a numerical procedure that involves, in concomitance, a Menétrey Willam-type yield criterion with cap in compression for concrete, a Tsai Wu-type yield criterion for the FRP strengthening reinforcement and the von Mises yield criterion for steel bars. The lack of associativity, postulated for concrete, to deal with its dilatancy, and for the FRP sheets, due to their compositive nature, has led to search for an upper and a lower bound to the peak load multiplier of the whole RC structural element. The former has been pursued by the LMM, the latter by the ECM. Both methods have been applied simultaneously to the three finite element types adopted to model concrete, FRP and steel; each type obeying the proper yield criterion. Large scale prototypes of a few FRP-strengthened RC beams, experimentally tested up to failure, have been numerically analyzed. The reliability and effectiveness of the proposed methodology has then been proved by comparison between experimental and numerical results showing also the capability of predicting, even if qualitatively, the failure mechanism type of the analyzed element.

References

1. Almusallam TH, Al-Salloum YA (2005) Use of glass FRP sheets as external flexure reinforcement in RC beams. Department of Civil Engineering, King Saud University, pp 1–15
2. American Concrete Institute ACI 440 (2008) Guide for the design and construction of externally bonded FRP systems for strengthening concrete structures, ACI 440.2R-08
3. Bažant ZP (1986) Mechanics of distributed cracking. Appl Mech Rev 39:675–705
4. Bresler B, Scordelis AC (1963) Shear strength of reinforced concrete beams. J Am Concr Inst 60(1):51–72
5. Chen WF (1982) Plasticity in reinforced concrete. McGraw-Hill
6. Drucker DC, Prager W (1952) Soil mechanics and plastic analysis or limit design. Q Appl Math 10:157–165
7. De Domenico D (2014) RC members strengthened with externally bonded FRP plates: a FE-based limit analysis approach. Compos Part B: Eng http://dx.doi.org/10.1016/j.compositesb.2014.11.013
8. De Domenico D, Pisano AA, Fuschi P (2014) A FE-based limit analysis approach for concrete elements reinforced with FRP bars. Compos Struct 107:594–603
9. FIB Bulletin 14 (2001) Externally bonded FRP reinforcement for RC structures. Task group 9.3. International Federation of Structural Concrete
10. Fuschi P (1999) Structural shakedown for elastic-plastic materials with hardening saturation surface. Int J Solids Struct 36:219–240
11. Le CV, Nguyen-Xuan H, Nguyen-Dang H (2010) Upper and lower bound limit analysis of plates using FEM and second-order cone programming. Comput Struct 88:65–73
12. Lee J, Fenves GL (1998) Plastic-damage model for cyclic loading of concrete structures. J Eng Mech ASCE 124:892–900
13. Li T, Crouch R (2010) A C_2 plasticity model for structural concrete. Comput Struct 88:1322–1332
14. Limam O, Foret G, Ehrlacher A (2003) RC two-way slabs strengthened with CFRP strips: experimental study and a limit analysis approach. Compos Struct 60:467–471
15. Lubliner J, Oliver J, Oller S, Oñate E (2010) A plastic-damage model for concrete. Int J Solids Struct 25(3):299–326
16. Mackenzie D, Boyle JT (1993) A method of estimating limit loads by iterative elastic analysis, Parts I, II, III. Int J Press Vessel Pip 77:77–142
17. Menétrey P, Willam KJ (1995) A triaxial failure criterion for concrete and its generalization. ACI Struct J 92:311–318
18. Pisano AA, Fuschi P (2007) A numerical approach for limit analysis of orthotropic composite laminates. Int J Numer Methods Eng 70:71–93
19. Pisano AA, Fuschi P, De Domenico D (2012) A layered limit analysis of pinned-joints composite laminates: numerical versus experimental findings. Compos: Part B 43:940–952

20. Pisano AA, Fuschi P, De Domenico D (2013) A kinematic approach for peak load evaluation of concrete elements. Comput Struct 119:125–139
21. Pisano AA, Fuschi P, De Domenico D (2013) Peak load prediction of multi-pin joints FRP laminates by limit analysis. Compos Struct 96:763–772
22. Pisano AA, Fuschi P, De Domenico D (2013) Peak loads and failure modes of steel-reinforced concrete beams: predictions by limit analysis. Eng Struct 56:477–488
23. Pisano AA, Fuschi P, De Domenico D (2014) Limit state evaluation of steel-reinforced concrete elements by von-Mises and Menétrey-Willam-type yield criteria. Int J Appl Mech 6(5):140058 (24 pages) Imperial College Press. doi:10.1142/S1758825114500586
24. Ponter ARS, Carter KF (1997) Limit state solutions, based upon linear elastic solutions with spatially varying elastic modulus. Comput Methods Appl Mech Eng 140:237–258
25. Ponter ARS, Fuschi P, Engelhardt M (2000) Limit analysis for a general class of yield conditions. Eur J Mech/A Solids 19:401–421
26. Prager W (1959) An introduction to plasticity. Addison-Wesley, Reading
27. Sloan SW (1988) Lower bound limit analysis using finite elements and linear programming. Int J Numer Anal Methods Geomech 12:61–77
28. Spiliopoulos K, Weichert D (2013) Direct methods for limit states in structures and materials. Springer Science+Business Media B.V., Dordrecht
29. Tsai SW, Wu EM (1971) A general theory of strength for anisotropic materials. J Compos Mater 5:58–80
30. Zhang J, Zhang Z, Chen C (2010) Yield criterion in plastic-damage models for concrete. Acta Mech Solida Sin 23(3):220–230

Shakedown Analysis Within the Framework of Strain Gradient Plasticity

Castrenze Polizzotto

Abstract A class of rate-independent material models is addressed within the framework of isotropic strain gradient plasticity. These models exhibit a size dependence through the *strengthening effects* (Hall–Petch effects), whereby the yield stress is related to the effective plastic strain by a suitable second-order partial differential equation with related boundary conditions. For a perfectly plastic material with strengthening effects, the classical concepts of plastic and shakedown limit analysis do hold, which lead to size dependent plastic and shakedown limit loads according to the dictum: *smaller is stronger*. In the perspective of a development of direct methods for applications to small-scale structures within micro/nano technologies, a shakedown theory for perfectly plastic materials with strengthening effects, previously elaborated by the present author [51], is presented and discussed. Apart from the inevitable mathematical complications carried in by the more complex constitutive behavior of the material herein considered, the overall conceptual architecture of the shakedown theorems remain within the classical Melan and Koiter theoretical framework. Further research efforts are needed to develop specific numerical procedures for the computation of the plastic and shakedown limit loads and the concomitant collapse mechanisms.

1 Introduction

Experimental observations show that, at micro/nano scales, crystalline materials exhibit a size dependent behavior whereby the yield stress, or strength, increases with the decreasing of grain size (the so-called Hall–Petch effects, or "strengthening"), see e.g. [10, 13, 25, 26, 60]. For an overview of the rich literature on this subject, see [1, 2, 18, 19, 21, 24, 27].

C. Polizzotto (✉)
Dipartimento di Ingegneria Civile Ambientale Aerospaziale e Dei Materiali,
Viale Delle Scienze, Università di Palermo, 90128 Palermo, Italy
e-mail: castrenze.polizzotto@unipa.it

P. Fuschi et al. (eds.), *Direct Methods for Limit and Shakedown Analysis of Structures*,
DOI 10.1007/978-3-319-12928-0_12

Fredriksson and Gudmundson [15] and, independently, Gurtin and Anand [19] employed similar power-law visco-plastic models in the purpose to describe strengthening effects, in such a way that, as the power law exponent tends to zero, a rate-independent plasticity behavior is approached. This method was improved for one-dimensional structures [5] and for pluri-dimensional ones obeying simplified constitutive laws [32]. Borg [6] extended the work by Gurtin and Anand [19] to crystal plasticity; Abu Al-Rub [1] addressed the influence of a boundary layer upon strengthening effects.

Within the framework of isotropic gradient plasticity, a different approach was employed by Polizzotto [50] to describe the strengthening effects. Namely, a strengthening potential was introduced, say $\psi_{st}(\kappa, \nabla\kappa)$, which is a degree-one positively homogeneous function of the effective plastic strain, κ, and of its gradient, such that the related thermodynamic forces (a scalar and a vector, called "primitive strengths") prove to be functions of the same type, but of degree zero. Consistent with thermodynamics principles, the primitive strengths provide the basis to the *strengthening law* of the material, in which the *strengthening stress* Y, measuring the increase (or decrease) of the yield strength produced by plastic deformation, is related to the effective plastic strain κ, through a relation having the shape of a second-order PDE (Partial Differential Equation) with higher order boundary conditions. Y is shown to depend on the current state of plastic deformation, but not on the past history of it.

Polizzotto [50] addressed the idealized case of a perfectly plastic material exhibiting strengthening effects and showed that the classical plastic limit analysis framework can be extended to such materials. The relevant collapse load problem was discussed and the solution uniqueness proved. Also, the extended forms of the lower bound and upper bound theorems were proved, together with the static-type maximum principle and kinematic-type minimum principles. With the aid of a simple numerical example, the collapse load was shown to increase with the decreasing of the structure's size. The resulting nonstandard plastic limit analysis proves to be mathematically more complex than the classical one, but suitable FEM-based procedures can lead to the evaluation of the relevant safety factor for micro/nano scale structures. Promising examples of evaluations of this kind were given in [5].

An extension of the above nonstandard plastic limit analysis, already achieved by [51], consists in a nonstandard shakedown theory for the considered class of strengthening materials. The interest in such a shakedown theory lies upon the need for suitable safety evaluation methods, or criteria, to be applied to micro/nano scale structures. This shakedown theory is reproposed in the present paper for informative purposes.

The proposed shakedown theory proves to be quite different from another one previously given by the author [47]. The latter theory comprises as particular cases a number of known gradient plasticity models of the literature, in which the gradient of plastic strain, or of another measure of plastic deformation, is explicitly involved, like in the gradient plasticity theories [1–3, 5, 14, 15, 18, 19, 45]. However, the proposed theory is able to describe the so-called energetic size effects, but not the strengthening effects. It was found, in fact, that the shakedown limit load is independent of the particular (gradient-type) hardening law of the material, hence it is size-independent.

On the contrary, in the present case of perfectly plastic material with strengthening effects, the plastic and shakedown limit loads are size dependent and exhibit a sort of Hall–Petch effects under the action of cyclic, or repeated, loads.

The outline of the paper is as follows. In Sect. 2 the constitutive model is described with emphasis on the strengthening potential and its main features. Suitable thermodynamic considerations are employed to derive the state equations and the plasticity evolution laws. The nature of the shakedown problem is addressed in Sect. 3, where some basic definitions of shakedown theory are recalled together with the relevant literature. Section 4 is devoted to the proof of the static shakedown theorem, Sect. 5 is analogously devoted to the proof of the kinematic theorem. Section 6 addresses the extension to strengthening materials of the lower bound and upper bound theorems; also, the complete set of equations governing the shakedown limit load problem in the presence of strengthening effects is established and discussed in some details. An application is reported in Sect. 7, where the shakedown analysis of a micro-scale two-bar system under cyclic thermal loading is accomplished. It is shown that the shakedown boundary expands with the decreasing of the sample size.

Notation. A compact notation is used, with boldface letters denoting vectors or tensors of any order. The scalar product between vectors or tensors is denoted with as many dots as the number of contracted index pairs. For instance, denoting by $\boldsymbol{u} = \{u_i\}$, $\boldsymbol{v} = \{v_i\}$, $\boldsymbol{\varepsilon} = \{\varepsilon_{ij}\}$, $\boldsymbol{\sigma} = \{\sigma_{ij}\}$, $\boldsymbol{\tau} = \{\tau_{ijk}\}$ and $\boldsymbol{A} = \{A_{ijkh}\}$ some vectors and tensors, one can write: $\boldsymbol{u} \cdot \boldsymbol{v} = u_i v_i$, $\boldsymbol{\sigma} : \boldsymbol{\varepsilon} = \sigma_{ij}\varepsilon_{ji}$, $\boldsymbol{A} : \boldsymbol{\varepsilon} = \{A_{ijkh}\varepsilon_{kh}\}$, $\boldsymbol{A} \vdots \boldsymbol{\tau} = \{A_{ijkh}\tau_{jkh}\}$, $\boldsymbol{A}^T \vdots \boldsymbol{\tau} = \{A_{ijkh}\tau_{kji}\}$. The summation rule for repeated indices holds and the subscripts denote components with respect to an orthogonal Cartesian co-ordinate system, say $\boldsymbol{x} = (x_1, x_2, x_3)$. An upper dot over a symbol denotes its time derivative, $\dot{\boldsymbol{u}} = \partial \boldsymbol{u}/\partial t$. The symbol ∇ denotes the spatial gradient operator, i.e. $\nabla \boldsymbol{u} = \{\partial_i u_j\}$. The symbol $\boldsymbol{e} = \{e_{ijk}\}$ indicates the alternating third order tensor. The symbol := means equality by definition. Other symbols will be defined in the text at their first appearance.

2 The Constitutive Model

The considered material is elastic-plastic and has a nonsimple nature. It exhibits no hardening, except for a fictitious isotropic one aimed at simulating *strengthening effects* (Hall–Petch effects), that is, the increase of the material *initial* (i.e. of the material being plastically undeformed) yield stress (or strength). The strengthening effects represent a form of size dependence of the material behavior, which in general is accompanied by other similar forms of size dependence, like for instance the size dependent hardening rate, but here we only consider a perfectly plastic material with strengthening effects for simplicity sake. The constitutive features of the latter material model were addressed in details by Polizzotto [50, 51]. For self-containment reasons they are briefly resumed in this section.

2.1 Thermodynamics Basis

Let the Helmholtz free energy potential of the material be of the form:

$$\psi = \frac{1}{2}\boldsymbol{\varepsilon}^{\mathrm{e}} : \boldsymbol{E} : \boldsymbol{\varepsilon}^{\mathrm{e}} + \psi_{\mathrm{st}}(\kappa, \nabla\kappa) \tag{1}$$

where $\boldsymbol{\varepsilon}^{\mathrm{e}} = \boldsymbol{\varepsilon} - \boldsymbol{\varepsilon}^{\mathrm{p}}$ is the elastic strain tensor, difference between the total and the plastic strain tensors; $\boldsymbol{E}$ denotes the fourth-order elasticity moduli tensor with its usual symmetries; κ, the effective plastic strain; whereas ψ_{st} is the *strengthening potential*. Isothermal conditions are considered for simplicity.

The potential ψ_{st} is a degree-one positively homogeneous function of κ and $\nabla\kappa$, κ being a field C^1-continuous in the volume V occupied by the material. It satisfies the condition $\psi_{\mathrm{st}} \to 0$ for $\nabla\kappa \to \mathbf{0}$, meaning that no strengthening effects arise in conditions of uniform plastic strain. The partial derivatives:

$$Y_0 := \frac{\partial\psi_{\mathrm{st}}}{\partial\kappa}, \quad \boldsymbol{Y}_1 := \frac{\partial\psi_{\mathrm{st}}}{\partial(\nabla\kappa)}, \tag{2}$$

here referred to as the *primitive strengths*, prove to be homogeneous functions like ψ_{st}, but of *degree zero*. As in the case of a the dissipation potential of flow plasticity [33, 36], the potential ψ_{st} subtends the existence of a deformation-theory plasticity-like law in which the primitive strengths Y_0 and $\boldsymbol{Y}_1$ are bounded by a *strengthening surface*, say $\phi = \phi(Y_0, \boldsymbol{Y}_1) \leq 0$, and are related to the strain quantities κ, $\nabla\kappa$ through the normality rule, that is:

$$\kappa = \mu\frac{\partial\phi}{\partial Y_0}, \quad \nabla\kappa = \mu\frac{\partial\phi}{\partial\boldsymbol{Y}_1} \tag{3}$$

where μ is a consistency coefficient satisfying the Kuhn–Tucker conditions:

$$\phi = \phi(Y_0, \boldsymbol{Y}_1) \leq 0, \ \ \mu \geq 0, \ \ \mu\phi(Y_0, \boldsymbol{Y}_1) = 0. \tag{4}$$

Primitive strengths below the strengthening limit, i.e. such that $\phi < 0$, pertain to the material being plastically undeformed, hence without strengthening effects. Primitive strengths at the strengthening limit ($\phi = 0$) pertain to the material being plastically deformed with plastic strains featured by the vector $(\kappa, \nabla\kappa)$ lying upon the outward normal to the surface $\phi = 0$ at the point $(Y_0, \boldsymbol{Y}_1)$. In turn, the latter conditions imply the existence of a degree-one homogeneous function $\psi_{\mathrm{st}}(\kappa, \nabla\kappa)$ satisfying (2).

In order to derive the thermodynamic restrictions on the constitutive equations for the material at hand, the so-called "residual-based strain gradient plasticity theory" is employed [7, 44–46, 48, 49, 52, 53], of which only the essential details are here given. The starting point is the Clausius–Duhem inequality which, in the present case of nonsimple material, reads:

$$D := \boldsymbol{\sigma} : \dot{\boldsymbol{\varepsilon}} - \dot{\psi} + P \geq 0 \ \text{ in } V. \tag{5}$$

where D denotes the dissipation density; $\boldsymbol{\sigma}$, the applied Cauchy stress; and P, the so-called (*nonlocality*) *energy residual*. In Eq. (5), P represents the long distance interaction energy density, vanishing only in the case of simple material.

The following constitutive assumptions hold:

- The *insulation condition*, which states that no long distance interactions occur between the material particles and the exterior ambient, hence (in the assumed absence of interfaces with surface energy) the volume integral of P must vanish, i.e.

$$\int_V P \, \mathrm{d}v = 0. \tag{6}$$

- The *bilinear dissipation condition*, consequence of the Onsager reciprocity principle assumed to hold also in the present context. It states that D exhibits a bilinear form in terms of independent plastic strain rates and related thermodynamic forces, that is (in the present case in which the nonlocality is confined to the strengthening behavior),

$$D := \boldsymbol{\sigma} : \dot{\boldsymbol{\varepsilon}}^{\mathrm{p}} - Y\dot{\kappa} \quad \text{in } V \tag{7}$$

  where Y is a scalar-valued force working through $\dot{\kappa}$.

- The *locality recovery condition*, which aims to guarantee that the material behaves as a simple one in the case of uniform plastic strain and requires that the residual P vanishes correspondingly.

On expanding the time derivative of ψ and with some mathematics, from (5) we can obtain the equality:

$$\begin{aligned} D = {} & (\boldsymbol{\sigma} - \boldsymbol{E} : \boldsymbol{\varepsilon}^{\mathrm{e}}) : \dot{\boldsymbol{\varepsilon}}^{\mathrm{e}} + \boldsymbol{\sigma} : \dot{\boldsymbol{\varepsilon}}^{\mathrm{p}} \\ & - (Y_0 - \nabla \cdot \boldsymbol{Y}_1)\,\dot{\kappa} - \nabla \cdot (\boldsymbol{Y}_1 \dot{\kappa}) + P \quad \text{in } V \end{aligned} \tag{8}$$

which, substituting D from (7), with an integration over V and taking into account (6), gives

$$\int_V (\boldsymbol{\sigma} - \boldsymbol{E} : \boldsymbol{\varepsilon}^{\mathrm{e}}) : \dot{\boldsymbol{\varepsilon}}^{\mathrm{e}} \mathrm{d}v + \int_V (Y - Y_0 + \nabla \cdot \boldsymbol{Y}_1)\dot{\kappa}\,\mathrm{d}v - \int_S \boldsymbol{n} \cdot \boldsymbol{Y}_1 \dot{\kappa} = 0. \tag{9}$$

Since the latter holds for whatever deformation mechanism and evolution law, the following relations are obtained as necessary and sufficient conditions, i.e.

$$\boldsymbol{\sigma} = \boldsymbol{E} : \boldsymbol{\varepsilon}^{\mathrm{e}} \quad \text{in } V \tag{10}$$

$$\left. \begin{aligned} & Y = Y_0 - \nabla \cdot \boldsymbol{Y}_1 \quad \text{in } V \\ & \dot{\kappa} = 0 \quad \text{on } S_c^{(1)}, \quad \boldsymbol{n} \cdot \boldsymbol{Y}_1 = 0 \quad \text{on } S_f^{(1)} \end{aligned} \right\} \tag{11}$$

where Y_0 and $\boldsymbol{Y}_1$ denote the primitive strengths in (2), whereas the residual is determined as

$$P = \nabla \cdot (\boldsymbol{Y}_1 \dot{\kappa}) \ \text{in}\ V. \tag{12}$$

Let us note that relations (10)–(12) make the dissipation D of (7) coincide with the one of (5).

Equation (10) is obviously the elasticity law and requires no comments. In $(11)_1$, Y represents the *strengthening stress*, i.e. the increase (or decrease) of the initial material strength, such that the plastically deformed material is characterized by an actual initial strength expressed as $\sigma_y = \sigma_0 + Y$ with Y depending on the primitive strengths. Equation $(11)_1$ represents a $Y - \kappa$ relation in the form of a second order PDE which is accompanied by the *higher order boundary conditions* in $(11)_2$, (which hold both in rate, as well as in time-finite form). The symbols $S_c^{(1)}$ and $S_f^{(1)}$ denote complementary parts of the boundary surface $S := \partial V$. In $S_c^{(1)}$, the onset of plastic strain is impeded and dislocations are forced to pile up; in $S_f^{(1)}$ dislocations can move freely and encounter no resistence. At points of V where $\phi < 0$ (plastically undeformed material), the indeterminate primitive strengths have to be set equal to zero, such that $Y = 0$ correspondingly. Equations (3), (4) and (11) govern the *strengthening law* obeyed by the material undergoing a strengthening process. The material strengthening depends on the current state of plastic deformation, not on its past history.

Equation (7) provides the (nonnegative) intrinsic dissipation power density, difference between the plastic power density, $\boldsymbol{\sigma} : \dot{\boldsymbol{\varepsilon}}^p$, and the dissipation power density pertaining to the strengthening process, $Y\dot{\kappa}$. The latter term appears here as a subtractive one because the strengthening effects are in the present model accounted for as fictitious (isotropic) hardening effects accompanied by the (irreversibly) stored energy $Y\dot{\kappa}$.

Equation (12) expresses the energy residual P as the divergence of the vector $\boldsymbol{Y}_1\dot{\kappa}$. It satisfies the *insulation condition*, i.e.

$$\int_V P \mathrm{d}v = \int_S \boldsymbol{n} \cdot \boldsymbol{Y}_1 \dot{\kappa} \mathrm{d}a = 0, \tag{13}$$

as a consequence of the higher order boundary conditions $(11)_2$.

Since for $\nabla\kappa \equiv \boldsymbol{0}$, hence κ uniform in V, it is $\boldsymbol{Y}_1 \equiv \boldsymbol{0}$ by $(2)_2$, and thus $P \equiv 0$ by (12), we can conclude that *the locality recovery condition is satisfied correspondingly*.

2.2 A Choice for the Strengthening Potential

Among many possible choices for ψ_{st}, for later use, let us choose a simple one in the form

$$\psi_{st} = \sigma_0 \, (p - \kappa), \qquad p := \sqrt{\kappa^2 + \ell^2 \|\nabla\kappa\|^2} \tag{14}$$

where σ_0 denotes the initial strength and ℓ is an internal length scale parameter. For $\ell \to 0$ (simple material) it is $\psi_{\rm st} = 0$ identically, which implies that no strengthening effects arise in a simple material for whatever plastic deformation. For $\ell \neq 0$, but $\|\nabla\kappa\| = 0$, again we have $\psi_{\rm st} = 0$, which means that in a nonsimple material under uniform strain no strengthening effects are allowed to occur.

The partial derivatives (2) take on the form

$$Y_0 = \sigma_0\left(\frac{\kappa}{p} - 1\right), \qquad \boldsymbol{Y}_1 = \ell^2\sigma_0\frac{\nabla\kappa}{p} \tag{15}$$

which, like (2), hold as long as plastic deformation is not trivially vanishing; otherwise Y_0 and $\boldsymbol{Y}_1$ are indeterminate, but below the strengthening limit, i.e. $\phi(Y_0, \boldsymbol{Y}_1) < 0$. We can obtain the equation $\phi = 0$ through a Legendre–Fenchel transformation [33] applied to (2), which is here implemented by rewriting (16) in the inverted form

$$\kappa = \frac{p}{\sigma_0}(Y_0 + \sigma_0), \qquad \nabla\kappa = \frac{1}{\ell^2}\frac{p}{\sigma_0}\boldsymbol{Y}_1, \tag{16}$$

then substituting the latter into the expression of p of (14) and dividing by p (assumed different from zero), we obtain

$$\phi := \sqrt{(Y_0 + \sigma_0)^2 + \frac{1}{\ell^2}\|\boldsymbol{Y}_1\|^2} - \sigma_0 = 0 \tag{17}$$

which is the desired strengthening surface. Then, by (3), we can write

$$\left.\begin{array}{ll} \kappa = \mu(Y_0 + \sigma_0)/\sigma_0, & \nabla\kappa = \frac{1}{\ell^2}\mu\boldsymbol{Y}_1/\sigma_0 \\ \mu = p \;\text{ for } \phi = 0, & \mu = 0 \;\text{ for } \phi < 0 \end{array}\right\} \tag{18}$$

Finally, Eq. (11) particularizes as

$$\left.\begin{array}{l} Y = \sigma_0\left[\frac{\kappa}{p} - \ell^2\,\nabla\cdot\left(\frac{\nabla\kappa}{p}\right) - 1\right] \;\text{ in } V \\ \dot{\kappa} = 0 \;\text{ on } S_c^{(1)}, \quad \ell^2\partial_n\kappa/p = 0 \;\text{ on } S_f^{(1)} \end{array}\right\} \tag{19}$$

which is the pertinent second-order differential-form $Y - \kappa$ relation. For $\ell = 0$ (simple material), obviously it is $Y \equiv 0$, $S_c^{(1)} = \emptyset$ and the higher order boundary conditions in (19) disappear.

The strengthening surface in (17) can be represented in the plane with co-ordinates $Y_0 + \sigma_0$, $\|\boldsymbol{Y}_1\|$. It has the shape of an ellipse centered at point $(-\sigma_0, 0)$ and with diameters $2\sigma_0$ and $2\ell\sigma_0$, Fig. 1. On letting ℓ increase from $\ell = 0$ to $\ell \to \infty$, the ellipse changes in shape. Namely, whereas the diameter along OA remains fixed, the diameter along OB increases from zero to ∞.

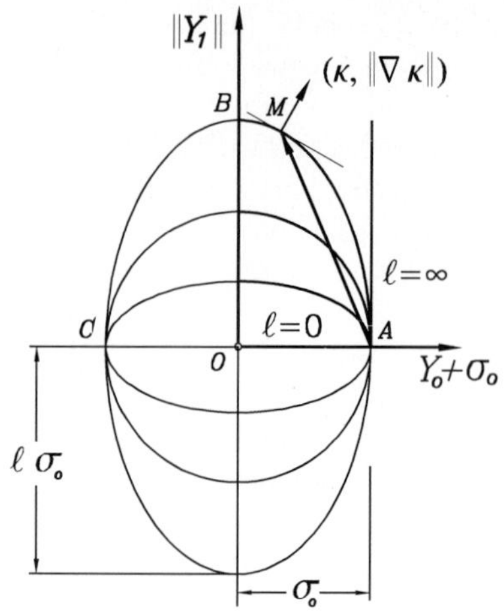

Fig. 1 Family of strengthening *ellipses* generated by varying the internal length parameter ℓ

2.3 Plasticity Evolution Laws

On the basis of the dissipation expression (7), the plasticity evolution laws of the material, in the assumed hypothesis of rate-independent associative plasticity, can be cast, by a standard notation, as follows:

$$\left.\begin{array}{l} f = f(\boldsymbol{\sigma}, Y) := f_0(\boldsymbol{\sigma}) - Y - \sigma_0 \leq 0 \\ \dot{\lambda} \geq 0, \quad \dot{\lambda} f(\boldsymbol{\sigma}, Y) = 0 \\ \dot{\boldsymbol{\varepsilon}}^{\mathrm{p}} = \dot{\lambda} f_{,\boldsymbol{\sigma}}, \quad \dot{\kappa} = -\dot{\lambda} f_{,Y} = \dot{\lambda} \end{array}\right\} \quad \text{in } V \tag{20}$$

where $\dot{\lambda}$ denotes the relevant consistency coefficient. It can be proved that, for whatever choice of the relevant variables, but complying with (20), the dissipation D of (7) proves to be nonnegative and therefore the Clausius–Duhem inequality (5) is satisfied correspondingly.

The yielding laws (20) obey a *maximum dissipation principle*, namely

$$\left.\begin{array}{l} D(\dot{\boldsymbol{\varepsilon}}^{\mathrm{p}}, \dot{\kappa}) = \max_{(\boldsymbol{\sigma}, Y)} (\boldsymbol{\sigma} : \dot{\boldsymbol{\varepsilon}}^{\mathrm{p}} - Y\dot{\kappa}) \\ \text{subject to} \quad f(\boldsymbol{\sigma}, Y) \leq 0. \end{array}\right\} \tag{21}$$

The latter problem provides the stress state $(\boldsymbol{\sigma}, Y)$ under which a *locally specified* plastic strain mechanism $(\dot{\boldsymbol{\varepsilon}}^{\mathrm{p}}, \dot{\kappa})$ is allowed to occur. The optimal objective function is the *dissipation function*, $D(\dot{\boldsymbol{\varepsilon}}^{\mathrm{p}}, \dot{\kappa})$, which satisfies the relations:

$$\boldsymbol{\sigma} = \frac{\partial D}{\partial \dot{\boldsymbol{\varepsilon}}^{\mathrm{p}}}, \qquad Y = -\frac{\partial D}{\partial \dot{\kappa}}. \tag{22}$$

These hold as long as the plastic mechanism is not trivially vanishing; otherwise, the stresses $(\boldsymbol{\sigma}, Y)$—constitutively determined through the state equations (10) and (11)—prove to be just plastically admissible, i.e. $f(\boldsymbol{\sigma}, Y) \leq 0$.

A consequence of the maximum dissipation principle is the *Druckerian inequality* [33, 36]:

$$(\boldsymbol{\sigma} - \boldsymbol{\sigma}^*) : \dot{\boldsymbol{\varepsilon}}^{\mathrm{p}} - (Y - Y^*)\dot{\kappa} \geq 0 \tag{23}$$

where the pairs $(\boldsymbol{\sigma}, Y)$ and $(\dot{\boldsymbol{\varepsilon}}^{\mathrm{p}}, \dot{\kappa})$ correspond to each other through the evolution laws (20), whereas the pair $(\boldsymbol{\sigma}^*, Y^*)$ is any plastically admissible stress state, i.e. $f(\boldsymbol{\sigma}^*, Y^*) \leq 0$. Another consequence is the doubled Druckerian inequality, which reads:

$$(\boldsymbol{\sigma}' - \boldsymbol{\sigma}'') : (\dot{\boldsymbol{\varepsilon}}^{\mathrm{p}'} - \dot{\boldsymbol{\varepsilon}}^{\mathrm{p}''}) - (Y' - Y'')(\dot{\kappa}' - \dot{\kappa}'') \geq 0, \tag{24}$$

where the two sets of primed and doubly primed quantities are arbitrary, but each satisfies the evolution equations (21) and (22).

Strong Druckerial inequality. For the purposes of the present paper, the *strong Druckerian inequality* is of paramount importance. This is obtained from (23) when the latter holds only under the *more stringent condition* $f(\boldsymbol{\sigma}^*, Y^*) < 0$. A consequence of the latter condition is that inequality (23) *holds as an equality if, and only if, the plastic deformation mechanism is trivially vanishing*, i.e. $\dot{\boldsymbol{\varepsilon}}^{\mathrm{p}} = \mathbf{0}$, $\dot{\kappa} = 0$.

Note that a maximum dissipation principle holds also for the strengthening laws (2) to (4), namely

$$\left.\begin{aligned} \psi_{\mathrm{st}}(\kappa, \nabla\kappa) = \max{}_{(Y_0, \boldsymbol{Y}_1)}\,(Y_0\kappa + \boldsymbol{Y}_1 \cdot \nabla\kappa) \\ \text{subject to } \ \phi(Y_0, \boldsymbol{Y}_1) \leq 0 \end{aligned}\right\} \tag{25}$$

Also note that the latter principle implies a Druckerian inequality as

$$(Y_0 - Y_0^*)\kappa - (\boldsymbol{Y}_1 - \boldsymbol{Y}_1^*) \cdot \nabla\kappa \geq 0 \tag{26}$$

where the pairs $(Y_0, \boldsymbol{Y}_1)$ and $(\kappa, \nabla\kappa)$ are related to each other through the strengthening laws (3) and (4), and the pair $(Y_0^*, \boldsymbol{Y}_1^*)$ does not exceed the strengthening limit, i.e. $\phi(Y_0^*, \boldsymbol{Y}_1^*) \leq 0$.

3 The Shakedown Problem

Although the shakedown problem is well understood in the literature (see e.g. [17, 30, 36, 61, 62]), some comments on the nature of this problem are believed to be helpful.

Let a (micro/nano scale) body of volume V and boundary surface $S := \partial V$ obey the constitutive equations of Sect. 2, namely Eqs. (3), (4), (10) and (11), as well as the plasticity evolution laws (20). The body is referred, in its undeformed state, to orthogonal Cartesian co-ordinates, say $\boldsymbol{x} = (x_1, x_2, x_3)$, is clamped at points of the portion S_c of S, free at point of the complementary part $S_f = S \setminus S_c$. The nonstandard surface partition $S = S_c^{(1)} \cup S_f^{(1)}$—not necessarily coincident with the standard one, i.e. $S = S_c \cup S_f$—specifies the portion ($S_c^{(1)}$) where some idealized substrate impedes the onset of plastic strains ($\dot{\kappa} = 0$) with consequent dislocations pileups, and the complementary portion ($S_f^{(1)}$) where dislocations can move freely, no dislocations pileups occur and thus the relevant Peach–Koehler force is vanishing ($\boldsymbol{n} \cdot \boldsymbol{Y}_1 = 0$). The latter partition of S is typical of micro/nano scale bodies, in which material inhomogeneities, or defects, here represented by dislocations, may influence substantially the overall body's response to the loads.

The body is subjected to body forces specified in V, tractions specified on S_f, imposed displacements specified on S_c. Thermal-like strains can also be imposed in V, but for simplicity sake we shall ignore them in the theoretical treatment. All these loads may act in a quasi-static manner, or even dynamically. We let these actions be collectively represented by means of the (fictitious) elastic stress response, say $\boldsymbol{\sigma}^E(\boldsymbol{x}, t)$. This denotes the stress response of the body to a specified load history, say $L(t), t \geq 0$, computed under the assumption of *purely elastic material behavior* ($\sigma_0 = \infty$) and *homogeneous initial conditions* (at $t = 0$).

A typical characteristics of the shakedown problem under discussion is that the load history $L(t)$ can be anyone within a specified set, say Π (*load domain*). Every *Admissible Load History*, (ALH), $L(t) \in \Pi$, represents a potentially active load history for the body. On letting $L(t)$ vary in all possible ways in Π, the relevant elastic stress response path $\boldsymbol{\sigma}^E(\boldsymbol{x}, t)$ at every $\boldsymbol{x} \in V$ belongs to a (finite) stress domain the convex hull of which is denoted $\Sigma(\boldsymbol{x})$, whereas in the whole V they belong to a (finite) stress domain the convex hull of which is Σ, the union of all $\Sigma(\boldsymbol{x})$. We can thus write:

$$\boldsymbol{\sigma}^E(\boldsymbol{x}, t) \in \Sigma(\boldsymbol{x}) \in \Sigma \qquad \forall t \geq 0,\ \forall \boldsymbol{x} \in V. \tag{27}$$

The boundary $\Gamma := \partial \Sigma$ denotes the *elastic stress response envelope*; $\Gamma(\boldsymbol{x}) := \partial \Sigma(\boldsymbol{x})$ is its projection at $\boldsymbol{x}$. The smallest subset of $\Gamma(\boldsymbol{x})$, say $\Gamma_B(\boldsymbol{x})$, the convex hull of which coincides with $\Sigma(\boldsymbol{x})$, is the *basic elastic stress response envelope* at $\boldsymbol{x}$.

With no loss in generality, we assume that the local stress domain $\Sigma(\boldsymbol{x})$ has the shape of a hyperpolyhedron of $N \geq 2$ vertices, denoted $\boldsymbol{\sigma}_i^E(\boldsymbol{x})$, ($i = 1, 2, \ldots, N$), and called the *basic elastic stresses at* $\boldsymbol{x}$, hence $\Gamma_B(\boldsymbol{x})$ has a *discrete form consisting in these N stress points*, i.e.

$$\Gamma_B(\boldsymbol{x}) = \{\boldsymbol{\sigma}_i^E(\boldsymbol{x}) \ \ \forall i \in I(N),\ \boldsymbol{x} \in V\} \tag{28}$$

whereas Γ_B is the union of all $\Gamma_B(\boldsymbol{x})$ in V.

The actual elastic-plastic response of the considered body to a generic ALH can in principle be computed making use of the set of constitutive equations established in Sect. 2, together with the equilibrium (or motion) and compatibility equations. By its own nature, the shakedown problem does not need for such (likely cumbersome, step-by-step) analyses, but rather it requires the application of suitable direct methods able to ascertain whether some shakedown safety criteria are satisfied by the considered structure/load system [34, 35].

Shakedown means that the structure subjected to variable repeated loads, after some initial period of limited plastic deformations, then responds to the subsequent loads elastically, without further plastic deformations (and thus, in the present context, without further strengthening effects). The shakedown theorems of the literature assess the pre-requisites of the structure/load system in order to guarantee that shakedown certainly occurs eventually during the deformation process. The latter theorems are well known for a wide class of constitutive laws (perfect plasticity, linear and nonlinear hardening, size dependent hardening, damage, shape memory, poroplasticity and temperature dependent yield stress), quasi-static and dynamic load conditions, geometric nonlinearities. It is not the purpose of the present paper to review the related literature, but it will suffice to quote a few representative papers and books with their reference lists: [8, 9, 11, 17, 22, 23, 28–31, 34–43, 54, 57–59, 61–63].

Shakedown theory was extended to nonlocal (integral) elasticity by Polizzotto et al. [55] and to gradient plasticity with size dependent hardening by Polizzotto [47]. Strengthening effects were addressed, as previously mentioned, in the paper by Polizzotto [51]. The content of the latter paper is herein summarized for seminal purposes.

4 Melan-Type Static Shakedown Theorem

Let $\bar{M} = \{\bar{\boldsymbol{\sigma}}^r, \bar{Y}, \bar{Y}_0, \bar{\boldsymbol{Y}}_1\}$ denote a set of load-independent stress-like space variables, collectively denoted *static shakedown parameters*, (SSPs). The set $\bar{M}$ is *admissible* if the relevant fields satisfy the following restrictions:

$$\left.\begin{array}{l} \nabla \cdot \bar{\boldsymbol{\sigma}}^r = \mathbf{0} \text{ in } V, \quad \boldsymbol{n} \cdot \bar{\boldsymbol{\sigma}}^r = \mathbf{0} \text{ on } S_f \\ \bar{Y} = \bar{Y}_0 - \nabla \cdot \bar{\boldsymbol{Y}}_1 \text{ in } V, \quad \boldsymbol{n} \cdot \bar{\boldsymbol{Y}}_1 = 0 \text{ on } S_f^{(1)} \\ \phi(\bar{Y}_0, \bar{\boldsymbol{Y}}_1) \le 0 \text{ in } V \end{array}\right\} \tag{29}$$

Namely, $\bar{\boldsymbol{\sigma}}^r$ is required to constitute a residual stress field, whereas $\bar{Y}$ must be some strengthening stress field associated to primitive strengths fields $\bar{Y}_0$ and $\bar{\boldsymbol{Y}}_1$ nowhere exceeding the strengthening surface and satisfying the static-type higher order boundary condition.

Let $\bar{\mathcal{M}}$ collect all sets $\bar{M}$ and $\bar{\mathcal{M}}_A$ all the admissible ones. With the just introduced notion of SSPs, we intend to prove the static, or Melan-type, theorem of shakedown theory for strengthening materials, in a unified form for either quasi-static, or dynamic, loads.

Static shakedown theorem
A necessary and sufficient condition in order that shakedown occurs in a (micro/nano scale) structure subjected to (either quasi-static, or dynamic) variable repeated loads, characterized by a discrete set Γ_B of basic elastic stresses, is that there exists a set of admissible SSPs, say $\bar{M} = \{\bar{\boldsymbol{\sigma}}^r, \bar{Y}, \bar{Y}_0, \bar{\boldsymbol{Y}}_1\} \in \bar{\mathcal{M}}_A$, such as to satisfy the following *static shakedown criterion*, namely:

$$\bar{f}_i := f(\boldsymbol{\sigma}_i^E + \bar{\boldsymbol{\sigma}}^r, \bar{Y}) \le 0 \quad \text{in } V, \quad \forall i \in I(N). \tag{30}$$

Proof A classical reasoning scheme is followed for this proof. For the *necessity* part of the theorem, let us assume that shakedown occurs. Then, plastic deformations promoted by any ALH, together with the consequent strengthening effects, must stop at a certain time, say t_a. The subsequent response is of purely elastic nature and has the form $\boldsymbol{\sigma} = \boldsymbol{\sigma}^E(\boldsymbol{x}, t) + \boldsymbol{\sigma}^r_a(\boldsymbol{x})$, $Y = Y_a(\boldsymbol{x})$, where the quantities $(.)_a$ denote values at time t_a. Obviously, the yield condition is satisfied, i.e.

$$f(\boldsymbol{\sigma}^E(\boldsymbol{x}, t) + \boldsymbol{\sigma}^r_a(\boldsymbol{x}), Y_a(\boldsymbol{x})) \leq 0 \quad \text{in } V, \quad \forall i \in I(N), \ \forall t \geq t_a \tag{31}$$

Since the latter inequality holds for whatever ALH, it is also satisfied within $\boldsymbol{\sigma}^E(\boldsymbol{x}, t)$ replaced by any stress point in $\Sigma(\boldsymbol{x})$ and in particular by anyone of the basic elastic stresses, $\boldsymbol{\sigma}^E_i$, $\forall i \in I(N)$. In other words, (31) implies the inequalities:

$$f(\boldsymbol{\sigma}^E_i + \boldsymbol{\sigma}^r_a, Y_a) \leq 0 \quad \text{in } V, \quad \forall i \in I(N). \tag{32}$$

This coincides with (30), but $\bar{\boldsymbol{\sigma}}^r \equiv \boldsymbol{\sigma}^r_a$ and $\bar{Y} \equiv Y_a$.

For the *sufficiency* part of the theorem, let us proceed in steps.

Step 1. Let us assume that the shakedown criterion (30) is satisfied, but in a *more stringent form* as:

$$\bar{f}_i := f(\boldsymbol{\sigma}^E_i + \bar{\boldsymbol{\sigma}}^r, \bar{Y}) \leq -\eta \quad \text{in } V, \quad \forall i \in I(N) \tag{33}$$

where η denotes an arbitrary (small) positive scalar. Also, let $L(t), t \geq 0$, be any ALH and let $\boldsymbol{u}, \boldsymbol{\sigma}, \ldots$, denote the related actual elastic-plastic response. Due to the convexity of f, inequalities (33) imply that:

$$f = f(\boldsymbol{\sigma}^E(\boldsymbol{x}, t) + \bar{\boldsymbol{\sigma}}^r(\boldsymbol{x}), \bar{Y}(\boldsymbol{x})) \leq -\eta \quad \text{in } V, \quad \forall t \geq 0 \tag{34}$$

where $\boldsymbol{\sigma}^E(\boldsymbol{x}, t)$ is the elastic stress response to the chosen ALH.

Step 2. With the above in mind, let us introduce the response functionals:

$$\begin{aligned} A_1(t) :=& \int_V \frac{1}{2}\rho(\dot{\boldsymbol{u}} - \dot{\boldsymbol{u}}^E) \cdot (\dot{\boldsymbol{u}} - \dot{\boldsymbol{u}}^E)\, \mathrm{d}v \\ &+ \int_V \int_0^t \mu_d(\dot{\boldsymbol{u}} - \dot{\boldsymbol{u}}^E) \cdot (\dot{\boldsymbol{u}} - \dot{\boldsymbol{u}}^E) \mathrm{d}t\, \mathrm{d}v \\ &+ \int_V \frac{1}{2}(\boldsymbol{\sigma} - \boldsymbol{\sigma}^E - \bar{\boldsymbol{\sigma}}^r) : \boldsymbol{E}^{-1} : (\boldsymbol{\sigma} - \boldsymbol{\sigma}^E - \bar{\boldsymbol{\sigma}}^r) \mathrm{d}v, \end{aligned} \tag{35}$$

$$A_2(t) := \int_V \left[\psi_{\mathrm{st}}(\kappa, \nabla\kappa) - \bar{Y}_0\kappa - \bar{\boldsymbol{Y}}_1 \cdot \nabla\kappa\right] \mathrm{d}v, \tag{36}$$

Both $A_1(t)$ and $A_2(t)$ depend on the chosen ALH by the related elastic response, as well as by the actual elastic-plastic response, herein described by the velocity $\dot{\boldsymbol{u}}$, the stress $\boldsymbol{\sigma}$ and the effective plastic strain κ; they also depend on the set $\bar{M}$ of SSPs, which we establish to coincide with the one entering the shakedown criterion (30). In (35) ρ and μ_d are the mass density and the damping coefficient, which must be taken equal to zero in the case of quasi-static loads.

Step 3. The functional $A(t) := A_1(t) + A_2(t)$ is *positive definite*. Obviously, this is true for $A_1(t)$. As for $A_2(t)$, let us note that the strengthening potential ψ_{st}, due to its degree-one homogeneity, can be written as

$$\psi_{st}(\kappa, \nabla\kappa) = Y_0\kappa + \boldsymbol{Y}_1 \cdot \nabla\kappa, \tag{37}$$

where Y_0 and $\boldsymbol{Y}_1$ are given by (1). Also, let us note that the integrand of (36) is nonnegative, since, by (37),

$$\begin{aligned} &\psi_{st}(\kappa, \nabla\kappa) - \bar{Y}_0\kappa - \bar{\boldsymbol{Y}}_1 \cdot \nabla\kappa \\ &\quad = (Y_0 - \bar{Y}_0)\kappa + (\boldsymbol{Y}_1 - \bar{\boldsymbol{Y}}_1) \cdot \nabla\kappa \geq 0 \quad \text{in } V \end{aligned} \tag{38}$$

as a consequence of the SSPs being admissible, i.e. $\bar{M} \in \bar{\mathscr{M}}_A$, as well as of the Druckerian inequality applied to the strengthening surface, Eq. (26).

Step 4. The time derivative of $A(t)$ reads as follows:

$$\begin{aligned} \dot{A}(t) = &\int_V \left[\rho(\ddot{\boldsymbol{u}} - \ddot{\boldsymbol{u}}^E) + \mu_d(\dot{\boldsymbol{u}} - \dot{\boldsymbol{u}}^E)\right] \cdot (\dot{\boldsymbol{u}} - \dot{\boldsymbol{u}}^E)\mathrm{d}v \\ &+ \int_V (\boldsymbol{\sigma} - \boldsymbol{\sigma}^E - \bar{\boldsymbol{\sigma}}^r) : \boldsymbol{E}^{-1} : (\dot{\boldsymbol{\sigma}} - \dot{\boldsymbol{\sigma}}^E)\mathrm{d}v \\ &+ \int_V (Y_0 - \bar{Y}_0)\dot{\kappa} + (\boldsymbol{Y}_1 - \bar{\boldsymbol{Y}}_1) \cdot \nabla\dot{\kappa}\mathrm{d}v. \end{aligned} \tag{39}$$

Next, using the principle of the virtual power and the divergence theorem applied to the last integral and considering the identity

$$\dot{\boldsymbol{\varepsilon}}^{\mathrm{p}} = \dot{\boldsymbol{\varepsilon}} - \dot{\boldsymbol{\varepsilon}}^E - \boldsymbol{E}^{-1} : (\dot{\boldsymbol{\sigma}} - \dot{\boldsymbol{\sigma}}^E) \tag{40}$$

along with the higher order boundary conditions $(11)_2$ and $(29)_2$, Eq. (39) becomes:

$$\dot{A}(t) = -\int_V \left[(\boldsymbol{\sigma} - \boldsymbol{\sigma}^E - \bar{\boldsymbol{\sigma}}^r) : \dot{\boldsymbol{\varepsilon}}^{\mathrm{p}} - (Y - \bar{Y})\dot{\kappa}\right]\mathrm{d}v. \tag{41}$$

Step 5. Since the stress pair $(\boldsymbol{\sigma}^E + \bar{\boldsymbol{\sigma}}^r, \bar{Y})$ satisfies (34), then the strong Druckerian inequality holds as far as the scalar η is positive, namely

$$(\sigma - \sigma^E - \bar{\sigma}^r) : \dot{\varepsilon}^{\mathrm{p}} - (Y - \bar{Y})\dot{\kappa} \geq 0 \quad \text{in } V, \quad \forall t \geq 0. \tag{42}$$

This implies $\dot{A}(t) < 0$, that is, $A(t)$ decreases at a nonzero rate as far as the plastic deformation process continues. Because $A(t)$ cannot become negative, it eventually must stop decreasing. A subsequent (likely finite) time t_a must exist, such that $\dot{A}(t) = 0, \forall t \geq t_a$, and thus inequality (42) is satisfied as an equality correspondingly. By the strong Druckerian inequality, such an occurrence implies that no more plastic deformations, nor strengthening effects, can further be produced. Since this result holds for whatever ALH, we can conclude that shakedown certainly occurs under condition (34). The same conclusion can be inferred on letting $\eta \to 0$, but we have to admit that at the limit the structure finds itself in a *shakedown limit state exposed to some impending inadaptation collapse mode*. The proof of the theorem is complete. □

In the absence of a strengthening potential, that is, if $\psi_{\mathrm{st}} = 0$, the above shakedown theorem coincides with the classical one for perfect plasticity [29].

As a *corollary* to the theorem above, we can prove that, under condition (34) not only shakedown certainly occurs, but the total plastic dissipation produced in the structure within the relevant deformation process proves to be bounded. This can be proved in a straightforward manner following the classical Koiter procedure [29]. Namely, let us again consider inequality (35) with some $\eta > 0$, but rewritten in the equivalent form

$$f(\beta(\sigma^E + \bar{\sigma}^r), \beta\bar{Y}) \leq 0 \quad \text{in } V, \quad \forall t \geq 0 \tag{43}$$

where $\beta > 1$ is suitably chosen. Via the well-known Koiter procedure, we easily arrive at the inequality:

$$D(\dot{\varepsilon}^{\mathrm{p}}, \dot{\kappa}) \leq \frac{\beta}{\beta - 1}\left[(\sigma - \sigma^E - \bar{\sigma}^r) : \dot{\varepsilon}^{\mathrm{p}} - (Y - \bar{Y})\dot{\kappa}\right]. \tag{44}$$

Then, with an integration over $V \times (0, t_1)$ and remembering (41), we can obtain:

$$\int_0^{t_1}\int_V D(\dot{\varepsilon}^{\mathrm{p}}, \dot{\kappa})\mathrm{d}v\,\mathrm{d}t < \frac{\beta}{\beta - 1}[A(0) - A(t_1)] < \frac{\beta}{\beta - 1}A(0) \tag{45}$$

which holds for arbitrary t_1, even for $t_1 \to \infty$. Assuming zero plastic deformation at $t = 0$, the (bounded) scalar $A(0)$ is given by

$$A(0) = \int_V \frac{1}{2}\bar{\sigma}^r : \boldsymbol{E}^{-1} : \bar{\sigma}^r \mathrm{d}v. \tag{46}$$

5 Koiter-Type Kinematic Shakedown Theorem

Let $\hat{M} := \{\hat{\boldsymbol{\varepsilon}}_1^{\rm p}, \hat{\kappa}_1, \ldots, \hat{\boldsymbol{\varepsilon}}_N^{\rm p}, \hat{\kappa}_N, \hat{\boldsymbol{\varepsilon}}^{\rm p}, \hat{\kappa}, \hat{\boldsymbol{u}}\}$ denote a set of plastic strain, effective plastic strain and displacement fields in space, which are collectively referred to as *kinematic shakedown parameters*, (KSPs). The pairs $(\hat{\boldsymbol{\varepsilon}}_i^{\rm p}, \hat{\kappa}_i)$, $i \in I(N)$, are associated to the basic elastic stress, $\boldsymbol{\sigma}_i^E$, respectively, but are not a consequence of the corresponding loads. The set $\hat{M}$ is *admissible* if the relevant KSPs satisfy the following restrictions:

$$\left.\begin{array}{l}\hat{\boldsymbol{\varepsilon}}^{\rm p} = \nabla^{\rm sym}\hat{\boldsymbol{u}} \ \text{ in } V, \quad \hat{\boldsymbol{u}} = \boldsymbol{0} \ \text{ on } S_c \\ \hat{\kappa}_i \geq 0 \text{ and } C^1\text{—continuous in } V, \quad \hat{\kappa}_i = 0 \text{ on } S_c^{(1)} \\ \hat{\boldsymbol{\varepsilon}}^{\rm p} = \sum_{i=1}^{i=N} \hat{\boldsymbol{\varepsilon}}_i^{\rm p}, \quad \hat{\kappa} = \sum_{i=1}^{i=N} \hat{\kappa}_i \quad \text{ in } V\end{array}\right\} \tag{47}$$

This means that the (fictitious) ratchet strain field $\hat{\boldsymbol{\varepsilon}}^{\rm p}$ is compatible with displacements $\hat{\boldsymbol{u}}$ vanishing on S_c, and that the effective plastic strains $\hat{\kappa}_i$ are differentiable and satisfy the kinematic-type higher order boundary condition on $S_c^{(1)}$. The scalar $\hat{\kappa}$ (*total effective plastic strain*) is unconstrained.

We let $\hat{\mathscr{M}}$ collect all the KSPs, $\hat{\mathscr{M}}_A$ all the admissible ones. Every set $\hat{M}$ of KSPs describes some kind of *impending inadaptation collapse mechanism*, or *kinematically admissible plastic strain cycle* after Koiter's language [29]. Equipped with the above notion of KSPs, we can prove the following theorem.

Kinematic shakedown theorem
A necessary and sufficient condition in order that shakedown certainly occurs in a (micro/nano scale) structure subjected to (either quasi-static, or dynamic) variable repeated loads, characterized by a discrete set Γ_B of basic elastic stresses, is that the *kinematic shakedown criterion*, i.e.

$$\mathscr{E}[\hat{M}] := \sum_{i=1}^{i=N} \int_V \left[D(\hat{\boldsymbol{\varepsilon}}_i^{\rm p}, \hat{\kappa}_i) - \boldsymbol{\sigma}_i^E : \hat{\boldsymbol{\varepsilon}}_i^{\rm p}\right] {\rm d}v + \int_V \psi_{\rm st}(\hat{\kappa}, \nabla\hat{\kappa}){\rm d}v \geq 0, \tag{48}$$

is satisfied for whatever admissible KSPs, $\hat{M} \in \hat{\mathscr{M}}_A$.

Proof For the *necessity* part of the theorem let us admit that shakedown occurs. Then, by the static shakedown theorem given for granted, there must exist some admissible SSPs, say $\bar{M} = \{\bar{\boldsymbol{\sigma}}^r, \bar{Y}, \bar{Y}_0, \bar{\boldsymbol{Y}}_1\}$, such that the yield condition $f(\boldsymbol{\sigma}^E + \bar{\boldsymbol{\sigma}}^r, \bar{Y}) \leq 0$ and the strengthening condition $\phi(\bar{Y}_0, \bar{\boldsymbol{Y}}_1) \leq 0$ are satisfied everywhere in V. Therefore, by the maximum dissipation principles, (21) and (25), applied to every set $\hat{M}$ of KSPs, we can write:

$$\left.\begin{array}{l}D(\hat{\boldsymbol{\varepsilon}}_i^{\rm p}, \hat{\kappa}_i) \geq (\boldsymbol{\sigma}_i^E + \bar{\boldsymbol{\sigma}}^r) : \hat{\boldsymbol{\varepsilon}}_i^{\rm p} - \bar{Y}\hat{\kappa}_i \quad \text{in } V, \ \forall i \in I(N) \\ \psi_{\rm st}(\hat{\kappa}, \nabla\hat{\kappa}) \geq \bar{Y}_0\hat{\kappa} + \bar{\boldsymbol{Y}}_1 \cdot \nabla\hat{\kappa} \quad \text{in } V\end{array}\right\} \tag{49}$$

Hence, by the aid of the latter inequalities, remembering $(47)_2$ and (48), and by the divergence theorem, we can write:

$$\mathscr{E}[\hat{M}] \geq \int_V \left(\bar{Y}_0\hat{\kappa} + \bar{\boldsymbol{Y}}_1 \cdot \nabla\hat{\kappa} - \bar{Y}\hat{\kappa}\right) \mathrm{d}v$$
$$= \int_V \left(\bar{Y}_0 - \nabla \cdot \bar{\boldsymbol{Y}}_1 - \bar{Y}\right) \hat{\kappa}\, \mathrm{d}v + \int_S \boldsymbol{n} \cdot \bar{\boldsymbol{Y}}_1 \hat{\kappa}\, \mathrm{d}a. \tag{50}$$

Since, by $(29)_2$ and $(47)_2$, the r.h.s. of (50) is vanishing, Eq. (50) can be recognize to coincide with the kinematic shakedown criterion (48).

As for the *sufficiency* part of the theorem, let inequality (48) be satisfied, but in a more stringent form, say

$$\mathscr{E}_\beta[\hat{M}] := \sum_{i=1}^{i=N} \int_V \left[D(\hat{\boldsymbol{\varepsilon}}_i^{\mathrm{p}}, \hat{\kappa}_i) - \beta \boldsymbol{\sigma}_i^E : \hat{\boldsymbol{\varepsilon}}_i^{\mathrm{p}}\right] \mathrm{d}v$$
$$+ \beta \int_V \psi_{\mathrm{st}}(\hat{\kappa}, \nabla\hat{\kappa}) \mathrm{d}v \geq 0 \quad \forall \hat{M} \in \hat{\mathscr{M}}_A, \tag{51}$$

where $\beta > 1$ is some scalar coefficient. As the set $\hat{\mathscr{M}}_A$ can be thought of to include the trivially vanishing KSPs, for which $\mathscr{E}_\beta = 0$, it follows that the minimum problem:

$$\min_{\hat{M}} \mathscr{E}_\beta[\hat{M}] \quad \text{s.t. } \hat{M} \in \hat{\mathscr{M}}_A \tag{52}$$

admits a solution (possibly also in correspondence to a nontrivial set of KSPs). This implies that the relevant Euler–Lagrange equations also admit a solution. The characteristic structure of the latter equations is thus fundamental to reveal the physical/mechanical implications of condition (51). We can obtain the mentioned equations with the Lagrange multiplier method. For this purpose, let the compatibility constraints in (47) be appended to (51) so obtaining the augmented functional

$$\mathscr{E}_\beta^* = \mathscr{E}_\beta + \int_V (\beta\hat{\boldsymbol{\sigma}}^r) : \left(\nabla^{\mathrm{sym}}\hat{\boldsymbol{u}} - \sum_{i=1}^{i=N} \hat{\boldsymbol{\varepsilon}}_i^{\mathrm{p}}\right) \mathrm{d}v - \int_{S_c} (\boldsymbol{n} \cdot \beta\hat{\boldsymbol{\sigma}}^r) \cdot \hat{\boldsymbol{u}}\, \mathrm{d}a \tag{53}$$

where $\beta\hat{\boldsymbol{\sigma}}^r$ denotes a (symmetric) stress-like Lagrange multiplier. The other constraints in (47) are left as side conditions to (53). The first variation of (53), after some mathematics, can be cast as follows:

$$\delta\mathscr{E}_\beta^* = \sum_{i=1}^{i=N} \int_V \delta\hat{\boldsymbol{\varepsilon}}_i^{\mathrm{p}} : \left[\frac{\partial D}{\partial \hat{\boldsymbol{\varepsilon}}_i^{\mathrm{p}}} - \beta(\boldsymbol{\sigma}_i^E + \hat{\boldsymbol{\sigma}}^r)\right] \mathrm{d}v + \beta \int_{S_f} \boldsymbol{n} \cdot \hat{\boldsymbol{\sigma}}^r \cdot \delta\hat{\boldsymbol{u}}\, \mathrm{d}a$$

$$+\sum_{i=1}^{i=N}\int_V \delta\hat{\kappa}_i\left[\frac{\partial D}{\partial\hat{\kappa}_i}+\beta\left(\frac{\partial\psi_{\mathrm{st}}}{\partial\hat{\kappa}}-\nabla\cdot\frac{\partial\psi_{\mathrm{st}}}{\partial(\nabla\hat{\kappa})}\right)\right]\mathrm{d}v$$

$$+\sum_{i=1}^{i=N}\int_S \boldsymbol{n}\cdot\frac{\partial\psi_{\mathrm{st}}}{\partial(\nabla\hat{\kappa})}\delta\hat{\kappa}_i\,\mathrm{d}a-\beta\int_V \nabla\cdot\hat{\boldsymbol{\sigma}}^r\cdot\delta\hat{\boldsymbol{u}}\,\mathrm{d}v$$

$$+\beta\int_V \delta\hat{\boldsymbol{\sigma}}^r:\left(\nabla^{\mathrm{sym}}\hat{\boldsymbol{u}}-\sum_{i=1}^{i=N}\hat{\boldsymbol{\varepsilon}}^{\mathrm{p}}\right)\mathrm{d}v+\beta\int_{S_f}\boldsymbol{n}\cdot\hat{\boldsymbol{\sigma}}^r\cdot\delta\hat{\boldsymbol{u}}\,\mathrm{d}a. \tag{54}$$

The identically vanishing of $\delta\mathscr{E}^*_\beta$ for arbitrary variations gives the desired Euler–Lagrange equations as in the following:

$$\nabla\cdot\hat{\boldsymbol{\sigma}}^r=\boldsymbol{0}\ \text{ in } V,\quad \boldsymbol{n}\cdot\hat{\boldsymbol{\sigma}}^r=\boldsymbol{0}\ \text{ on } S_f \tag{55}$$

$$\hat{Y}_0:=\frac{\partial\psi_{\mathrm{st}}}{\partial\hat{\kappa}},\quad \hat{\boldsymbol{Y}}_1:=\frac{\partial\psi_{\mathrm{st}}}{\partial(\nabla\hat{\kappa})}\quad\text{in } V \tag{56}$$

$$\hat{Y}:=\hat{Y}_0-\nabla\cdot\hat{\boldsymbol{Y}}_1\quad\text{in } V \tag{57}$$

$$\boldsymbol{n}\cdot\hat{\boldsymbol{Y}}_1=0\quad\text{on } S_f^{(1)} \tag{58}$$

$$\beta(\boldsymbol{\sigma}_i^E+\hat{\boldsymbol{\sigma}}^r)=\frac{\partial D}{\partial\hat{\boldsymbol{\varepsilon}}_i^{\mathrm{p}}}\quad\text{in } V,\ \forall i\in I(N) \tag{59}$$

$$\beta\hat{Y}=-\frac{\partial D}{\partial\hat{\kappa}_i}\quad\text{in } V,\ \forall i\in I(N) \tag{60}$$

besides Eq. $(47)_2$ and $(47)_3$.

Equation (55) states that the Lagrange multiplier $\hat{\boldsymbol{\sigma}}^r$ proves to be a residual stress field. Equations (56)–(58) state that the primitive strengths $\hat{Y}_0, \hat{\boldsymbol{Y}}_1$, obtained as partial derivatives of $\psi_{\mathrm{st}}(\hat{\kappa},\nabla\hat{\kappa})$, provide a strengthening stress $\hat{Y}$ complying with the strengthening law, such that the set $\hat{M}=\{\hat{\boldsymbol{\sigma}}^r,\hat{Y},\hat{Y}_0,\hat{\boldsymbol{Y}}_1\}$ is a set of admissible SSPs, i.e. $\hat{M}\in\bar{\mathscr{M}}_A$. Moreover, the stress pairs $\{\beta(\boldsymbol{\sigma}_i^E+\hat{\boldsymbol{\sigma}}^r),\beta\hat{Y}\}, \forall i\in I(N)$, being each derived from the dissipation function D, satisfy the yield conditions, i.e.

$$f(\beta(\boldsymbol{\sigma}_i^E+\hat{\boldsymbol{\sigma}}^r),\beta\hat{Y})\le 0\quad\text{in } V,\ \forall i\in I(N) \tag{61}$$

Since $\beta>1$, the latter inequalities amount to

$$f(\boldsymbol{\sigma}_i^E+\hat{\boldsymbol{\sigma}}^r,\hat{Y})<0\quad\text{in } V,\ \forall i\in I(N) \tag{62}$$

which, by the static shakedown theorem, implies that shakedown certainly occurs. We can thus conclude that, under assumption of validity of the kinematic shakedown criterion (51) with $\beta > 1$, shakedown occurs. The same conclusion can be admitted to occur even at the limit for $\beta \to 1$, but then the structure finds itself in a shakedown limit state exposed to some kind of impending inadaptation collapse mode. The proof of the theorem is complete. □

In the absence of a strengthening potential, that is, if $\psi_{st} = 0$, the above shakedown theorem coincides—except for the proof technique—with the classical counterpart one for perfect plasticity [29].

6 Lower Bound and Upper Bound Theorems

The lower bound and upper bound theorems of classical shakedown theory [29] can be extended to strengthening materials. For this purpose, let the applied load be amplified by a factor $s > 0$, such that the related elastic stress response can be denoted with the symbol $s\boldsymbol{\sigma}^E(\boldsymbol{x}, t)$, hence $s\boldsymbol{\sigma}_i^E(\boldsymbol{x})$ denote the basic elastic stresses. We call s the "load multiplier".

The considered structure/load system is featured by a particular value of s, say s_c, such that shakedown certainly occurs under loads amplified by $s < s_c$, does not if $s > s_c$. s_c is called *shakedown limit load multiplier*. Its value is of obvious interest for applications. Under loads amplified by s_c, shakedown can still be thought of to occur, but the structure finds itself, correspondingly, in a shakedown limit state with an impending inadaptation collapse mode, either alternating plasticity, or ratchetting, also of practical interest.

6.1 Extended Lower Bound Theorem

Let us call *statically and plastically admissible* a load multiplier , say $\bar{s}$, one which can be associated to a set of admissible SSPs, say $\bar{M} = \{\bar{\boldsymbol{\sigma}}^r, \bar{Y}, \bar{Y}_0, \bar{\boldsymbol{Y}}_1\} \in \mathscr{M}_A$, such as to satisfy the static shakedown criterion (30), which here takes on the form:

$$\bar{f}_i := f(\bar{s}\boldsymbol{\sigma}_i^E + \bar{\boldsymbol{\sigma}}^r, \bar{Y}) \leq 0 \ \text{ in } V, \quad \forall i \in I(N). \tag{63}$$

By the static shakedown theorem of Sect. 4, this definition implies that shakedown certainly occurs under a load amplified by $\bar{s}$. The value of $s_c := \max \bar{s}$ is thus the maximum of s for which shakedown occurs. Formally we can write:

$$\left.\begin{array}{l} s_c = \max \ \bar{s} \qquad \text{subject to :} \\ \text{constraints in Eqs. (31) and (72)} \end{array}\right\} \tag{64}$$

It results that $\bar{s} \leq s_c$ and thus *any statically and plastically admissible load multiplier is a lower bound to* s_c. This is the (extended) lower bound theorem.

6.2 Extended Upper Bound Theorem

Let us consider a set of admissible KSPs, say $\hat{M} = \{\hat{\boldsymbol{\varepsilon}}^{\mathrm{p}}_1, \hat{\kappa}_1, \ldots, \hat{\boldsymbol{\varepsilon}}^{\mathrm{p}}_N, \hat{\kappa}_N, \hat{\boldsymbol{\varepsilon}}^{\mathrm{p}}, \hat{\kappa}, \hat{\boldsymbol{u}}\} \in \hat{\mathscr{M}}_A$ satisfying the *normalization condition*, that is

$$\sum_{i=1}^{i=N} \int_V \boldsymbol{\sigma}_i^E : \hat{\boldsymbol{\varepsilon}}_i^{\mathrm{p}} \,\mathrm{d}v = 1. \tag{65}$$

Then, we call *kinematically admissible* the load multiplier, say $\hat{s}$, given by

$$\hat{s} := \sum_{i=1}^{i=N} \int_V D(\hat{\boldsymbol{\varepsilon}}_i^{\mathrm{p}}, \hat{\kappa}_i) \,\mathrm{d}v + \int_V \psi_{\mathrm{st}}(\hat{\kappa}, \nabla\hat{\kappa}) \,\mathrm{d}v. \tag{66}$$

The upper bound theorem states that any such $\hat{s}$ cannot be smaller than s_c, the shakedown load multiplier.

In order to prove the latter statement, let us rewrite (66) in the equivalent form:

$$\begin{aligned} \mathscr{E}[\hat{M}] = \sum_{i=1}^{i=N} \int_V \left[D(\hat{\boldsymbol{\varepsilon}}_i^{\mathrm{p}}, \hat{\kappa}_i) \,\mathrm{d}v - s_c \boldsymbol{\sigma}_i^E : \hat{\boldsymbol{\varepsilon}}_i^{\mathrm{p}} \right] \mathrm{d}v \\ + \int_V \psi_{\mathrm{st}}(\hat{\kappa}, \nabla\hat{\kappa}) \,\mathrm{d}v = \hat{s} - s_c \end{aligned} \tag{67}$$

which is obtained from (66) by subtracting s_c from both sides and using (65). Since shakedown occurs under loads amplified by s_c, by the kinematic shakedown theorem the kinematic shakedown criterion functional $\mathscr{E}[\hat{M}]$ proves to be nonnegative for whatever KSPs, such that we have from (67) the inequality $\hat{s} \geq s_c$.

The shakedown load multiplier can alternatively be computed as the minimum upper bound value, i.e.

$$\left.\begin{aligned} s_c = \min \left(\sum_{i=1}^{i=N} \int_V D(\hat{\boldsymbol{\varepsilon}}_i^{\mathrm{p}}, \hat{\kappa}_i) \,\mathrm{d}v + \int_V \psi_{\mathrm{st}}(\hat{\kappa}, \nabla\hat{\kappa}) \,\mathrm{d}v \right) \\ \text{s.t. the constraints in Eqs. (47) and (65)} \end{aligned}\right\} \tag{68}$$

6.3 The Shakedown Limit Load Problem

In the previous subsections the shakedown limit load multiplier, s_c, has been formally expressed either as the maximum lower bound value, or as the minimum upper bound value. In principle, these maximum and minimum problems can be addressed for the numerical evaluation of s_c. But this problem proves to be quite more complex than in classical shakedown theory as a consequence of the strengthening effects to be accounted for.

Another way to address the above problem consists in writing out the complete set of equations governing the shakedown limit state of the structure, which is generated by the load being amplified by s_c. Under this load, the structure is exposed to an impending inadaptation (or non-instantaneous) collapse mode (either alternating plasticity, or ratchetting), which becomes an *actual inadaptation collapse mode* as soon as the shakedown multiplier slightly exceeds the limit value s_c. Addressing this equation set by means of FEM techniques may be a useful way to solve the shakedown load problem.

The equation set under discussion can be derived as the Euler–Lagrange equations associated either to problem (64), or (68). Let us proceed with the former. On appending the pertinent constraint equations to the negative objective function, we have the augmented functional:

$$\mathscr{L} = -s + \sum_{i=1}^{i=N} \int_V \kappa_i \left[f_0(s\boldsymbol{\sigma}_i^E + \boldsymbol{\sigma}^r) - Y - \sigma_0 \right] \mathrm{d}v + \int_V \kappa (Y - Y_0 + \nabla \cdot \boldsymbol{Y}_1)\, \mathrm{d}v - \int_{S_f^{(1)}} \kappa \boldsymbol{n} \cdot \boldsymbol{Y}_1 \kappa \, \mathrm{d}a + \int_V \nabla \cdot \boldsymbol{\sigma}^r \cdot \boldsymbol{u}\, \mathrm{d}v - \int_{S_f} \boldsymbol{n} \cdot \boldsymbol{\sigma}^r \cdot \boldsymbol{u}\, \mathrm{d}a + \int_V \mu \phi(Y_0, \boldsymbol{Y}_1)\, \mathrm{d}v \tag{69}$$

where the upper bars have been eliminated for simplicity. The fields $\boldsymbol{u}$, $\kappa_i \geq 0$, κ, $\mu \geq 0$ are suitable Lagrange multipliers. Writing the first variation of (69) and with some straightforward mathematics (details are skipped for brevity), we can arrive at the pertinent Euler–Lagrange equations. These read:

$$\nabla \cdot \boldsymbol{\sigma}^r = \boldsymbol{0} \text{ in } V, \quad \boldsymbol{n} \cdot \boldsymbol{\sigma}^r = \boldsymbol{0} \text{ on } S_f \quad \text{(self stresses)} \tag{70}$$

$$\boldsymbol{\sigma}_i^M := s_c \boldsymbol{\sigma}_i^E + \boldsymbol{\sigma}^r \quad \text{in } V,\ \forall i \in I(N) \quad \text{(Melan stresses)} \tag{71}$$

$$f_i := f_0(\boldsymbol{\sigma}_i^M) - Y - \sigma_0 \leq 0,\ \kappa_i \geq 0,\ \kappa_i f_i = 0 \quad \text{in } V,\ \forall i \in I(N) \tag{72}$$

$$\boldsymbol{\varepsilon}_i^{\mathrm{p}} = \kappa_i \frac{\partial f_i}{\partial \boldsymbol{\sigma}_i^M} \quad \text{in } V,\ \forall i \in I(N) \tag{73}$$

$$\left.\begin{array}{l} \boldsymbol{\varepsilon}^{\mathrm{p}} = \sum_{i=1}^{i=N} \boldsymbol{\varepsilon}_i^{\mathrm{p}}, \quad \text{in } V \\ \boldsymbol{\varepsilon}^{\mathrm{p}} = \nabla^{\mathrm{sym}} \boldsymbol{u} \ \text{ in } V, \quad \boldsymbol{u} = \boldsymbol{0} \ \text{ on } S_c \end{array}\right\} \quad \text{(compatibility conds.)} \tag{74}$$

$$\sum_{i=1}^{i=N} \int_V \boldsymbol{\sigma}_i^E : \boldsymbol{\varepsilon}_i^{\mathrm{p}} \,\mathrm{d}v = 1 \quad \text{(normalization condition)} \tag{75}$$

$$\kappa = \sum_{i=1}^{i=N} \kappa_i \quad \text{in } V \quad \text{(total effective plastic strain)} \tag{76}$$

$$\left.\begin{array}{l} \phi(Y_0, \boldsymbol{Y}_1) \leq 0, \;\; \mu \geq 0, \;\; \mu\phi = 0 \\ \kappa = \dfrac{\partial \phi}{\partial Y_0}, \quad \nabla\kappa = \dfrac{\partial \phi}{\partial \boldsymbol{Y}_1} \end{array}\right\} \quad \text{in } V \tag{77}$$

$$\left.\begin{array}{l} Y = Y_0 - \nabla \cdot \boldsymbol{Y}_1 \quad \text{in } V \\ \kappa = 0 \;\; \text{on } S_c^{(1)}, \quad \boldsymbol{n} \cdot \boldsymbol{Y}_1 \quad \text{on } S_f^{(1)} \end{array}\right\} \quad \text{(strengthening law)} \tag{78}$$

The following can be remarked:

1. The vector-valued Lagrange multiplier $\boldsymbol{u}$ takes on the meaning of displacement field, the analogous scalars, κ_i and μ take the meanings of consistency coefficients in the yield conditions (72) and strengthening conditions (77), whereas κ has the meaning of total effective plastic strain, Eq. (76). Equation (73) provides the relationship between κ_i and $\boldsymbol{\varepsilon}_i^{\mathrm{p}}$. The stresses in (71) are conventionally called "Melan stresses".
2. Equations (70)–(74) are typical of classical shakedown, except for the presence of the strengthening stress, Y, in the yield conditions (72). The type of impending collapse mode is revealed by the strain ratchet, $\boldsymbol{\varepsilon}^{\mathrm{p}}$, according to whether it is nonvanishing, at least somewhere in V, (in which case we have *ratchetting* behavior), or instead $\boldsymbol{\varepsilon}^{\mathrm{p}} = \boldsymbol{0}$ everywhere in V, hence $\boldsymbol{u} \equiv \boldsymbol{0}$, (in which case we have *alternating plasticity* behavior, or *plastic shakedown*).
3. The presence of the strengthening effects manifests itself also through the additional equation set (76)–(78), which describes the strengthened state of the material and its coupling with plastic deformations. Since, by (77), Y_0 and $\boldsymbol{Y}_1$ prove to be some specific functions of κ and $\nabla\kappa$ like in (2), then Eq. $(78)_1$ constitutes a second order PDE relating Y to the total effective plastic strain κ. This PDE is accompanied by the higher order boundary conditions $(78)_2$. Equation (76) relate the mentioned κ to the plastic strains $\boldsymbol{\varepsilon}_i^{\mathrm{p}}$ through (73).
4. The maximum and minimum problems (64) and (68) represent true variational principles associated to the shakedown limit load problem (70)–(78). It can be proved that *the shakedown limit load problem has a unique solution* (if it exists)—except perhaps for the residual stresses $\boldsymbol{\sigma}^r$ in the region (if any) of V where the material remains plastically undeformed (see [51] for details).
5. The shakedown limit load problem (70)–(78) is not an evolution problem, since in fact no stress redistribution is demanded. The (fictitious) plastic strains associated to every basic load sum up with one another to give rise to the *impending inadaptation collapse mechanism* described by the strain field $\boldsymbol{\varepsilon}^{\mathrm{p}}$, *compatible* with the displacements $\boldsymbol{u}$ vanishing on S_c. Since the above plastic strains, $\boldsymbol{\varepsilon}^{\mathrm{p}}$, are

formed up by contributions occurring in different times during the load cycle, the collapse mechanism cannot be an instantaneous one. The load can be further amplified causing the structure to exit from the shakedown domain and enter one in which actual plastic strains are produced under every load cycle. Was the load proportional ($N = 1$), the collapse mechanism would be instantaneous and the shakedown limit state would be a plastic collapse state.

6. The latter features are qualitatively the same as for a material without strengthening effects. These effects just cause an increase (or decrease, possibly) of the shakedown limit load multiplier, in more or less extent depending on the size of the considered body. The knowledge of the law in which the shakedown limit load multiplier varies with the body's size is of fundamental importance for applications.

7 Application

For an application of the theory expounded in the preceding sections, a simple structural scheme typical of classical shakedown theory is considered, but in a miniaturized configuration. The structural system is formed by two vertical bars of lengths L_1 and $L_2 = cL_1$ and cross sections $A_1 = A$, $A_2 = \omega A$. The bar upper ends are fixed, the lower ones are connected to a same vertically movable rigid block, as depicted in Fig. 2.

The geometry of the sample is so scaled such as to make its greatest linear dimension (L_1) comparable with the material internal length parameter, ℓ. Each bar has its extremes clamped to substrates impeding the onset of plastic strains (hard boundaries), hence *no uniform plastic strain distributions are permitted in each bar*. This bar model is elastic perfectly plastic and exhibits strengthening effects; it

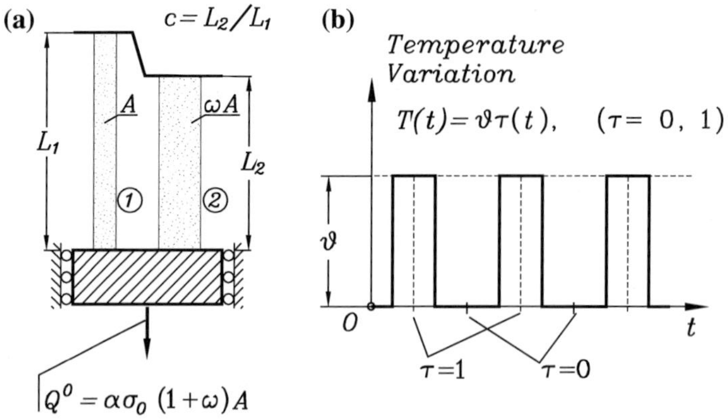

Fig. 2 Micro-scale two–bar system subjected to thermal cycles in bar 2 and to a permanent load Q^0 upon the rigid block: **a** Geometrical sketch; **b** Temperature variation as a function of time

can be considered the homogenized equivalent of a series of crystals with interfaces (similar to one used by Aifantis et al. [4], although for a different purpose).

Bar 2 is subjected to cyclically varying temperature, $T(t) = \theta\tau(t)$, with $\theta > 0$ and $\tau(t)$ taking cyclically the values 0 and 1. The rigid block is subjected to a permanent load, say $Q^0 = \alpha\sigma_0(1+\omega)A$, where α is a scalar parameter. The problem to solve consists in finding the maximum value of θ such that the structural system shakes down under the action of the thermal cycles, while the permanent load in taken fixed. In doing this, the ability of the material to strengthen has to be considered. This task can be achieved either addressing the maximum problem (64), or the minimum problem (68), or even the equation set (70)–(78). Here the latter way is followed, with the proviso that the solution is searched for within the framework of the particular case presented in Sect. 2.2.

The thermo-elastic stresses in the bars prove to be:

$$\sigma_1^E(t) = \sigma^\theta \tau(t) + \alpha\sigma_0 m_1, \quad \sigma_2^E = -(\sigma^\theta/\omega)\tau(t) + \alpha\sigma_0 m_2, \quad \forall t \geq 0 \tag{79}$$

where

$$\sigma^\theta := \frac{\alpha_T \theta E \omega c}{\omega + c}, \quad m_1 := \frac{(1+\omega)c}{\omega + c}, \quad m_2 := m_1/c; \tag{80}$$

moreover, α_T denotes the thermal expansion coefficient; E, the Young modulus.

Noting that the residual stresses in (70) can be written in terms of an arbitrary scalar R as

$$\sigma_1^r = R\sigma_0, \qquad \sigma_2^r = -R\sigma_0/\omega, \tag{81}$$

and with the position $\beta := \sigma^\theta/\sigma_0$, the Melan stresses in (71) become:

$$\left.\begin{array}{ll} \sigma^M_{1(1)}/\sigma_0 = \beta + \alpha m_1 + R, \quad \sigma^M_{2(1)}/\sigma_0 = -\frac{\beta}{\omega} + \alpha m_2 - \frac{R}{\omega}, & (\tau = 1) \\ \sigma^M_{1(0)}/\sigma_0 = \alpha m_1 + R, \quad \sigma^M_{2(0)}/\sigma_0 = \alpha m_2 - \frac{R}{\omega}, & (\tau = 0) \end{array}\right\} \tag{82}$$

The yield conditions (72) read, with the notation $Y_i + \sigma_0 = \sigma_y := \sigma_0\varphi_i$, $(i = 1, 2)$:

$$|\sigma^M_{i(\tau)}| \leq \sigma_0\varphi_i, \quad (i = 1, 2; \; \tau = 1, 0). \tag{83}$$

Obviously, $\max\beta \to \max\theta$. The plot of $\max\beta$ (the latter being identified with the shakedown limit s_c) as a function of α is the shakedown contour line in the (α, β)-plane (the so-called Bree-like diagram). On the basis of previous experience on shakedown analysis, we can envision two distinct situations as follows.

7.1 Small α Values: Alternating Plasticity

For small α values, the shakedown limit state of the system is mainly conditioned by the cyclic nature of the thermal loading; it in fact exhibits an alternating plasticity behavior, in which bar 1 suffers reverse plasticity, whereas bar 2 remains elastic.

Then, inequalities (83) written for $i = 1$ and enforced as equalities can be written as

$$\left.\begin{aligned} \sigma^M_{1(1)}/\sigma_0 &= \beta + \alpha m_1 + R = \varphi_1 \\ \sigma^M_{1(0)}/\sigma_0 &= \alpha m_1 + R = -\varphi_1 \end{aligned}\right\} \tag{84}$$

from where we get

$$\beta = \beta_{al} := 2\varphi_1, \qquad R = -\varphi_1 - \alpha m_1. \tag{85}$$

This result holds as far as bar 2 remains elastic, i.e., by (83) written for $i = 2$,

$$\sigma^M_{2(0)}/\sigma_0 = \alpha m_2 - \frac{R}{\omega} \leq \varphi_2, \tag{86}$$

which, eliminating R by means of $(85)_2$, yields

$$\alpha \leq \hat{\alpha} := \frac{\omega}{1+\omega}\varphi_2 - \frac{1}{1+\omega}\varphi_1 \geq 0. \tag{87}$$

The condition $\sigma^M_{2(1)}/\sigma_0 \geq -\varphi_2$ is always satisfied for $\alpha > 0$, as it can be easily verified.

Since $\varepsilon^{\mathrm{p}}_{1(1)} = -\varepsilon^{\mathrm{p}}_{1(0)}$ all along bar 1, by (74) we have $\varepsilon^{\mathrm{p}}_1 = \varepsilon^{\mathrm{p}}_{1(1)} + \varepsilon^{\mathrm{p}}_{1(0)} = 0$, $\varepsilon^{\mathrm{p}}_2 = \varepsilon^{\mathrm{p}}_{2(1)} = \varepsilon^{\mathrm{p}}_{2(0)} = 0$ (alternating plasticity), hence $u_b = 0$ (no ratchet displacement of the rigid block). The total accumulated plastic strain is $\kappa_1 = 2\varepsilon^{\mathrm{p}}_{1(1)}$ in bar 1 and $\kappa_2 = 0$ in bar 2.

For the evaluation of the plastic strain profile in bar 1, i.e. the function $\varepsilon^{\mathrm{p}}_{1(1)}(\eta)$, where $\eta := y/L_1$ is the adimensional abscissa over bar 1, $-1/2 \leq \eta \leq 1/2$, we follow [50]. Remembering $(19)_1$, Eq. $(85)_1$ proves to be the differential equation:

$$\varphi_1(\kappa_1, \nabla\kappa_1) = \frac{\xi_1 \varepsilon^{\mathrm{p}}_{1(1)}}{p_1} - \frac{1}{\xi_1}\left(\frac{\varepsilon^{\mathrm{p}'}_{1(1)}}{p_1}\right)' = s, \qquad (s := \beta_{\mathrm{al}}/2) \tag{88}$$

where $(\cdot)' = \partial_\eta$, $\xi_1 := L_1/\ell$ (*size ratio*), and

$$p_1 = \sqrt{(\xi_1 \varepsilon^{\mathrm{p}}_{1(1)})^2 + (\varepsilon^{\mathrm{p}'}_{1(1)})^2}. \tag{89}$$

Due to the symmetry of the plastic strain profile, the differential equation (88) can be integrated in the interval $0 \leq \eta \leq 1/2$, with the higher order boundary conditions $\varepsilon^{\mathrm{p}'}_{1(1)}(0) = 0$ and $\varepsilon^{\mathrm{p}}_{1(1)}(1/2) = 0$. With the same numerical procedure used in [50] for the plastic limit analysis problem, let us pose:

$$\varepsilon^{\mathrm{p}'}_{1(1)}/\varepsilon^{\mathrm{p}}_{1(1)} = -\xi_1 \tan\Theta(\eta) \tag{90}$$

where $\Theta(\eta)$ is an unknown function required to satisfy the boundary conditions $\Theta(0) = 0$ and $\Theta(1/2) = \pi/2$. By (90), Eq. (88) can be transformed into the integral equation

$$\Theta(\eta) = \frac{\pi + \xi_1}{2} \frac{\int_0^\eta \sec\Theta(t)\mathrm{d}t}{\int_0^{1/2} \sec\Theta(t)\mathrm{d}t} - \xi_1\eta \tag{91}$$

which satisfies the mentioned boundary conditions on Θ and s is given by

$$s = \beta_{al}/2 = \frac{1 + \pi/\xi_1}{2\int_0^{1/2} \sec\Theta(t)\mathrm{d}t}. \tag{92}$$

Equation (91) has been solved by means of an iterative procedure of the type:

$$\Theta_n(\eta) = \frac{\pi + \xi_1}{2} \frac{\int_0^\eta \sec\Theta_{n-1}(t)\mathrm{d}t}{\int_0^{1/2} \sec\Theta_{n-1}(t)\mathrm{d}t} - \xi_1\eta. \tag{93}$$

Next, Eq. (90), by an integration, gives

$$\varepsilon^{\mathrm{p}}_{1(1)}(\eta) = C\exp\left[-\xi_1 \int_0^\eta \tan\Theta(t)\mathrm{d}t\right], \quad (0 \le \eta \le 1/2) \tag{94}$$

which represents the plastic strain profile of bar 1 in the thermal loading period and satisfies the higher order boundary condition $\varepsilon^{\mathrm{p}}_{1(1)}(1/2) = 0$.

Let us note that the normalization condition (75) involves only the thermoelastic stresses written for $\beta = 1$ and $\alpha = 0$, that is the stresses:

$$\tilde{\sigma}^E_{1(1)} = \sigma_0, \quad \tilde{\sigma}^E_{2(1)} = -\frac{\sigma_0}{\omega}, \quad \tilde{\sigma}^E_{1(0)} = \tilde{\sigma}^E_{2(0)} = 0. \tag{95}$$

Since only bar 1 deforms plastically in the shakedown limit state being considered, the normalization condition reduces to:

$$\begin{aligned} 2AL_1 \int_0^{1/2} \tilde{\sigma}_{1(1)}\varepsilon^{\mathrm{p}}_{1(1)}\mathrm{d}\eta &= \sigma_0 AL_1\langle\varepsilon^{\mathrm{p}}_{1(1)}\rangle \\ &= 2\sigma_0 AL_1 C \int_0^{1/2} \exp[-\xi_1 \int_0^\eta \tan\Theta(t)\mathrm{d}t]\mathrm{d}\eta = 1[\text{work unit}] \end{aligned} \tag{96}$$

where $\langle\varepsilon^{\mathrm{p}}_{1(1)}\rangle$ denotes the mean value of $\varepsilon^{\mathrm{p}}_{1(1)}$ over the bar length, L_1. Equation (96) then gives

$$C = \frac{1[\text{work unit}]}{2\sigma_{0AL_1}\int_0^{1/2} \exp\left[-\xi_1 \int_0^\eta \tan\Theta(t)\mathrm{d}t\right]\mathrm{d}\eta}. \tag{97}$$

7.2 Higher α Values: Ratchetting

Equation (87) gives the value $\hat{\alpha}$ of the permanent load at which the transition from alternating plasticity to ratchetting occurs. For $\alpha > \hat{\alpha}$, the shakedown limit state exhibits a ratchetting behavior, namely, both bars exceed—in the two half cycles, respectively—the plastic limit in traction, but not in compression. Thus, enforcing the yield conditions (83) written for $i = 1, \tau = 1$ and then for $i = 2, \tau = 0$, we can write:

$$\left.\begin{aligned} \sigma^M_{1(1)}/\sigma_0 &= \beta + \alpha m_1 + R = \varphi_1 \\ \sigma^M_{2(0)}/\sigma_0 &= \alpha m_2 - \tfrac{R}{\omega} = \varphi_2 \end{aligned}\right\} \tag{98}$$

from where we have, remembering $(80)_2$ and $(80)_3$,

$$\frac{\beta}{1+\omega} + \alpha = \frac{1}{1+\omega}\varphi_1 + \frac{\omega}{1+\omega}\varphi_2. \tag{99}$$

This equation represents a straight line of the (α, β)-plane, intersecting the α axis at the point of abscissa

$$\alpha = \alpha_L := \frac{1}{1+\omega}\varphi_1 + \frac{\omega}{1+\omega}\varphi_2. \tag{100}$$

The scalar α_L specifies the plastic collapse load of the system subjected to the permanent load only.

By (74) we have $\varepsilon^p_1 = \varepsilon^p_{1(1)}, \varepsilon^p_2 = \varepsilon^p_{2(0)}, \varepsilon^p_{1(0)} = \varepsilon^p_{2(1)} = 0$ (ratchetting), and thus $\kappa_1 = \varepsilon^p_{1(1)}, \kappa_2 = \varepsilon^p_{2(0)}$. In order to evaluate the $\varepsilon^p_{1(1)}$ and $\varepsilon^p_{2(2)}$ profiles, by $(19)_1$ and (98) we can write, respectively,

$$\varphi_1(\kappa_1, \nabla\kappa_1) = \frac{\xi_1 \varepsilon^p_{1(1)}}{p_1} - \frac{1}{\xi_1}\left(\frac{\varepsilon^{p'}_{1(1)}}{p_1}\right)' = s_1 \tag{101}$$

$$\varphi_2(\kappa_2, \nabla\kappa_2) = \frac{\xi_2 \varepsilon^p_{2(0)}}{p_2} - \frac{1}{\xi_2}\left(\frac{\varepsilon^{p'}_{2(0)}}{p_2}\right)' = s_2 \tag{102}$$

where

$$p_1 = \sqrt{(\xi_1 \varepsilon^p_{1(1)})^2 + (\varepsilon^{p'}_{1(1)})^2}, \quad p_2 = \sqrt{(\xi_2 \varepsilon^p_{2(0)})^2 + (\varepsilon^{p'}_{2(0)})^2}. \tag{103}$$

The differential equations (101) and (102) are both similar to (88) and can be solved in a similar way. We can thus write, correspondingly:

$$\Theta_i(\eta) = \frac{\pi + \xi_i}{2} \frac{\int_0^\eta \sec\Theta_i(t)\mathrm{d}t}{\int_0^{1/2} \sec\Theta_i(t)\mathrm{d}t} - \xi_i \eta, \quad (i = 1, 2) \tag{104}$$

$$s_i = \frac{1 + \pi/\xi_i}{2\int_0^{1/2} \sec \Theta_i(t)\mathrm{d}t}, \quad (i = 1, 2) \tag{105}$$

$$\varepsilon^{\mathrm{p}}_{1(1)} = C_1 \exp\left[-\xi_1 \int_0^{\eta} \tan \Theta_1(t)\mathrm{d}t\right], \tag{106}$$

$$\varepsilon^{\mathrm{p}}_{2(0)} = C_2 \exp\left[-\xi_2 \int_0^{\eta} \tan \Theta_2(t)\mathrm{d}t\right], \tag{107}$$

and thus, by (87), (101) and (102), it is

$$\hat{\alpha} = \frac{1}{1+\omega}(\omega s_2 - s_1), \quad \alpha_L = \frac{1}{1+\omega}(\omega s_2 + s_1). \tag{108}$$

Noting that the normalization condition is formally similar to (96), we can write for C_1 a formula similar to (97), i.e.

$$C_1 = \frac{1[\text{work unit}]}{2\sigma_0 A L_1 \int_0^{1/2} \exp\left[-\xi_1 \int_0^{\eta} \tan \Theta_1(t)\mathrm{d}t\right]\mathrm{d}\eta}. \tag{109}$$

Substituting from the latter into (106), the mean value of $\varepsilon^{\mathrm{p}}_{1(1)}$ can be expressed as

$$\langle \varepsilon^{\mathrm{p}}_{1(1)} \rangle = \frac{1}{\sigma_0 A L_1} = \frac{u_b}{L_1} \tag{110}$$

where u_b is the (nonvanishing) ratchet displacement of the rigid block. Analogously, the mean value of $\varepsilon^{\mathrm{p}}_{2(0)}$ of (107) in L_2 can be expressed as

$$\langle \varepsilon^{\mathrm{p}}_{2(0)} \rangle = 2C_2 \int_0^{1/2} \exp\left[-\xi_2 \int_0^{\eta} \tan \Theta_2(t)\,\mathrm{d}t\right]\mathrm{d}\eta = \frac{u_b}{L_2} \tag{111}$$

which, by the aid of (110), gives

$$\begin{aligned} C_2 &= \frac{u_b}{2L_2 \int_0^{1/2} \exp\left[-\xi_2 \int_0^{\eta} \tan \Theta_2(t)\mathrm{d}t\right]\mathrm{d}\eta} \\ &= \frac{1}{2\sigma_0 A L_2 \int_0^{1/2} \exp\left[-\xi_2 \int_0^{\eta} \tan \Theta_2(t)\mathrm{d}t\right]\mathrm{d}\eta} \end{aligned} \tag{112}$$

from where we obtain C_2 and thus, by (107), the plastic strain profile in bar 2.

7.3 Results and Comments

On the basis of the numerical evaluations described above, the solution of the shakedown problem at hand can be summarized as in the following.

In the (α, β)-plane, the shakedown contour line is a bilateral segment $A - B - C$, with co-ordinates $A(\alpha_L, 0)$, $B(\hat{\alpha}, \beta_{al})$, $C(0, \beta_{al})$. The position of this line depends on the value of ℓ. The line position closest to the axes origin corresponds to $\ell = 0$ (simple material), for which $\varphi_1 = \varphi_2 = 1$. The Bree-like diagrams of Fig. 3 looked like the classical one (see e.g. [16, 56]). For any load below to this curve, shakedown certainly occurs. At points upon the curve, the structure finds itself in a shakedown limit state, subjected to an impending inadaptation collapse mode. The latter is an alternating plasticity collapse mode for loads upon the "upper plateau" BC (where the outward unit normal is parallel to the β-axis), but it is a ratchetting collapse mode for loads upon the discendent branch AB (where the outward unit normal has a nonzero α-component). What is interesting in the present result is that the shakedown contour line ABC changes location with changing the ratio ζ. For $\zeta = 0$ (simple material), the contour line lies upon its standard position $A_0B_0C_0$, whereas it expands with increasing ζ, from $A_0B_0C_0$ to $A_1B_1C_1$ for $\zeta = 1$.

The plots of Fig. 4 provide the ways the values of $\beta_{\rm al}$ (the alternating plasticity limit load multiplier) and α_L (plastic collapse limit load for $\beta = 0$) vary with varying the ratio ζ. An almost linear variation law is recognizable. For ζ increasing from zero to one, the load multiplier increases of almost three times.

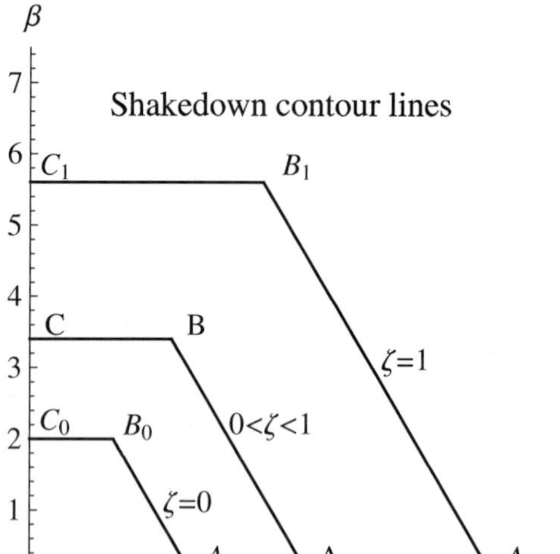

Fig. 3 Bree-like diagrams in the (α, β)-plane, showing the shakedown contour lines locations at increasing values of the ratio $\zeta = 1/\xi = \ell/L_1$, from $A_0B_0C_0$ at $\zeta = 0$ (simple material) to $A_1B_1C_1$ for $\zeta = 1$ (micro-scale sizes)

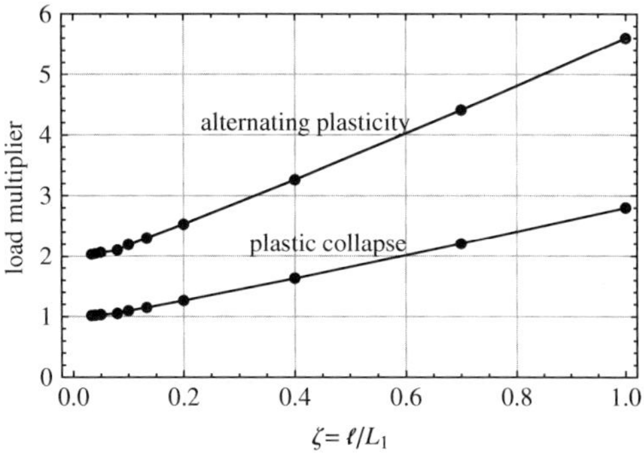

Fig. 4 Plots representing the alternating plasticity multiplier β_{al} and the plastic collapse multiplier α_L as functions of the ratio $\zeta = 1/\xi_1 = \ell/L_1$. An almost linear law is recognizable with an increase of the load multiplier of almost three times as ζ increases from 0 to 1

8 Conclusions

An elastic perfectly plastic material model with strengthening effects (Hall–Petch effects) has been considered and the classical shakedown theorems have been extended to such materials. A micro-scale specimen subjected to cyclic thermal loads combined with a permanent load clearly indicates that the shakedown limit load increases with decreasing size and that the shakedown load boundary expands correspondingly, in a way in which the Hall–Petch effects are classically known to manifest themselves (see the literature quoted in Sect. 1). A main result of the present work is that the gradient plasticity model advanced in [50] is able to predict Hall–Pech type effects also within a contest of cyclic-type loadings.

The present work is founded on the assumptions that (i) the cyclic nature of loadings, typical of shakedown analysis, does not modify the Hall–Petch effects in their own way to manifest themselves under monotonic loads, and (ii) no interaction between these effects and temperature exists, at least within certain limits of interest. Although these assumptions appear to be quite reasonable, specific laboratory experiments to probe them are hoped for. There is additionally a need for more efforts in order to derive, from the present theoretical study, suitable computational methods for practical applications to micro/nano technologies. These issues are left open to future research.

References

1. Abu Al-Rub RK (2008) Interfacial gradient plasticity governs scale-dependent yield strength and strain hardening rates in micro/nano structural metals. Int J Plast 24:1277–1306
2. Abu Al-Rub RK, Voyiadjis GZ, Aifantis EC (2009) On the thermodynamics of higher-order gradient plasticity for size effects at the micron and submicron length scale. Int J Mater Prod Technol 34:172–187
3. Aifantis EC (2003) Update on a class of gradient theories. Mech Mater 35:259–280
4. Aifantis KE, Soer WA, De Hosson JThM, Willis JR (2006) Interfaces within strain gradient plasticity: theory and experiments. Acta Materialia 54:5077–5085

5. Anand L, Gurtin ME, Lele SP, Gething C (2005) A one-dimensional theory of strain-gradient plasticity: formulation, analysis, numerical results. J Mech Phys Solids 53:1789–1826
6. Borg U (2007) A strain gradient crystal plasticity analysis of strain gradient effects in polycrystals. Eur J Mech A/Solids 26:313–324
7. Borino G, Polizzotto C (2007) A thermodynamically consistent gradient plasticity theory and comparison with other formulations. Model Mater Sci Eng 15:23–35
8. Ceradini G (1980) Dynamic shakedown in elastic-plastic bodies. J Eng Mech Div, ASCE 106:481–499
9. Débordes O, Nayroles B (1976) Sur la théorie et le calcul à l'adaptation des structures élastoplastiques. J Mécanique 15:1–53
10. Espinosa HD, Prorok BC, Peng B (2004) Plasticity size effects in free-standing submicron polycrystalline FCC films subjected to pure tension. J Mech Phys Solids 52:667–689
11. Feng X-Q, Sun Q (2007) Shakedown analysis of shape memory structures. Int J Plast 23:183–206
12. Fleck NA, Hutchinson JW (1997) Strain gradient plasticity. Adv Appl Mech 33:295–361
13. Fleck NA, Muller GM, Ashby MF, Hutchinson JW (1994) Strain gradient plasticity: theory and experiments. Acta Metall Mater 42:475–487
14. Fleck NA, Willis JR (2008) A mathematical basis for strain-gradient plasticity theory—Part I: scalar plastic multiplier. J Mech Phys Solids 57:161–177
15. Fredriksson P, Gudmundson P (2005) Size-dependent yield strength of thin films. Int J Plast 21:1834–1854
16. Fuschi P, Polizzotto C (1995) The shakedown load boundary of an elastic-perfectly plastic structure. Meccanica 30:155–174
17. Gokhfeld DA, Cherniavsky DF (1980) Limit analysis of structures at thermal cycling. Sijthoff and Noordhoff, Alphen aan der Rijn
18. Gudmundson P (2004) A unified treatment of strain gradient plasticity. J Mech Phys Solids 52:1377–1406
19. Gurtin ME, Anand L (2005) A theory of strain-gradient plasticity for isotropic, plastically irrotational materials. Part I: small deformation. J Mech Phys Solids 53:1624–1649
20. Gurtin ME, Anand L, Lele SP (2007) Gradient single-crystal plasticity with free energy dependent on dislocation density. J Mech Phys Solids 55:1853–1878
21. Gurtin ME, Fried E, Anand L (2010) The mechanics and thermodynamics of continua. Cambridge University Press, Cambridge
22. Halphen B (1979) Adaptation of elasto-visco-plastic structures. In: Matériax et Structures Sous Chargement Cyclique Association Amicale des Ingénieurs Ancien Elèves. Ponts et Chaussés, Paris, pp 203–229
23. Halphen B, Nguyen QS (1979) Sur les matériaux standards dénéralisés. J Mécanique 14:39–63
24. Hansen N (2004) Hall-Petch relation and boundary strengthening. Scripta Mater 51:801–806
25. Haque MA, Saif MTA (2003) Strain gradient effects in nanoscale thin films. Acta Mater 51:3053–3061
26. Huang H, Spaepan F (2000) Tensile testing of free-standing Cu, Ag and Al thin films and Ag/Cu multilayers. Acta Mater 48:3261–3269
27. Hutchinson JW (2000) Plasticity at micron scale. Int J Solids Struct 37:225–238
28. Kamenjarzh J, Merzljakov A (1994) On kinematic method in shakedown theory: I. Duality of extremum problems; II. Modified kinematic method. Int J Plast 10:363–392
29. Koiter WT (1960) General theorems for elastic-plastic solids. In: Hill R, Sneddon I (eds) Progress in solid mechanics, I. North-Holland, The Netherlands, pp 167–221
30. König JA (1987) Shakedown of elastic plastic structures. Elsevier, Amsterdam
31. König JA, Maier G (1981) Shakedown analysis of elastoplastic structures: a review of recent developments. Nucl Eng Des 66:81–95
32. Lele SP, Anand L (2008) A small-deformation strain-gradient theory for isotropic viscoplastic materials. Phil Mag 88:3655–3689
33. Lemaitre J, Chaboche J-L (1990) Mechanics of solid materials. Cambridge University Press, Cambridge

34. Maier G (2001) On some issues of shakedown analysis. J Appl Mech, ASME 68:799–808
35. Maier G, Carvelli V, Cocchetti G (2000) On direct methods for shakedown and limit analysis. Eur J Mech A/Solids 19:79–100
36. Martin JB (1975) Plasticity: fundamentals and general results. MTI Press, Cambridge
37. Mróz Z, Weichert D, Dorosz S (eds) (1995) Inelastic behaviour of structures under repeated loads. Kluwer Academic Publishers, Dordrecht
38. Pham DC (1996) Dynamic shakedown and a reduced kinematic theorem. Int J Plast 12:1055–1068
39. Pham DC (2003) Shakedown theory for elastic-perfectly plastic bodies revisited. Int J Mech Sci 45:1011–1027
40. Pham DC (2005) Shakedown static and kinematic theorems for elastic-plastic limited linear kinematic hardening solids. Eur J Mech A/Solids 24:35–45
41. Polizzotto C (1982) A unified treatment of shakedown theory and related bounding techniques. Solid Mech Arch 7:19–75
42. Polizzotto C (1984) On shakedown of structures under dynamic agencies. In: Sawczuk A, Polizzotto C (eds) Inelastic analysis under variable repeated loads. Cogras, Palermo
43. Polizzotto C (1993) On the conditions to prevent plastic shakedown: Part I-Theory; Part II-The plastic shakedown limit load. J Appl Mech, ASME 60:15–25
44. Polizzotto C (2003) Unified thermodynamic framework for nonlocal/gradient continuum theories. Eur J Mech A/Solids 22:651–668
45. Polizzotto C (2007) Strain-gradient elastic-plastic material models and assessment of the higher order boundary conditions. Eur J Mech A/Solids 26:189–211
46. Polizzotto C (2008) Thermodynamics-based gradient plasticity theories with an application to interface models. Int J Solids Struct 45:4820–4834
47. Polizzotto C (2008) Shakedown theorems for elastic-plastic solids in the framework of gradient plasticity. Int J Plast 24:218–241
48. Polizzotto C (2009) Interfacial energy effects in the framework of strain gradient plasticity. Int J Solids Struct 46:1685–1694
49. Polizzotto C (2009) A link between the residual-based gradient plasticity theory and the analogous theories based on the virtual work principle. Int J Plast 25:2169–2180
50. Polizzotto C (2010) Strain gradient plasticity, strengthening effects and plastic limit analysis. Int J Solids Struct 47:100–112
51. Polizzotto C (2010) Shakedown analysis for a class of strengthening materials within the framework of gradient plasticity. Int J Plast 26:1050–1069
52. Polizzotto C (2011) A unified residual-based thermodynamic framework for strain gradient theories of plasticity. Int J Plast 27:388–413
53. Polizzotto C, Borino G (1998) A thermodynamics-based formulation of gradient dependent plasticity. Eur J Mech A/Solids 17:741–761
54. Polizzotto C, Borino G (2014) Shakedown under thermo-mechanical loads. In: Hetnarski RB (ed) Encyclopedia of thermal stresses, vol 8. Springer, Dordrecht, pp 4317–4333
55. Polizzotto C, Borino G, Fuschi P (2000) Shakedown of cracked bodies with nonlocal elasticity. In: CD Proceedings of ECCOMAS, Barcelona, September 2000
56. Ponter ARS (1983) Shakedown and ratchetting below the creep range commission of the European communities, Report EUR 8702 EN, Brussels, Belgium
57. Ponter ARS (2002) A linear matching method for shakedown analysis. In: Weichert D, Maier G (eds) Inelastic behaviour of structures under variable repeated loads, Springer, Wien
58. Ponter ARS, Karadenitz S (1985) An extended shakedown theory for structures that suffer cyclic thermal loadings. Part I-Theory. J Appl Mech 52:877–882
59. Stein E, Zhang G, König JA (1992) Shakedown with nonlinear strain hardening including structural computation using finite element method. Int J Plast 8:1–31
60. Stölken JS, Evans AG (1998) A microbend test method for measuring the plasticity length-scale. Acta Mater 46:5109–5115
61. Weichert D, Maier G (eds) (2000) Inelastic analysis of structures under variable repeated loads, theory and engineering applications. Kluwer, Dordrecht

62. Weichert D, Maier G (eds) (2002) Inelastic behaviour of structures under variable repeated loads. Springer, Wien
63. Zarka J, Casier J (1979) Cyclic loading on elastic-plastic structures. In: Nemat-Nasser S (ed) Mechanics to-day. Pergamon Press, Oxford, pp 93–198

Shakedown Analysis of 3D Frames with an Effective Treatment of the Load Combinations

Giovanni Garcea, Leonardo Leonetti and Raffaele Casciaro

Abstract Using the Melan static theorem and an algorithm based on dual decomposition, a formulation for the shakedown analysis of 3D frames is proposed. An efficient treatment of the load combinations and an accurate and simple definition of the cross-section yield function are employed to increase effectiveness and to make shakedown analysis an affordable design tool. The section yield function, obtained by its support function values associated with presso-flexural mechanisms, is defined as the Minkowski sum of ellipsoids. The return mapping process, resulting from the dual decomposition, is solved at the element level by means of an algorithm based again on the dual decomposition. It allows the separation of the problem at the ellipsoid level and the use of a simple and inexpensive radial return mapping process for its solution. A series of numerical tests are presented to show both the accuracy and the effectiveness of the proposed formulation.

1 Introduction

Structures are subjected at different times to different load actions including dead loads and variable natural and anthropic variable loads. Building codes fix the range and extension of load variability through combination formulas.

Shakedown analysis furnishes, in a direct and elegant manner, a reliable safety factor against plastic collapse, loss in functionality due to excessive deformation (ratcheting) or collapse due to low cycle fatigue (plastic shakedown) [1] and also provides valuable information about the internal stress redistribution due to plastic adaptation phenomenon. The latter plays an important role in reducing tip stresses

G. Garcea (✉) · L. Leonetti · R. Casciaro
Dipartimento di Modellistica per L'Ingegneria, Università della Calabria, Cosenza, Italy
e-mail: giovanni.garcea@unical.it

L. Leonetti
e-mail: leonardo.leonetti@unical.it

R. Casciaro
e-mail: raffaele.casciaro@unical.it

P. Fuschi et al. (eds.), *Direct Methods for Limit and Shakedown Analysis of Structures*,
DOI 10.1007/978-3-319-12928-0_13

coming from the elastic solutions and allows to refer to smoother stress fields, better suited for design purposes to improve both the elements dimensioning and the reinforcements design. The advantages can be noticeable in R.C. structures when the rebars design is produced by fully automatic procedures.

Due to its importance as a tool for designers [2–5], in the last few years a notable effort has been made to propose efficient numerical algorithms of analysis. The interest in direct methods for limit and shakedown analysis has been encouraged by the availability of new and efficient optimization algorithms [6–8] such as the Interior Point method which is employed for solving very large non-linear problems, [9–11], like those obtained in the Finite Element (FE) discretization of real-scale engineering structures. An alternative approach to evaluate the shakedown safety factors in an FE context of analysis, is represented by the specialized direct method proposed in [12–14]. It is based on a strain-driven strategy of analysis hinged on closest point projection return mapping schemes and Riks arc-length solution techniques. It can be seen as the application of the proximal point algorithm to the static shakedown theorem and in the solution of the resulting problem by means of a dual decomposition strategy [14]. When applied to the static limit analysis theorem, the method coincides with a standard strain driven incremental elastoplastic analysis in which the stresses are obtained by a Backward Euler integration process and the equilibrium path is evaluated by means of a Riks arc-length strategy [15]. For its relation to standard elastoplastic analysis, to which it reduces in the limit analysis case, it will be named pseudo elastoplastic analysis and denoted as SD-CPP (Strain Driven—Closest Point Projection).

Despite its important implications and the development of new efficient numerical methods shakedown analysis still seems confined to the research community instead of being a common tool of structural design. This is, in part, still due to a series of problems regarding the efficiency and the robustness of the algorithms of analysis used. In particular, in the case of 3D frames, when considering standard rules as the Eurocodes ones, to be used by designers, shakedown analysis requires a preliminary fine tuning of two important aspects: (i) a suitable treatment of the usually large number of load combinations; (ii) an accurate and simple definition of the yield function which defines the nonlinear behavior of the generic cross-section. This last aspect is also important when using standard path following elasto-plastic analyses.

Regarding the first point, we know that the number of different load conditions entering into the combination sensibly affects the computational costs especially when an approach based on the static theorem is used. For instance, in the very simple case of load domain defined as a combination of basic actions varying between a minimum and a maximum value, the *elastic envelope*, that is the set of elastic stresses due to all possible loads in the combinations, becomes a convex polytope with 2^p vertexes p being the number of basic loads. For typical values of p (of order of ten) to impose the plastic admissibility for all these vertexes strongly affects the efficiency of the analysis. The difficulties further increase when the load domain definition is not so trivial, as in the case of the actual design rules adopted in Eurocodes.

The plastic admissibility condition has, however to be checked at the finite element or Gauss point level that we will call from now on the *local level* of analysis. As the larger part of the elastic stresses associated with the load domain vertexes could be in the interior of the elastic envelope, in order to improve efficiency without affecting the accuracy, we can use only the convex hull vertexes to check plastic admissibility. Standard convex hull algorithms, however, require a preliminary evaluation of all the points of the set before selecting, from these, those belonging to the convex hull so they become very expensive when the number of load combinations is large. For these reasons we present an approximate evaluation of the convex hull that retains only the significant vertexes with respect to a piecewise external linearization of the yield surface. Increasing the number of hyperplanes of the yield surface linearization allows the evaluation of the true convex hull of the elastic domain. It will be shown, however, that this simplified evaluation of the convex hull, also using a small number of hyperplanes, does not produce errors that are meaningful from a practical point of view.

Regarding the second point, while the yield function of 3D frames is usually evaluated considering only presso–flexural failures, so in a somewhat oversimplified form, to computationally define accurate yield surfaces combining axial force and two bending moments is not an easy task. This aspect is important not only for shakedown analysis but also for limit and standard incremental elastoplastic analyses and, for this reason, has received increasing attention in the literature [2, 4, 5, 16]. A piecewise linearization often requires a large number of polyhedral facets to obtain a sufficiently accurate approximation, which can have an important effect on the quality of the estimated bounds [5], but also on the efficiency of the algorithm. Since the yield criterion has to be verified at a large number of points throughout the whole structure, a compromise between accuracy and computational efficiency is required in the case of large-scale problems [12, 13].

Recently a strategy for approximating the true nonlinear yield surfaces by using a Minkowski sum of ellipsoids (MSE), has been proposed for limit analysis problems in [4] and the resulting Second Order Cone Programming problem has been solved with the commercial code `MOSEK`. The MSE allows us to accurately describe the section elastic domain using only a few analytical functions but it produces a noticeable increasing of the number of variables of the problems, being the stresses parameters now expressed as sums of ellipsoidal contributions, so penalizing the analysis in terms of both efficiency and robustness especially in the shakedown case. In this work with the aim of implementing an effective and reliable algorithm, we explore the possibility of using the MSE, within the SD-CPP approach.

The decomposition strategy allows us to decompose the global optimization problem at the element level where we have to solve a smaller optimization subproblem. Unfortunately the MSE representation of the elastic domain and of the stresses don't allow the use of standard return mapping algorithms commonly adopted in strain driven analysis. In the paper we propose a solution scheme which exploits the dual decomposition algorithm even at the local level and the exploitation of standard formulation based on the elastic predictor/plastic corrector scheme and so making the algorithm robust and effective. The approach is general, simple and can be effectively employed also in standard path–following elasto-plastic analysis of a 3D frame requiring few code modifications.

In order to evaluate the effectiveness and accuracy of our proposal a series of numerical tests are presented regarding both the accuracy and effectiveness of: (i) the yield function approximation using the MSE and the consequent decomposed CPP solution; (ii) the effect of using an approximate evaluation of the elastic envelope.

2 The 3D Beam Model

In the following the beam model and its discrete finite element version is briefly presented. We refer to [17, 18] for more details.

2.1 Beam Kinematics and Statics

Let us consider a cylinder occupying a reference configuration $\mathscr{B}$ of length ℓ confined by the lateral boundary denoted by $\partial\mathscr{B}$ and two terminal bases Ω_0 and Ω_ℓ. The cylinder is referred to a Cartesian frame $(\mathscr{O}, x_1 \equiv s, x_2, x_3)$ with unit vectors $\{\mathbf{e}_1, \mathbf{e}_2, \mathbf{e}_3\}$ and $\mathbf{e}_1$ aligned with the cylinder axis. In this system, see Fig. 1, we denote with $\mathbf{X} = \mathbf{X}_0 + \mathbf{x}$ the position of a point P, $\mathbf{X}_0 = s\mathbf{e}_1$ is the position of P with respect to the beam axis, s being an abscissa which identifies the generic cross-section $\Omega[s]$ of the beam, while $\mathbf{x} = x_2\mathbf{e}_2 + x_3\mathbf{e}_3$ is the position of P inside $\Omega[s]$.

The displacement field $\mathbf{u}[\mathbf{X}]$ of the model is expressed, as usual, as a section rigid motion

$$\mathbf{u}[\mathbf{X}] = \mathbf{u}_0[s] + \boldsymbol{\varphi}[s] \wedge \mathbf{x} \tag{1}$$

where $\mathbf{u}_0[s]$ and $\boldsymbol{\varphi}[s]$ are the mean translation and rotation of the section, $\wedge$ denotes the cross product. The kinematics assumed in Eq. (1) allows us to evaluate, using a standard linear 3D Cauchy continuum, the stress strain work $\mathscr{W}$ in terms of the generalized strains and stresses on the section (see [17]) as

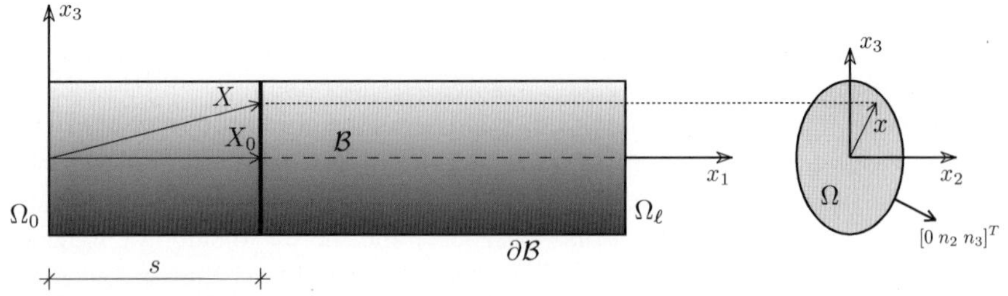

Fig. 1 The cylindrical solid

$$\mathscr{W} \equiv \int_{\ell} (\mathbf{N}[s] \cdot \boldsymbol{\varepsilon}[s] + \mathbf{M}[s] \cdot \boldsymbol{\chi}[s]) ds = \int_{\ell} \mathbf{t}_{\sigma}[s]^T \boldsymbol{\rho}[s] ds \tag{2}$$

where $\boldsymbol{\rho}[s] = \{\boldsymbol{\varepsilon}[s], \boldsymbol{\chi}[s]\}$ collects the generalized strain parameters $\boldsymbol{\varepsilon}$ and $\boldsymbol{\chi}$ defined as

$$\boldsymbol{\varepsilon}[s] = \mathbf{u}_{0},_{s}[s] + \mathbf{e}_1 \wedge \boldsymbol{\varphi}[s], \quad \boldsymbol{\chi} = \boldsymbol{\varphi}[s],_{s} \tag{3}$$

where, from now on, a comma stands for derivative and $\mathbf{t}_{\sigma}[s] = \{\mathbf{N}[s], \mathbf{M}[s]\}$ are the resultant force $\mathbf{N}[s] = \{N_1, N_2, N_3\}$ and moment $\mathbf{M}[s] = \{M_1, M_2, M_3\}$ defined as

$$\mathbf{N}[s] = \int_{\Omega} \mathbf{s} d\Omega, \quad \mathbf{M}[s] = \int_{\Omega} \mathbf{x} \wedge \mathbf{s} d\Omega. \tag{4}$$

and $\mathbf{s} = \boldsymbol{\sigma}\mathbf{e}_1$ is the traction applied to the generic cross section $\boldsymbol{\sigma}$ being the Cauchy stress tensor. Finally the elastic constitutive laws are expressed as

$$\boldsymbol{\rho}[s] = \mathbf{H}\mathbf{t}_{\sigma}[s], \quad \mathbf{H}[s] = \begin{bmatrix} \mathbf{H}_{NN} & \mathbf{H}_{NM} \\ \mathbf{H}_{NM}^T & \mathbf{H}_{MM} \end{bmatrix} \tag{5}$$

where the coefficients of the cross-section compliance matrix $\mathbf{H}$ can be obtained as in [17, 19, 20] (see also [21, 22] for the extension to generic anisotropic materials).

Furthermore we assume an elastic perfectly plastic material and the beam section yield function can be defined in terms of the normal actions only $\mathbf{t}[s] := \{N_1, M_2, M_3\}$ extracted from $\mathbf{t}_{\sigma}[s]$. In particular denoting the section elastic domain with $\mathbb{E}[s]$ and with $\boldsymbol{\Phi}[s, \mathbf{t}[s]]$ the convex yield function of the section s, that we will define explicitly in the next sections, we have

$$\mathbb{E}[s] = \{\mathbf{t}[s] : \boldsymbol{\Phi}[s, \mathbf{t}[s]] \leq \mathbf{0}\}. \tag{6}$$

2.2 The Finite Element for the Beam

The beam equilibrium equation for zero body forces

$$\mathbf{N},_{s} = \mathbf{0}, \qquad \mathbf{M},_{s} + \mathbf{e}_1 \wedge \mathbf{N} = \mathbf{0} \tag{7}$$

states that $\mathbf{N}$ and the torsional moment component M_1 are constant, while the two flexural components $M_2[s]$ and $M_3[s]$ of $\mathbf{M}[s]$ are linear with s and linked to the shear resultants so that $N_2\ell = -(M_3[\ell] - M_3[0])$ and $N_3\ell = (M_2[\ell] - M_2[0])$. The internal work then becomes

$$\mathscr{W} \equiv \mathbf{N} \cdot (\mathbf{u}_0[\ell] - \mathbf{u}_0[0]) + \mathbf{M}[\ell] \cdot \boldsymbol{\varphi}[\ell] - \mathbf{M}[0] \cdot \boldsymbol{\varphi}[0] = \mathbf{d}_e^T \mathbf{Q}_e^T \boldsymbol{\beta}_e \tag{8}$$

so allowing us to directly obtain the discrete form of $\mathscr{W}$ without the need to use any FEM interpolation for the kinematic variables. In Eq. (8) the vectors collecting the finite element generalized parameters and the compatibility operator $\mathbf{Q}_e$ are

$$\boldsymbol{\beta}_e = \begin{bmatrix} N \\ M_2[0] \\ M_3[0] \\ M_2[\ell] \\ M_3[\ell] \\ M_1 \end{bmatrix}, \quad \mathbf{d}_e = \begin{bmatrix} \mathbf{u}_0[0] \\ \boldsymbol{\varphi}[0] \\ \mathbf{u}_0[\ell] \\ \boldsymbol{\varphi}[\ell] \end{bmatrix}, \quad \mathbf{Q}_e = \frac{1}{\ell} \begin{bmatrix} -\ell\,\mathbf{e}_1^T & \mathbf{0} & \ell\,\mathbf{e}_1^T & \mathbf{0} \\ \mathbf{e}_3^T & -\ell\,\mathbf{e}_2^T & -\mathbf{e}_3^T & \mathbf{0} \\ -\mathbf{e}_2^T & -\ell\,\mathbf{e}_3 & \mathbf{e}_2^T & \mathbf{0} \\ -\mathbf{e}_3^T & \mathbf{0} & \mathbf{e}_3^T & \ell\,\mathbf{e}_2^T \\ \mathbf{e}_2^T & \mathbf{0} & -\mathbf{e}_2^T & \ell\,\mathbf{e}_3^T \\ \mathbf{0} & -\ell\,\mathbf{e}_1^T & \mathbf{0} & \ell\,\mathbf{e}_1^T \end{bmatrix} \tag{9}$$

Equation (9) allows us to write the discrete form of the equilibrium equations as

$$\mathbf{Q}^T\boldsymbol{\beta} - \lambda\mathbf{p} = \mathbf{0} \quad \text{with} \quad \mathbf{Q}^T\boldsymbol{\beta} = \mathscr{A}_e\{\mathbf{Q}_e{}^T\boldsymbol{\beta}_e\} \tag{10}$$

where $\boldsymbol{\beta}$ denotes the global vector collecting all the stress parameters $\boldsymbol{\beta}_e$ and $\mathbf{p}$ is the load vector. The global equilibrium matrix $\mathbf{Q}^T$ is obtained as usual by means of the contribution of each finite element and $\mathscr{A}_e$ is the standard assembling operator which takes into account the inter–element continuity conditions on $\mathbf{u}$ and $\boldsymbol{\varphi}$. From now on a subscript e denotes the finite element counterpart of a global vector or matrix.

Letting $\mathbf{t}_\sigma[s] = \mathbf{D}_t[s]\boldsymbol{\beta}_e$ be the finite element interpolation for the generalized stress the element elastic compliance matrix, is obtained from the equivalence

$$\int_\ell \mathbf{t}_\sigma[s]\cdot\mathbf{H}\mathbf{t}_\sigma[s]ds = \boldsymbol{\beta}_e^T\mathbf{H}_e\boldsymbol{\beta}_e \qquad \mathbf{H}_e = \int_\ell \mathbf{D}_t[s]^T\mathbf{H}\mathbf{D}_t[s]\,ds \tag{11}$$

We express the plastic admissibility of the beam finite element in terms of the plastic admissibility of its end section normal actions $\mathbf{t}[s]$ obtained from the generalized stress element vector $\boldsymbol{\beta}_e$ as $\mathbf{t}[s] = \mathbf{S}[s]\,\boldsymbol{\beta}_e$ with $s = 0, \ell$ that is

$$\boldsymbol{\Phi}_e[\boldsymbol{\beta}_e] := \begin{bmatrix} \boldsymbol{\Phi}[0, \mathbf{t}[0]] \\ \boldsymbol{\Phi}[\ell, \mathbf{t}[\ell]] \end{bmatrix} \tag{12}$$

where $\boldsymbol{\Phi}_e[\boldsymbol{\beta}_e]$ represents the element yield function and, as usual, inequalities are intended in componentwise. Note that, both sections $s = 0$ and $s = \ell$ share the same value of the normal action N_1 as prescribed by the beam equilibrium Eq. (7), so the plastic admissibility for the two sections is coupled.

From now on we denote with a subscript the dependence of the quantities on the section while a vector column will be represented between curly brackets in the body of the text.

3 Shakedown Analysis Based on the Proximal Point Method and Dual Decomposition

In this section the approach proposed in [14] is particularized to the shakedown analysis of a 3D frame.

3.1 The Envelope of Elastic Stresses

We assume that the external actions $\mathbf{p}[t]$, variable with the time t, are expressed as a combination of p basic loads $\mathbf{p}_i$ belonging to the convex admissible *load domain* $\mathbb{P}$. This is defined according to the Eurocode rules [23] which prescribe p combination of different load conditions, each of them obtained by considering one of the load conditions $\mathbf{p}_k$ as the leading one and the others $\mathbf{p}_j$, $j \neq k$ as accompanying. All of these are affected by a safety factor α_j varying from α_j^{min} to α_j^{max} and by a combination factor $\psi_{kj} \leq 1$ ($\psi_{kk} = 1$). So the load domain is defined by

$$\mathbb{P} \equiv \bigcup_{k=1}^{p} \mathbb{P}^{(k)}, \quad \mathbb{P}^{(k)} \equiv \left\{ \mathbf{p} \equiv \sum_{i=1}^{p} \psi_{ki}\alpha_i \mathbf{p}_i \; : \; \alpha_i^{min} \leq \alpha_i \leq \alpha_i^{max} \right\}. \tag{13}$$

Denoting with $\hat{\mathbf{t}}[s, t] = \{\hat{N}_1[s, t], \; \hat{M}_2[s, t], \hat{M}_3[s, t]\}$ the set of the elastic normal actions produced by each load path contained in $\mathbb{P}$ we define the *elastic envelope* $\mathbb{S}[s]$ of the generic beam section s

$$\mathbb{S} = \bigcup_{k=1}^{p} \mathbb{S}^{(k)}, \quad \mathbb{S}^{(k)} := \left\{ \hat{\mathbf{t}} \equiv \sum_{i=1}^{p} \psi_{ki}\alpha_i \hat{\mathbf{t}}_i \; : \; \alpha_i^{min} \leq \alpha_i \leq \alpha_i^{max} \right\} \tag{14}$$

where $\hat{\mathbf{t}}_i$ are the normal actions elastic solution for $\mathbf{p}_i$ while the dependence on s will be omitted, from now on, for easerer writing.

Finally if the external loads increase by a real number λ, called the *load domain multiplier*, the elastic envelope becomes $\lambda\mathbb{S} := \left\{ \lambda\hat{\mathbf{t}} : \; \hat{\mathbf{t}} \in \mathbb{S} \right\}$.

3.2 The Plastic Admissibility Conditions for Shakedown

Shakedown requires that all the stresses contained in the amplified elastic envelopes $\lambda\, \mathbb{S}[s]$, translated by $\mathbf{t}[s]$, will be plastically admissible. Due to the convexity of $\boldsymbol{\Phi}[s, \mathbf{t}[s]]$, this is easily expressed in terms of plastic admissibility of the N_v stress vertexes of the convex hull of $\mathbb{S}$. This means that we can directly substitute $\mathbb{S}$ with its convex hull, i.e. a convex polytope whose vertex will be, in general, a subset of the

$p \cdot 2^p$ stresses corresponding to the vertexes of $\mathbb{P}$. With this substitution, each $\hat{\mathbf{t}} \in \mathbb{S}$ can be expressed as a convex combination of the N_v of the convex hull vertex $\hat{\mathbf{t}}^{\alpha}$:

$$\hat{\mathbf{t}} = \sum_{\alpha=1}^{N_v} \xi^{\alpha}\, \hat{\mathbf{t}}^{\alpha}, \quad \xi^{\alpha} \geq 0, \quad \sum_{\alpha=1}^{N_v} \xi^{\alpha} = 1 \tag{15}$$

where, from now on, a Greek superscript denotes a vertex of $\mathbb{S}[s]$.

Letting $\lambda\, \hat{\mathbf{t}}^{\alpha}[s] + \mathbf{t}[s]$ be the vertexes of the amplified and translated hull, the plastic admissibility condition for all stresses in $\lambda\mathbb{S}[s] \oplus \{\mathbf{t}[s]\}$ becomes

$$\lambda\mathbb{S}[s] \oplus \mathbf{t}[s] \in \mathbb{E}[s] \Longleftrightarrow \lambda\, \hat{\mathbf{t}}^{\alpha}[s] + \mathbf{t}[s] \in \mathbb{E} \quad \forall\alpha \tag{16}$$

Defining shakedown yield functions for the generic section s as $\boldsymbol{\Phi}_s[\lambda, \mathbf{t}[s]] \equiv \{\boldsymbol{\Phi}_s[\lambda\hat{\mathbf{t}}^1 + \mathbf{t}[s]], \ldots, \boldsymbol{\Phi}_s[\lambda\hat{\mathbf{t}}^{N_v} + \mathbf{t}[s]]\}$ we obtain the shakedown yield function for the element as

$$\boldsymbol{\Phi}_e[\lambda, \boldsymbol{\beta}_e] := \begin{bmatrix} \boldsymbol{\Phi}_0[\lambda, \mathbf{t}[0]] \\ \boldsymbol{\Phi}_\ell[\lambda, \mathbf{t}[\ell]] \end{bmatrix} \tag{17}$$

Note that the number of vertexes N_v in the convex hull depends on the section considered.

3.3 The Pseudo-elastoplastic Step for Shakedown Analysis

Shakedown analysis is performed using the algorithm proposed in [12–14] (see also [24]), which corresponds to the application of the proximal point method to the Melan static theorem. A sequence of subproblems or *steps* are obtained by adding a quadratic positive term, using the element compliance matrix $\mathbf{H}_e$ in Eq. (11), to the objective function of the Melan static theorem

$$\begin{aligned} \text{maximize} \quad & \Delta\xi^{(n)}\lambda^{(n)} - \frac{1}{2}\sum_e \Delta\boldsymbol{\beta}_e{}^T \mathbf{H}_e \Delta\boldsymbol{\beta}_e \\ \text{subject to} \quad & \mathbf{Q}^T\boldsymbol{\beta}^{(n)} - \lambda^{(n)}\mathbf{p}_0 = \mathbf{0} \\ & \boldsymbol{\Phi}_e[\lambda, \boldsymbol{\beta}_e] \leq \mathbf{0}, \qquad \forall e \end{aligned} \tag{18}$$

where the superscript $(\cdot)^{(n)}$ will denote quantities evaluated in the nth step, the symbol $\Delta(\cdot) = (\cdot)^{(n)} - (\cdot)^{(n-1)}$ is the increment of a quantity from the previous step and $\Delta\xi^{(n)} > 0$ is an assigned real positive number. The load $\mathbf{p}_0 \equiv \mathbf{Q}^T\hat{\boldsymbol{\beta}}^0$ associated with the equilibrium equation to a reference stress $\hat{\boldsymbol{\beta}}^0$ is used to simply recast the formulation in Eq. (18) with that of limit analysis (see [14] for further details).

3.3.1 First Order Conditions

Introducing the dual multipliers $\Delta\mathbf{d}$ and $\Delta\boldsymbol{\kappa}$ associated with the equality and inequality constraints of (18) respectively, the finite step equations are defined by the first order conditions of the following Lagrangian $\mathscr{L}^{(n)}$

$$\begin{aligned}\mathscr{L}^{(n)} &= \Delta\xi^{(n)}\lambda^{(n)} - \frac{1}{2}\sum_e \Delta\boldsymbol{\beta}_e{}^T\mathbf{H}_e\Delta\boldsymbol{\beta}_e + \Delta\mathbf{d}^T(\mathbf{Q}^T\boldsymbol{\beta}^{(n)} - \lambda^{(n)}\mathbf{p}_0)\\ &\quad -\sum_e \Delta\boldsymbol{\kappa}_e{}^T\boldsymbol{\Phi}_e[\boldsymbol{\beta}_e{}^{(n)}, \lambda^{(n)}]\end{aligned} \tag{19}$$

In order to simplify the notation the superscript (n) will be omitted from now on.

In particular, from the stationary condition of (19) with respect to $\boldsymbol{\beta}$ and $\Delta\boldsymbol{\kappa}$ and remembering Eq. (10) we obtain the finite step form of the constitutive law, i.e. the plastic admissibility and plastic consistence conditions for shakedown

$$\begin{cases}\mathbf{r}_\sigma \equiv -\mathbf{H}_e\Delta\boldsymbol{\beta}_e - \mathbf{A}_e[\boldsymbol{\beta}_e, \lambda]\Delta\boldsymbol{\kappa}_e + \mathbf{Q}_e\Delta\mathbf{d}_e = \mathbf{0}\\ \mathbf{r}_\mu \equiv \boldsymbol{\Phi}_e[\boldsymbol{\beta}_e, \lambda] \le 0, \quad \boldsymbol{\kappa}_e \ge 0, \quad \boldsymbol{\kappa}_e{}^T\boldsymbol{\Phi}_e[\boldsymbol{\beta}_e, \lambda] = 0,\end{cases} \tag{20a}$$

where $\mathbf{A}_e[\boldsymbol{\beta}_e, \lambda] := \left(\frac{\partial\boldsymbol{\Phi}_e[\boldsymbol{\beta}_e,\lambda]}{\partial\boldsymbol{\beta}_e}\right)^T$. Note that, when $\Delta\mathbf{d}_e$ and λ are assigned, the admissibility conditions (20a) are expressed in terms of element quantities alone. For this reason they will be denoted, from now on, as *local equations* and $\boldsymbol{\beta}_e$ and $\boldsymbol{\kappa}_e$ will be denoted as *local variables*. Following [24] a task that uses only local variables and equations will be said to be at the *local level*.

In the same way the stationary condition with respect to $\Delta\mathbf{d}$ and λ furnishes the equilibrium equations and the normalization condition, coupling the global variables of the problem and defining the *global level* of the analysis,

$$\begin{cases}\mathbf{r}_u \equiv \mathbf{Q}^T\boldsymbol{\beta} - \lambda\mathbf{p}_0 = \mathbf{0}\\ r_\lambda \equiv \Delta\xi - \Delta\mathbf{d}^T\mathbf{p}_0 - \sum_e \Delta\boldsymbol{\kappa}_e{}^T\boldsymbol{\Phi}_{e,\lambda} = 0\end{cases} \tag{20b}$$

where $\boldsymbol{\Phi}_{e,\lambda} := \left(\frac{\partial\boldsymbol{\Phi}_e[\boldsymbol{\beta}_e,\lambda]}{\partial\lambda}\right)$. Equation (20b) will be denoted, from now on, as *global equations* while $\mathbf{d}$ and λ will be denoted as *global variables*.

In fixed load cases, that is for a single vertex of the load domain, Eq. (20a) coincide with the backward-Euler integration of the elasto-plastic constitutive equations while Eq. (20a, 20b) exactly correspond to a step of the arc-length algorithm used to solve the incremental elastoplastic problem [12, 13, 15]. Due to its meaning in the case of fixed loads we call this kind of analysis *pseudo elastoplastic*. $\Delta\mathbf{d}$ and $\Delta\boldsymbol{\kappa}_e$ assume the meaning of displacements and plastic multipliers of the problem.

As for elastic perfectly plastic structures the limit load can be evaluated by recovering the complete equilibrium path by means of path-following algorithms, in the same fashion the skakedown multiplier can be obtained by evaluating a sequence of states, $\mathbf{z}^{(n)} := \{\lambda^{(n)}, \boldsymbol{\beta}^{(k)}, \mathbf{d}^{(n)}, \Delta\boldsymbol{\kappa}_e{}^{(n)}\}$ by solving a series of problems (18), i.e. defining a pseudo-elastoplastic equilibrium curve [12]. In [14] it has been shown that starting from the known elastic limit $\mathbf{z}^{(0)}$, the sequence $\mathbf{z}^{(n)}$ generated in this way is safe in the sense of the static theorem and monotonously increasing in $\lambda^{(n)}$. In the case $\lambda^{(n)} = \lambda^{(n-1)}$ with $\Delta\mathbf{d} \neq \mathbf{0}$ it is simple to show (see [14]) that we have the convergence at the desired shakedown multiplier.

3.4 The Dual Decomposition Solution of the Pseudo Elastoplastic Step

The similarity of Eq. (20a, 20b) with standard strain driven path-following elastoplastic analysis suggests that the same method of solution can also be used. This is the approach followed in [12, 13] and it is based on an exact solution of the local conditions in (20a) for an assigned value of $\Delta\mathbf{d}_e$ and $\Delta\lambda$, so expressing $\boldsymbol{\beta}_e$ and $\boldsymbol{\kappa}_e$ as implicit functions of the displacements and of the load multiplier. This step is performed at the local level by using a return mapping by closest point projection process. Equation (20a) are, in fact, the first order conditions of the following problem

$$\begin{cases} \min\limits_{\boldsymbol{\beta}_e} & \dfrac{1}{2}\left(\boldsymbol{\beta}_e - \boldsymbol{\beta}_e{}^*\right)^T \mathbf{H}\left(\boldsymbol{\beta}_e - \boldsymbol{\beta}_e{}^*\right), \\ \text{subject to:} & \boldsymbol{\Phi}_e[\boldsymbol{\beta}_e, \lambda] \leq 0. \end{cases} \tag{21}$$

that is the convex projection of the *trial stress* (or the elastic predictor) $\boldsymbol{\beta}_e{}^*$, defined by

$$\boldsymbol{\beta}_e{}^* = \boldsymbol{\beta}_e{}^{(n-1)} + \mathbf{H}^{-1}\mathbf{Q}_e\Delta\mathbf{d}_e, \tag{22}$$

onto the elastic shakedown domain bounded by the convex function $\boldsymbol{\Phi}_e[\boldsymbol{\beta}_e, \lambda]$.

From the closest point projection (CPP) in Eq. (21) we have the stresses and the plastic multipliers as a function of $\Delta\mathbf{d}$ and λ

$$\boldsymbol{\beta} = \boldsymbol{\beta}[\mathbf{z}^{(n-1)}, \Delta\mathbf{d}, \lambda], \quad \Delta\boldsymbol{\kappa} = \Delta\boldsymbol{\kappa}[\mathbf{z}^{(n-1)}, \Delta\mathbf{d}, \lambda]$$

and of the known initial step quantities collected in $\mathbf{z}^{(n-1)}$.

Omitting the dependence on $\mathbf{z}^{(n-1)}$ the global Eq. (20b) can be rewritten, in terms of $\Delta\mathbf{d}$ and λ, as

$$\begin{cases} \mathbf{r}_u \equiv \mathbf{Q}^T\boldsymbol{\beta}[\Delta\mathbf{d}, \lambda] - \lambda\mathbf{p}_0 = \mathbf{0} \\ r_\lambda \equiv \Delta\xi - \Delta\mathbf{d}^T\mathbf{p}_0 - \boldsymbol{\Phi},_\lambda^T \Delta\boldsymbol{\kappa}[\Delta\mathbf{d}, \lambda] = 0 \end{cases} \tag{23}$$

If the nonlinear system (23) is solved by means of a Newton iteration as in [12, 14], we obtain:

$$\begin{cases} \Delta\mathbf{d}^{j+1} = \Delta\mathbf{d}^{j} + \dot{\mathbf{d}}, \\ \Delta\lambda^{j+1} = \Delta\lambda^{j} + \dot{\lambda}, \end{cases} \quad \text{with} \quad \begin{cases} \mathbf{K}^{j}\dot{\mathbf{d}} - \dot{\lambda}\mathbf{y}^{j} = -\mathbf{r}_{u}^{j}, \\ -\mathbf{y}^{j\,T}\dot{\mathbf{d}} + \dot{\lambda}\, h_{\lambda\lambda}^{j} = -r_{\lambda}^{j}, \end{cases} \tag{24}$$

where $\mathbf{K}^j \equiv \partial\mathbf{r}_u/\partial\mathbf{d}$ is the algorithmic tangent matrix while r_λ^j and $\mathbf{r}_u^j$ are the residuals defined in Eq. (23) evaluated in $(\mathbf{d}^j, \lambda^j)$ after performing the return mapping process (21) to evaluate $\boldsymbol{\beta}[\Delta\mathbf{d}^j, \lambda^j]$ and $\Delta\boldsymbol{\kappa}[\Delta\mathbf{d}^j, \lambda^j]$, while

$$\mathbf{y}^j \equiv \left.\frac{\partial\mathbf{r}_u}{\partial\lambda}\right|_{(\lambda^j,\mathbf{d}^j)} = \left.\frac{\partial r_\lambda}{\partial\mathbf{d}}\right|_{(\lambda^j,\mathbf{d}^j)} \qquad h_{\lambda\lambda}^j \equiv \left.\frac{\partial r_\lambda}{\partial\lambda}\right|_{(\lambda^j,\mathbf{d}^j)}.$$

We have convergence to a new equilibrium point when the norm of $\mathbf{r}_u^j$ become sufficiently small, in this case we set $\mathbf{z}^{(n)} = \mathbf{z}^j$. We recall that the use of a modified Newton method that uses the initial elastic stiffness matrix assures global, even if simply linear, convergence [12].

4 The Elastic Domain of the Beam Section

In this section we briefly present the construction of the 3D elastic domain for each beam section using the approach recently proposed by Bleyer and De Buhan [4] (see also [5] and [2]). Then we propose a simple way to use the Minkowski sum of ellispoids in the return mapping process of the strain driven analysis previously described.

The algorithm we propose is general and can be used not only for shakedown but also for any other analysis based on return mapping processes like those currently employed for the reconstruction of the equilibrium path of elastoplastic structures.

4.1 Evaluation of the Support Functions of the Beam Elastic Domain

We assume that the beam section domain Ω is the union of n_d sub domain Ω_i in which the material is homogeneous (see Fig. 2). For each Ω_i, the plastic admissibility condition is expressed in terms of normal stress only as $-\sigma_{ci} \le \sigma_{11} \le \sigma_{ti}$ where σ_{ti} is the ultimate normal stress in tension (positive) and σ_{ci} in compression (negative). This corresponds to assume, as is usual for technical applications involving slender beams, infinitely resistant frame members with respect to shear effects as well as torsion. Hence the yield surface will be drawn in the 3D space involving axial force N_1 and bending moments M_2 and M_3 that is it depends on vector $\mathbf{t}[s] = \{N_1, M_2, M_3\}$ only.

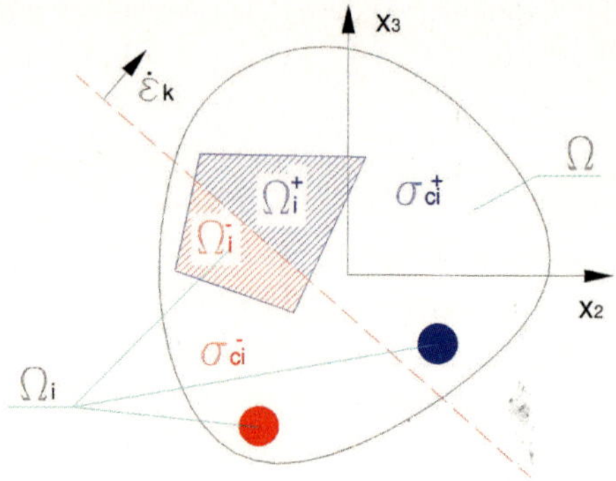

Fig. 2 Beam section with different material

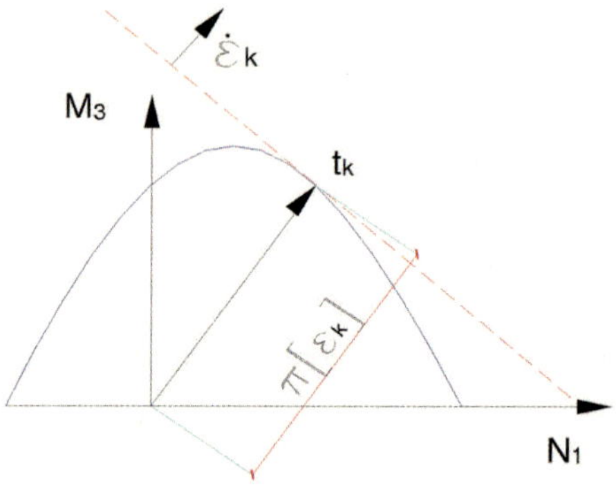

Fig. 3 Support function for the the elastic domain

Due to the section rigid motion hypotheses a generic *section collapse mechanism* will be defined by the position of the neutral axis

$$\varepsilon_1 + x_3\ \chi_2 - x_2\ \chi_3 = 0 \tag{25}$$

Denoting with $\dot{\boldsymbol{\varepsilon}} = \{\varepsilon_1, \chi_2, \chi_3\}$ the vector collecting the collapse mechanism parameters we have

$$\pi_{\mathbb{E}}[\dot{\boldsymbol{\varepsilon}}] = \max\{\dot{\boldsymbol{\varepsilon}}^T\mathbf{t} : \mathbf{t} \in \mathbb{E}\} \tag{26}$$

i.e. $\pi_{\mathbb{E}}[\dot{\boldsymbol{\varepsilon}}]$ is the signed distance with respect to the origin of the hyperplane tangent to $\mathbb{E}$ and with normal $\dot{\boldsymbol{\varepsilon}}$. $\pi_{\mathbb{E}}[\dot{\boldsymbol{\varepsilon}}]$ is then *the support function* of $\mathbb{E}$ (see Fig. 3).

The vector $\mathbf{t}_y \in \mathbb{E}$ collecting the generalized section resultants associated with $\dot{\boldsymbol{\varepsilon}}$ by the max condition in Eq. (26), can be evaluated by the expression

$$\mathbf{t}_y = \begin{bmatrix} N_{y1} \\ M_{y2} \\ M_{y3} \end{bmatrix} \quad \text{with} \quad \begin{cases} N_{y1} = \sum_i \left(\int_{\Omega_i^+} \sigma_{ti} d\Omega_i - \int_{\Omega_i^-} \sigma_{ci} d\Omega_i \right) \\ M_{y2} = \sum_i \left(\int_{\Omega_i^+} x_3 \sigma_{ti} d\Omega_i - \int_{\Omega_i^-} x_3 \sigma_{ci} d\Omega_i \right) \\ M_{y3} = \sum_i \left(\int_{\Omega_i^+} x_2 \sigma_{ci} d\Omega_i - \int_{\Omega_i^-} x_2 \sigma_{ti} d\Omega_i \right) \end{cases} \tag{27}$$

where σ_{ti} and σ_{ci} are the yield normal stress in traction and in compression, respectively and Ω_i^+ and Ω_i^- are the portion of dS_i in traction or compression. The integrals can be easily evaluated by the numerical procedure described in [5].

This definition states that, for each position of the neutral axis defined by $\dot{\boldsymbol{\varepsilon}}_k$ the corresponding generalized stress $\mathbf{t}_{yk}$ on the boundary of the section elastic domain $\mathbb{E}[s]$ is obtained by considering uniaxial stress fields reaching their maximum strength capacity in each region, either in tension or in compression.

4.2 The Approximation of $\mathbb{E}$ Using a Minkowski Sum of Ellipsoids

Once the support function values $\pi_{\mathbb{E}}[\dot{\boldsymbol{\varepsilon}}_k]$ for a series of N_p direction $\dot{\boldsymbol{\varepsilon}}_k$ have been obtained, we use them to approximate $\mathbb{E}$ by means of a Minkowski sum of $N_{\mathscr{E}}$ ellipsoids using the approach proposed in [4], to which we refer for further details. The Ith ellipsoid is defined by the following equation

$$\mathscr{E}_I[\mathbf{C}_I, \mathbf{c}_I] = \left\{ \mathbf{t} : \ \|\mathbf{J}_I^{-1}(\mathbf{t} - \mathbf{c}_I)\| - 1 \leq 0 \right\} \tag{28}$$

where $\|\cdot\|$ stands for the Euclidean norm, the symmetric and definite positive matrix $\mathbf{C}_I = \mathbf{C}_I^T = \mathbf{J}_I^T \mathbf{J}_I$ gives the shape and orientation and $\mathbf{c}_I$ the origin of the ellipsoid. The support function of $\mathscr{E}_I$ is

$$\pi_{\mathscr{E}_I}[\mathbf{n}] = \|\mathbf{J}_I \mathbf{n}\| + \mathbf{n}^T \mathbf{c}_I \tag{29}$$

As the support function of the Minkowski sum of convex sets is the sum of their support functions [4] we have that

$$\pi_{\mathscr{E}}[\mathbf{n}] = \sum_{I=1}^{N_{\mathscr{E}}} \pi_{\mathscr{E}_I}[\mathbf{n}] = \sum_{I=1}^{N_{\mathscr{E}}} \|\mathbf{J}_I \mathbf{n}\| + \mathbf{n}^T \mathbf{c} \tag{30}$$

where $\mathbf{c} = \sum_I \mathbf{c}_I$. From now on, capital romans letters are used to denote the Ith ellipsoid.

With the procedure adopted in [2, 4] we obtain, from the values of $\pi_{\mathbb{E}}[\dot{\boldsymbol{\varepsilon}}_k]$ evaluated for N_p directions $\dot{\boldsymbol{\varepsilon}}_k$, the unknowns that define $\mathbf{J}_I$ (six for each ellipsoid) and $\mathbf{c}$ using the following minimization problem

$$\min_{(\mathbf{J}_I, \mathbf{c})} \sum_{k=1}^{N_p} (\pi_{\mathbb{E}}[\dot{\boldsymbol{\varepsilon}}_k] - \pi_{\mathscr{E}}[\dot{\boldsymbol{\varepsilon}}_k])^2 \tag{31}$$

Eq. (30) allows us to approximate the elastic domain as the Minkowski sum of n ellipsoids and of a singleton $\mathbf{c}$, as

$$\mathbb{E}[s] \equiv \left(\bigoplus_{I=1}^{N_{\mathscr{E}}} \mathscr{E}_I \right) \oplus \mathbf{c}, \quad \mathscr{E}_I \equiv \mathscr{E}[\mathbf{C}_I, \mathbf{0}] \tag{32}$$

Denoting with a subscript the dependence of the quantities on the section s Eq. (32) means that the condition for static variable $\mathbf{t}_s \equiv \mathbf{t}[s]$ to be admissible is that it has to be expressed as the sum of $N_{\mathscr{E}} + 1$ contributions

$$\mathbf{t}[s] = \left(\mathbf{c}_s + \sum_{I=1}^{N_{\mathscr{E}}} \mathbf{t}_{sI} \right) \in \mathbb{E}[s] \Longleftrightarrow \mathbf{t}_{sI} \in \mathscr{E}_I \tag{33}$$

Introducing the ellipsoid vector $\mathbf{t}_{s\mathscr{E}} = \{\mathbf{t}_{s1}, \ldots, \mathbf{t}_{sN_{\mathscr{E}}}\}$ the plastic admissibility of the stress $\mathbf{t}_s$ in terms of the ellipsoid decomposition of the yield functions $\phi_{sI}[\mathbf{t}_{sI}] \equiv \|\mathbf{J}_{sI}^{-1} \mathbf{t}_{sI}\| - 1$ of the section in s rewrites as

$$\begin{cases} \mathbf{t}[s] = \mathbf{c}_s + \boldsymbol{\Sigma}_s \mathbf{t}_{s\mathscr{E}} \\ \boldsymbol{\Phi}_{s\mathscr{E}}[\mathbf{t}_{s\mathscr{E}}] \leq \mathbf{0} \end{cases} \tag{34}$$

where $\boldsymbol{\Phi}_{s\mathscr{E}}[\mathbf{t}_{s\mathscr{E}}] := \{\phi_{s1}[\mathbf{t}_{s1}], \ldots, \phi_{sN_{\mathscr{E}}}[\mathbf{t}_{sN_{\mathscr{E}}}]\}$ and the matrix $\boldsymbol{\Sigma}_s$ is implicitly defined by Eq. (33).

4.2.1 The Shakedown Yield Function as MSE

The shakedown yield function is defined in terms of the translated by $\mathbf{t}_s$ vertexes of $\mathbb{S}$ that is defining the admissibility of each $\mathbf{t}_s^\alpha = \lambda \hat{\mathbf{t}}_s^\alpha + \mathbf{t}_s$. Expressing $\mathbf{t}_s^\alpha = \mathbf{c}_s + \boldsymbol{\Sigma}_s \mathbf{t}_{s\mathscr{E}}^\alpha$ as MSE the plastic admissibility condition in Eq. (34) rewrites for shakedown as

$$\begin{cases} \mathbf{c}_s + \boldsymbol{\Sigma}_s \mathbf{t}_{s\mathscr{E}}^{\alpha} = \lambda \hat{\mathbf{t}}_s^{\alpha} + \mathbf{t}_s \\ \boldsymbol{\Phi}_{s\mathscr{E}}[\mathbf{t}_{s\mathscr{E}}^{\alpha}] \leq \mathbf{0} \end{cases} \quad \forall \alpha \tag{35}$$

Introducing the vectors $\mathbf{p}_{s\mathscr{E}} = \{\lambda \hat{\mathbf{t}}_s^1 - \mathbf{c}_s, \ldots, \lambda \hat{\mathbf{t}}_s^{N_v} - \mathbf{c}_s\}$ and the shakedown yield function of the section $\boldsymbol{\Phi}_{s\mathscr{E}}[\lambda, \mathbf{t}_{s\mathscr{E}}] := \{\boldsymbol{\Phi}_{s\mathscr{E}}[\mathbf{t}_{s\mathscr{E}}^1], \ldots, \boldsymbol{\Phi}_{s\mathscr{E}}[\mathbf{t}_{s\mathscr{E}}^{N_v}]\}$ Eq. (35) rewrites as

$$\begin{cases} -\mathbf{I}_{s\alpha} \mathbf{t}_s + \boldsymbol{\Sigma}_{s\mathscr{E}} \mathbf{t}_{s\mathscr{E}} = \mathbf{p}_{s\mathscr{E}} \\ \boldsymbol{\Phi}_{s\mathscr{E}}[\lambda, \mathbf{t}_{s\mathscr{E}}] \leq \mathbf{0} \end{cases} \tag{36}$$

where matrices $\mathbf{I}_{s\alpha}$ and $\boldsymbol{\Sigma}_{s\mathscr{E}}$ are implicitly defined by Eq. (36).

4.3 Simplified Evaluation of the Elastic Envelope

Following the König [25] theorem a general rule adopted in numerical shakedown analyses consist in considering only the convex hull of the loading history in $\mathbb{P}$ i.e., in our case, the $p \cdot 2^p$ *load vertexes* of the polytope in Eq. (13). Adopting this approach the elastic stresses we have to check for plastic admissibility are those corresponding to load vertexes.

However, as described in Sect. 3.2, using the convexity of the elastic domain $\mathbb{E}$, the control of the plastic admissibility can be performed by referring to the convex hull of $\mathbb{S}$. It involves only local quantities defined on the finite element (in our proposal the normal actions of the beam end sections). When the value of p increases it is possible that part of the elastic stresses associated with the load domain vertexes, are in the interior of the local elastic envelope. This means that, while at the global level we have to retain all the load vertexes, at the local level only a part of these will define the convex hull of $\mathbb{S}$ and then will be significant to impose the plastic admissibility condition. This is more clear if we consider the simple example of a truss structure in which the plastic admissibility condition for the normal resultant N_1 is defined by $-N_y \leq N_1 \leq N_y$, with N_y a reference yield value. In this case the convex hull of each bar is defined only by the two vertexes corresponding to the maximum and minimum value achieved, within all load combinations, by the normal resultant N_1, i.e. we have two vertexes for each bar, instead of $p \cdot 2^p$, independently from the value of p.

Then, the use of the vertexes of the convex hull of $\mathbb{S}$ produces, in general, a reduction of the number of constraints of the optimization problem to be solved. For this reason it is always convenient, independently from the method of solution adopted, with respect to both effectiveness and robustness and does not *introduce any approximation*. It is worth mentioning as this aspect is not considered in shakedown algorithms due both to (i) the limited number of load conditions usually analyzed, (ii) a not clearly distinction between the global and the local level of the analysis.

An other important problem regards the evaluation of the convex hull of $\mathbb{S}$. The use of standard algorithms presents some inconvenient as they require: (i) first to evaluate the $2^p p$ vertexes corresponding to the load combinations; (ii) then to select the N_v of these defining the convex hull of $\mathbb{S}$.

To overcome these difficulties in the following, we propose a fast, simplified procedure that directly provides the N_v significant vertexes in $\mathbb{S}$ as the most restrictive with respect to the *supporting hyperplanes* of the section yield function $\mathbb{E}[s]$, defined in Eq. (26). In fact, shakedown yield condition for the section can be rewritten as

$$f[\mathbf{t}, \lambda, \mathbf{n}] := \left\{ \mathbf{n}^T (\mathbf{t} + \lambda\, \hat{\mathbf{t}}) - \pi_{\mathbb{E}}[\mathbf{n}] \right\} \leq 0, \quad \forall \hat{\mathbf{t}} \in \mathbb{S} \tag{37}$$

for all unitary vectors $\mathbf{n}$. Obviously, only vertexes $\hat{\mathbf{t}}^{\beta} \in \mathbb{S}$ satisfying the condition

$$b[\mathbf{n}] := \mathbf{n}^T \hat{\mathbf{t}} = \max \tag{38}$$

for some $\mathbf{n}$ and are then significant in Eq. (37). A suitable approximation for this *selection rule* is obtained by considering only a certain number N_h of relevant directions $\mathbf{n}_h$ and select vertex $\hat{\mathbf{t}}^h$ associated with $\mathbf{n}_h$ by maximizing the projection $b[\mathbf{n}_h]$ for all possible $\hat{\mathbf{t}} \in \mathbb{S}$. Recalling the definition (14) of $\mathbb{S}$, this condition rewrites as

$$b_h := \max_{k=1\cdots p} \{b_{hk}\}, \quad b_{hk} = \sum_{j=1}^{p} \gamma_{kj}\, \mathbf{n}_k^T\, \hat{\mathbf{t}}_j, \quad \gamma_{kj} := \begin{cases} \psi_{kj}\, \alpha_j^{min} & \text{if } \mathbf{n}_k^T \hat{\mathbf{t}}_j < 0 \\ \psi_{kj}\, \alpha_j^{max} & \text{if } \mathbf{n}_k^T \hat{\mathbf{t}}_j \geq 0 \end{cases} \tag{39}$$

We obtain $\hat{\mathbf{t}}^h = \sum_{i=1}^{p} \gamma_{ki} \hat{\mathbf{t}}_i$ using the same kth combination that gives b_h. The use of this selection rule allows us to associate one vertex $\hat{\mathbf{t}}^h$ with each direction $\mathbf{n}_h$ and then to drastically reduce the number N_v of vertexes to be considered. In fact, we will have $N_v \leq N_h$. In particular note that $N_v < N_h$ if different directions are associated with the same vertex as usually occurs.

The filter rule previously described corresponds to retain, as vertexes of $\mathbb{S}$, only those that would be used when a piecewise external linearization of the elastic domain (see [13]) is adopted. As will be shown in Sect. 5, usually few control directions are sufficient to give very accurate results. In any case, as this selection is performed once and for all at the beginning of the analysis, it is also possible to use a large number of directions in order to select all the vertexes of the convex hull of $\mathbb{S}$. It is worth to note that the filtering procedure it is only used to select the convex hull vertexes while the shakedown analysis always uses, a yield function described by ellipsoids.

The proposed algorithm presents further advantages: (i) it directly provides the relevant vertexes of the elastic envelope without needing of the preliminary evaluation of all the vertexes corresponding to the load domain (2^p for each $\mathbb{S}^k$ using Eq. (13)) and the successive selection of those characterizing the convex hull of the domain; (ii) moreover, it is also suitable for load combinations defined in a more complex fashion with respect to that presented in Eq. (13) like, for instance, the combinations including contribution of dynamical modes through SRSS or CQC combination rules which are used in seismic analysis.

4.4 The Closest Point Projection Problem in Terms of Minkowski Sums of Ellipsoids

The CPP scheme in Eq. (21) has now to be expressed in terms of the MSE in which we write the plastic admissibility condition in Eqs. (12) and (36). To build up an efficient and accurate algorithm it has to be solved efficiently and in a robust fashion. In particular remembering that $\mathbf{t}_s = \mathbf{S}_s \boldsymbol{\beta}_e$ for $s = 0, \ell$ and introducing the following vector quantities

$$\boldsymbol{\beta}_{\mathscr{E}} = \begin{bmatrix} \boldsymbol{\beta}_e \\ \mathbf{t}_{0\mathscr{E}} \\ \mathbf{t}_{\ell\mathscr{E}} \end{bmatrix}, \quad \mathbf{p}_{\mathscr{E}} = \begin{bmatrix} \mathbf{p}_{0\mathscr{E}} \\ \mathbf{p}_{\ell\mathscr{E}} \end{bmatrix} \quad \mathbf{Q}_{\mathscr{E}}^T = \begin{bmatrix} -\mathbf{I}_{0\alpha}\mathbf{S}_0 & \boldsymbol{\Sigma}_{0\mathscr{E}} & \mathbf{0} \\ -\mathbf{I}_{\ell\alpha}\mathbf{S}_\ell & \mathbf{0} & \boldsymbol{\Sigma}_{\ell\mathscr{E}} \end{bmatrix},$$

the CPP problem rewrites as

$$\begin{cases} \text{minimize} & \frac{1}{2}\boldsymbol{\beta}_e{}^T \mathbf{H}_e \boldsymbol{\beta}_e - \boldsymbol{\beta}_e{}^T \mathbf{H}_e \boldsymbol{\beta}_e{}^* \\ \text{subject to} & \mathbf{Q}_{\mathscr{E}}^T \boldsymbol{\beta}_{\mathscr{E}} = \mathbf{p}_{\mathscr{E}} \\ & \boldsymbol{\Phi}_{\mathscr{E}}[\boldsymbol{\beta}_{\mathscr{E}}] \leq \mathbf{0} \end{cases} \tag{40}$$

where $\boldsymbol{\Phi}_{\mathscr{E}}[\boldsymbol{\beta}_{\mathscr{E}}] = \{\boldsymbol{\Phi}_{0\mathscr{E}}[\mathbf{t}_{0\mathscr{E}}], \boldsymbol{\Phi}_{\ell\mathscr{E}}[\mathbf{t}_{\ell\mathscr{E}}]\}$.

4.4.1 The Return Mapping Solution of the Closest Point Projection Problem

The CPP problem defined in Eq. (40) over the element can be solved using an approach similar to that used for solving the global problem in Eq. (18).

In particular the CPP problem has again the equality constraints as *complicating constraints* that couple all the ellipsoids contribution and various dual decomposition techniques can be applied for its solution. In the following we briefly describe one of these to illustrate how the dual decomposition and the proximal point algorithms can be applied to this problem.

We again add quadratic terms to the objective function of problem (40) so obtaining again a sequence of steps n as for Eq. (18). The nth of these steps is defined by the problem

$$\begin{cases} \text{minimize} & \frac{1}{2}\Delta\boldsymbol{\beta}_{\mathscr{E}}^T \mathbf{H}_{\mathscr{E}} \Delta\boldsymbol{\beta}_{\mathscr{E}} \\ \text{subject to} & \mathbf{Q}_{\mathscr{E}}^T \boldsymbol{\beta}_{\mathscr{E}} = \mathbf{p}_{\mathscr{E}} \\ & \boldsymbol{\Phi}_{\mathscr{E}}[\boldsymbol{\beta}_{\mathscr{E}}] \leq \mathbf{0} \end{cases} \tag{41}$$

where

$$\mathbf{H}_{\mathscr{E}} = \text{diag}\left[\mathbf{H},\, \mathbf{H}_I^1,\, \ldots,\, \mathbf{H}_I^{\alpha} \ldots\right], \qquad \Delta\boldsymbol{\beta}_{\mathscr{E}} = \boldsymbol{\beta}_{\mathscr{E}} - \boldsymbol{\beta}_{0\mathscr{E}} := \begin{bmatrix} \boldsymbol{\beta}_e - \boldsymbol{\beta}_e^* \\ \Delta\mathbf{t}_{0\mathscr{E}} \\ \Delta\mathbf{t}_{\ell\mathscr{E}} \end{bmatrix}$$

The positive definite matrices $\mathbf{H}_I^{\alpha}$ can be appropriately selected to simplify the solution process.

Letting $\Delta\mathbf{v}$ be the Lagrangian multipliers of the equality constraints and $\Delta\boldsymbol{\kappa}_{\mathscr{E}}^T$ be that of the inequality ones we obtain the following first order condition for the problem (41)

$$\begin{aligned} &\mathbf{r}_{\mathbf{v}} := \mathbf{Q}_{\mathscr{E}}^T\boldsymbol{\beta}_{\mathscr{E}} - \mathbf{p}_{\mathscr{E}} = \mathbf{0} \\ &\mathbf{r}_{\boldsymbol{\beta}} := \mathbf{H}_{\mathscr{E}}(\boldsymbol{\beta}_{\mathscr{E}} - \boldsymbol{\beta}_{0\mathscr{E}}) - \mathbf{Q}_{\mathscr{E}}\Delta\mathbf{v} + \mathbf{A}\boldsymbol{\kappa}_{\mathscr{E}} = \mathbf{0} \\ &\boldsymbol{\Phi}_{e\mathscr{E}}[\boldsymbol{\beta}_{\mathscr{E}}] \leq \mathbf{0}, \quad \Delta\boldsymbol{\kappa}_{\mathscr{E}} \geq 0, \quad \Delta\boldsymbol{\kappa}_{\mathscr{E}}^T\boldsymbol{\Phi}_{e\mathscr{E}} = \mathbf{0} \end{aligned} \tag{42}$$

are solved using a dual decomposition technique

- *Elastic predicor*
  Assuming $\Delta\boldsymbol{\kappa} = \mathbf{0}$ in the last equations (42) we have

$$\boldsymbol{\beta}_{\mathscr{E}}^* = \boldsymbol{\beta}_{0\mathscr{E}} + \mathbf{H}_{\mathscr{E}}^{-1}(\mathbf{Q}_{\mathscr{E}}\Delta\mathbf{v}) \tag{43}$$

  that represents the ellipsoids decomposition of *trial stress* $\boldsymbol{\beta}_{\mathscr{E}}^*$. The initial value of the process is obtained by substituting the so evaluated $\boldsymbol{\beta}_{\mathscr{E}}^*$ in the first of Eqs. (42) so obtaining

$$\mathbf{K}_{\mathscr{E}}\Delta\mathbf{v} = \bar{\mathbf{p}}_{\mathscr{E}}$$

  where

$$\mathbf{K}_{\mathscr{E}} = \mathbf{Q}_{\mathscr{E}}^T\mathbf{H}_{\mathscr{E}}^{-1}\mathbf{Q}_{\mathscr{E}}, \qquad \bar{\mathbf{p}}_{\mathscr{E}} = \mathbf{p}_{\mathscr{E}} - \mathbf{Q}_{\mathscr{E}}^T\boldsymbol{\beta}_{0\mathscr{E}}$$

- *Return mapping corrector*
  With fixed $\Delta\mathbf{v}$ the following problem is solved

$$\begin{cases} \text{minimize} & \frac{1}{2}(\boldsymbol{\beta}_{\mathscr{E}} - \boldsymbol{\beta}_{\mathscr{E}}^*)^T\mathbf{H}_{\mathscr{E}}(\boldsymbol{\beta}_{\mathscr{E}} - \boldsymbol{\beta}_{\mathscr{E}}^*) \\ & \boldsymbol{\Phi}_{\mathscr{E}}[\boldsymbol{\beta}_{\mathscr{E}}] \leq \mathbf{0} \end{cases} \tag{44}$$

  that represents the CPP problem for the stress decomposed as MSE. Due to the block nature of matrix $\mathbf{H}_{\mathscr{E}}$ and of the yield function $\boldsymbol{\Phi}_{\mathscr{E}}[\boldsymbol{\beta}_{\mathscr{E}}]$ it can be separated for each ellipsoid component as follows

$$
\begin{cases} \min\limits_{\beta_{eI}^{\alpha}} & \frac{1}{2}\left(\mathbf{t}_{sI}^{\alpha}-\mathbf{t}_{sI}^{\alpha*}\right)^T \mathbf{H}_{eI}^{\alpha}\left(\mathbf{t}_{sI}^{\alpha}-\mathbf{t}_{sI}^{\alpha*}\right), \\ \text{subject to:} & \phi_I[s,\mathbf{t}_{sI}^{\alpha}] \le 0. \end{cases} \tag{45}
$$

where $\mathbf{t}_{sI}^{\alpha*}$ is the component of $\boldsymbol{\beta}_{\mathscr{E}}^{*}$ in Eq. (43) that corresponds to $\mathbf{t}_{sI}^{\alpha}$. The solution of this problem, using the predictor corrector scheme is really very simple when we select for $\mathbf{H}_{eI}^{\alpha}$ the same matrix $\mathbf{C}_I$ that defines the corresponding constraint, obtaining in this way a standard *radial return mapping process*.

- *Equality constraints solution*
  At the element level we have to solve the equality constraints

$$
\mathbf{Q}_{\mathscr{E}}^T\boldsymbol{\beta}_{\mathscr{E}}[\mathbf{v}]-\mathbf{p}_{\mathscr{E}}=\mathbf{0} \tag{46}
$$

As in the previous algorithm we use a Newton scheme that letting

$$
\begin{cases} \Delta\mathbf{v}_{j+1}=\Delta\mathbf{v}_j+\dot{\mathbf{v}} \\ \mathbf{K}_{t\mathscr{E}}\dot{\mathbf{v}}=-\mathbf{r}_{\mathbf{v}j}, \end{cases}
$$

where the algorithmic tangent matrix $\mathbf{K}_{t\mathscr{E}}=\mathbf{Q}_{\mathscr{E}}^T\mathbf{C}_{t\mathscr{E}}\mathbf{Q}_{\mathscr{E}}$ can easily be updated starting from the initial valued $\mathbf{C}_{\mathscr{E}}=\mathbf{H}_{\mathscr{E}}^{-1}$. In fact we obtain for radial return mapping that

$$
\mathbf{C}_{t\mathscr{E}}=\operatorname{diag}\{\cdots\frac{\mathbf{H}_{eI}^{\alpha}}{1-c_{sI}^{\alpha}},\cdots\}
$$

where $0\le c_{sI}^{\alpha}\le 1$ is a scalar function of the plastic multiplier.

5 Numerical Results

In this section we present some tests regarding the accuracy and the efficiency of the new proposal in the shakedown analysis of 3D reinforced concrete frames.

5.1 Yield Function and Elastic Envelope Vertexes for an L-Shaped Section

In Fig. 4 the yield function approximations for L shaped reinforced concrete (see also [5]) are reported on the basis of 1, 3 and 5 ellipsoids. The points in blue correspond

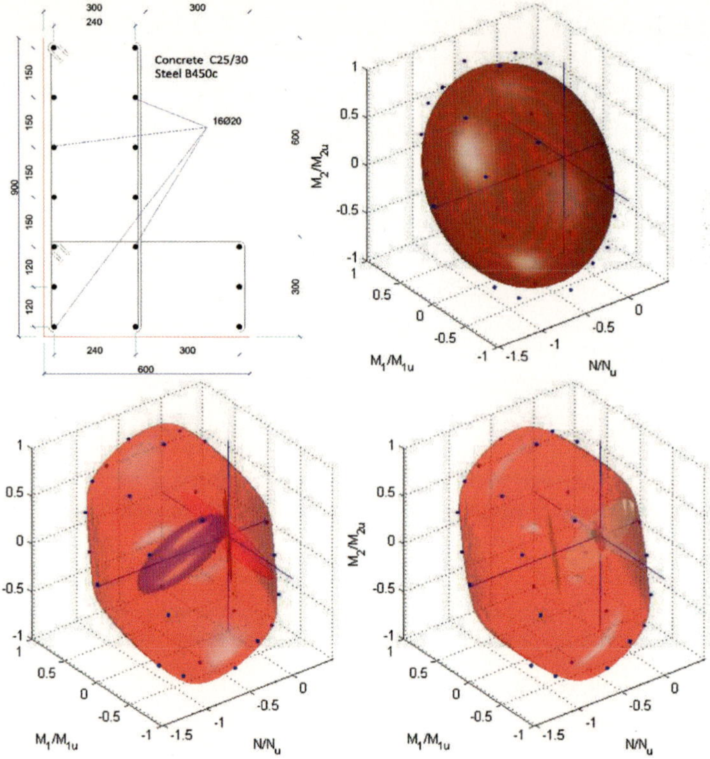

Fig. 4 Yield function approximations with 1, 3 and 5 ellipsoids for the considered cross-section

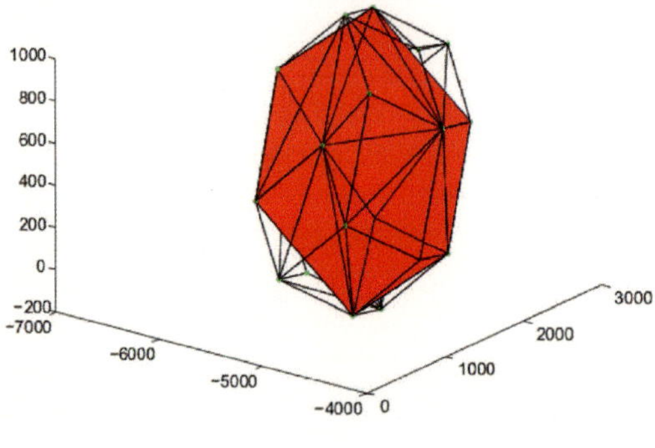

	Basic elastic actions				
	$\hat{N}_1$	$\hat{M}_2$	$\hat{M}_3$	α_j^{min}	α_j^{max}
$\hat{\mathbf{t}}_1$	407.6	-829.9	-412.4	-0.716	0.311
$\hat{\mathbf{t}}_2$	1174.1	-456.7	566.8	-0.929	-0.156
$\hat{\mathbf{t}}_3$	-5380.2	96.7	550.7	0.698	0.831
$\hat{\mathbf{t}}_4$	237.9	943.4	-17.6	0.584	0.868
$\hat{\mathbf{t}}_5$	-1127.1	958.6	361.7	0.357	0.919

Fig. 5 Convex hull evaluation, the approximated one in *red*

to the true values of $\mathbf{t}_{yk}$. The increase in accuracy is evident in the case of more ellipsoids.

In Fig. 5 we report, for the same section, the convex hull vertexes of $\mathbb{S}$ for a sequence of 5 assigned $\hat{\mathbf{t}}_j$ with $0 \leq \alpha_j \leq 1$. The Figure corresponds to the exact and approximate evaluation of $\mathbb{S}$ using the 26 filtering hyperplanes proposed in [5]. It is worth noting as the true convex hull of $\mathbb{S}_e$ is defined by $22 \leq 2^5 = 32$ vertexes while the filtering procedure only select 18 vertexes.

5.2 Shakedown Analysis of a Continuous Beam

The first numerical test on shakedown analysis refers to the multibay simply supported beam shown in Fig. 6, with constant section characterized by ultimate strength M_0 and already analyzed in [5]. Two elements are used to describe each horizontal span. External loads are defined by the simple rule

$$p[t] := \alpha_1[t]P_1 + \alpha_2[t]P_2, \quad \begin{cases} 0 \leq \alpha_1[t] \leq 1 \\ 0 \leq \alpha_2[t] \leq 1 \end{cases}$$

The values regarding the elastic multiplier and the elastic and plastic shakedown multipliers are shown in Table 1 for the assigned ratio $P_2 = 2\ P_1$ and using a different number of ellipsoids and the exact or approximate evaluation of the convex hull.

In particular, in this simple case, the use of the simplified evaluation of the convex hull furnishes exactly, for each section, the true convex hull vertexes also using only the 26 filtering hyperplanes in [5]. Differences in the evaluation of the multipliers can be observed by changing the number of the ellipsoids used to approximate the elastic domain. In particular, the elastic limit λ_e and the shakedown multiplier λ_a present the larger differences due to the worst approximation of the elastic domain in the zone with $N_1 \approx 0$ that characterizes the elastic solution. A better match is obtained for the plastic multiplier λ_p (see [14] for its definition). On the contrary the shakedown multiplier λ_a presents negligible differences as the stresses corresponding to the ratcheting mechanism are in the zone of the maximum flexural moment, that is in the zone of $\mathbb{E}$ better approximated also with a single ellipsoid (see Fig. 7).

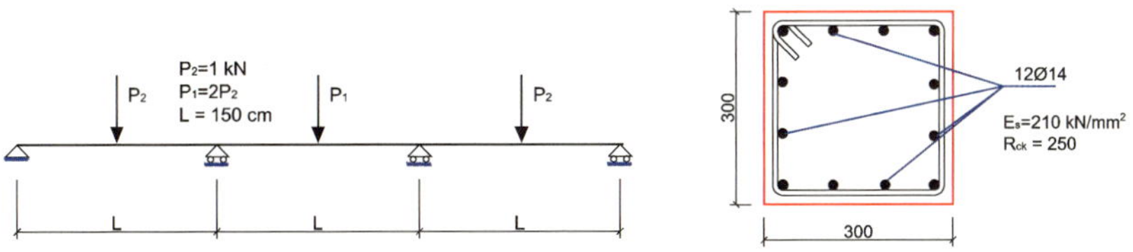

Fig. 6 Loads and geometry for the continuous beam

Table 1 Shakedown multipliers

	1 Ellipsoid		3 Ellipsoids	
	$N_h = 26$	Exact	$N_h = 26$	Exact
λ_e	82.205	82.205	75.822	75.822
λ_s	124.57	124.57	128.97	128.97
λ_p	168.39	168.39	174.33	174.33

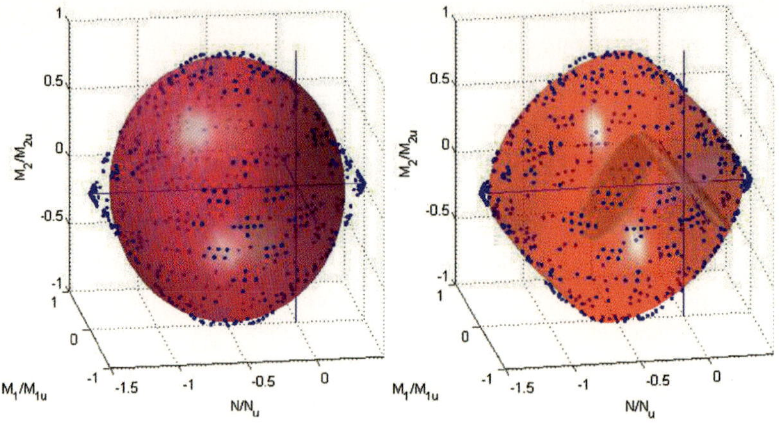

Fig. 7 Continuous beam: yield domain approximation for the *rectangular* section

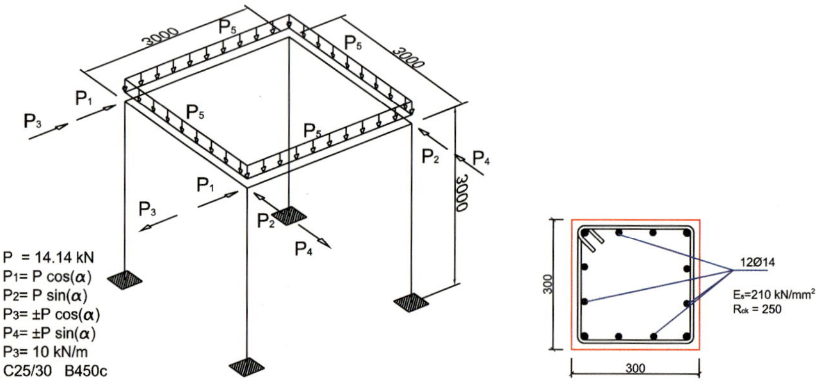

Fig. 8 Loads and geometry

5.3 Shakedown Analysis of a 3D Frame

The second test refers to the four storey 3D frame shown in Fig. 8 and already analyzed in [5]. The structure is under the action of varying horizontal concentrated loads $P_1 = P\cos(\alpha)$, $P_2 = P\sin(\alpha)$, $P_3 = \pm P_1$, $P_4 = \pm P_2$. These forces act in the presence of distributed vertical loads P_5 applied on the beams and defined in Table 2. The results refer to a counterclockwise angle of $\alpha = \pi/4$ with respect to the direction of P_1. The load domain is defined by

$$p[t] = \sum_{i=1}^{5} \alpha_i[t] P_i, \qquad \begin{cases} -1 \leq \alpha_1[t], \alpha_2[t] \leq 1 \\ -1 \leq \alpha_3[t], \alpha_4[t] \leq 1 \\ 0.9 \leq \alpha_5[t] \leq 1 \end{cases}$$

Table 2 3D Frame: Shakedown multipliers

	Single ellipsoid		Three ellipsoids	
	$N_h = 26$	Exact	$N_h = 26$	Exact
λ_e	40.765	40.765	38.610	38.610
λ_s	41.205	41.205	40.005	39.925
λ_p	47.206	47.206	47.330	47.330

The simplified evaluation of $\mathbb{S}$ has been performed using the 26 hyperplanes defined in [5]. In this case the selection rule described in Sect. 4.3 don't furnishes exactly the same vertexes as the exact convex hull of $\mathbb{S}$ for each section. However the results are coincident (or practically coincident) with those obtained using the complete convex hull, as can be observed in Table 2 where the values of the elastic λ_e, plastic λ_p and shakedown λ_a multipliers are reported.

Some differences, in the evaluation of the multipliers can be observed by changing the number of the ellipsoids used to approximate the elastic domain.

6 Conclusions

An FEM procedure for the shakedown analysis of 3D frames subject to complex loading conditions has been presented, based on the iterative method proposed in [12, 14]. Only flexural plastic mechanisms have been considered while the yield surface of the section has been obtained by approximating the true values obtained by exploiting the support function concept, by means of a Minkowski sum of ellipsoids. The return mapping by closest point projection process, coming from the dual decomposition algorithm, is further decomposed with a dual technique. In this way a simple and effective radial return mapping process can be used for its solution.

With the aim of reducing the number of vertexes of the elastic envelope of each section an approximate convex hull is evaluated retaining only the significant vertexes with respect to a piecewise external linearization of the yield surface. Increasing the number of hyperplanes we can evaluate the true convex hull of the elastic domain. However the numerical tests show how the error, also when using a limited number of hyperplanes, is not meaningful from a practical point of view.

The algorithm proposed can be effectively used for the design and the analysis of reinforced concrete 3D frames subject to complex and large load combinations.

Acknowledgments The research leading to these results has received regional funding from the European Communitys Seventh Framework Programme FP7-FESR: "PIA Pacchetti Integrati di Agevolazione industria, artigianato e servizi" in collaboration with the Newsoft s.a.s. (www.newsoft-eng.it).

References

1. Zouain N (2004) Shakedown and safety assessment. Encyclopedia of computational mechanics, vol 2. Wiley, New York, pp 291–334
2. Skordeli M-A, Bisbos C (2010) Limit and shakedown analysis of 3D steel frames via approximate ellipsoidal yield surfaces. Eng Struct 32(6):1556–1567
3. Nguyen A, Hachemi A, Weichert D (2008) Application of the interior-point method to shakedown analysis of pavements. Int J Numer Methods Eng 75(4):414–439
4. Bleyer J, De Buhan P (2013) Yield surface approximation for lower and upper bound yield design of 3d composite frame structures. Comput Struct 129:86–98
5. Malena M, Casciaro R (2008) Finite element shakedown analysis of reinforced concrete 3d frames. Comput Struct 86(11–12):1176–1188
6. Zouain N, Borges L, Silveira JL (2002) An algorithm for shakedown analysis with nonlinear yield functions. Comput Methods Appl Mech Eng 191(2324):2463–2481. doi:10.1016/S0045-7825(01)00374-7
7. Vu D, Yan A, Nguyen-Dang H (2004) A primal-dual algorithm for shakedown analysis of structures. Comput Methods Appl Mech Eng 193(42–44):4663–4674
8. Tran T, Liu G, Nguyen-Xuan H, Nguyen-Thoi T (2010) An edge-based smoothed finite element method for primal-dual shakedown analysis of structures. Int J Numer Methods Eng 82(7):917–938
9. Simon J-W, Kreimeier M, Weichert D (2013) A selective strategy for shakedown analysis of engineering structures. Int J Numer Methods Eng 94:985–1014
10. Simon J-W, Weichert D (2012) Shakedown analysis of engineering structures with limited kinematical hardening. Int J Solids Struct 49(15–16):2177–2186
11. Simon J-W, Weichert D (2012) Shakedown analysis with multidimensional loading spaces. Comput Mech 49(4):477–485
12. Casciaro R, Garcea G (2002) An iterative method for shakedown analysis. Comput Methods Appl Mech Eng 191(49–50):5761–5792
13. Garcea G, Armentano G, Petrolo S, Casciaro R (2005) Finite element shakedown analysis of two-dimensional structures. Int J Numer Methods Eng 63(8):1174–1202. doi:10.1002/nme.1316
14. Garcea G, Leonetti L (2011) A unified mathematical programming formulation of strain driven and interior point algorithms for shakedown and limit analysis. Int J Numer Methods Eng 88:1085–1111
15. Bilotta A, Leonetti L, Garcea G (2011) Three field finite elements for the elastoplastic analysis of 2D continua. Finite Elem Anal Des 47(10):1119–1130
16. Rosati L, Marmo F, Serpieri R (2008) Enhanced solution strategies for the ultimate strength analysis of composite steel-concrete sections subject to axial force and biaxial bending. Comput Methods Appl Mech Eng 197(9–12):1033–1055
17. Genoese A, Genoese A, Bilotta A, Garcea G (2013) A mixed beam model with non-uniform warpings derived from the Saint Venànt rod. Comput Struct 121:87–98
18. Garcea G, Madeo A, Zagari G, Casciaro R (2009) Asymptotic post-buckling Fem analysis using corotational formulation. Int J Solids Struct 46(2):377–397
19. Petrolo A, Casciaro R (2004) 3D beam element based on Saint Venant's rod theory. Comput Struct 82(29–30):2471–2481
20. Genoese A, Genoese A, Bilotta A, Garcea G (2014) A geometrically exact beam model with non-uniform warping coherently derived from the Saint Venant rod. Eng Struct 68(0):33–46. doi:10.1016/j.engstruct.2014.02.024, http://www.sciencedirect.com/science/article/pii/S0141029614001175
21. Genoese A, Genoese A, Bilotta A, Garcea G (2014) A generalized model for heterogeneous and anisotropic beams including section distortions. Thin-Walled Struct 74:85–103
22. Genoese A, Genoese A, Bilotta A, Garcea G (2014) A composite beam model including variable warping effects derived from a generalized Saint Venant solution. Compos Struct 110:140–151

23. European Union (1991) EN 1991—Eurocode 1: Actions on structures
24. Bilotta A, Leonetti L, Garcea G (2012) An algorithm for incremental elastoplastic analysis using equality constrained sequential quadratic programming. Comput Struct 102–103:97–107
25. König J (1987) Shakedown of elastic-plastic structures. Elsevier, Amsterdam

Uncertainty Multimode Failure and Shakedown Analysis of Shells

Thanh Ngọc Trần and M. Staat

Abstract This paper presents a numerical procedure for reliability analysis of thin plates and shells with respect to plastic collapse or to inadaptation. The procedure involves a deterministic shakedown analysis for each probabilistic iteration, which is based on the upper bound approach and the use of the exact Ilyushin yield surface. Probabilistic shakedown analysis deals with uncertainties originated from the loads, material strength and thickness of the shell. Based on a direct definition of the limit state function, the calculation of the failure probability may be efficiently solved by using the First and Second Order Reliability Methods (FORM and SORM). The problem of reliability of structural systems (series systems) is handled by the application of a special technique which permits to find all the design points corresponding to all the failure modes. Studies show, in this case, that it improves considerably the FORM and SORM results.

Keywords Limit analysis · Shakedown analysis · Reliability analysis · Multimode failure · Non-linear optimization

1 Introduction

The plastic collapse limit and the shakedown limit which define the load-carrying capacity of structures are important in assessing the structural integrity. Due to the high expenses of experimental setups and the time consuming full elastic-plastic cyclic loading analysis, the determination of these limits by means of numerical direct plasticity methods has been of great interest to many designers. Moreover, a

T.N. Trần
Chair of Mechanics and Robotics, University of Duisburg-Essen, Lotharstr. 1, 47057 Duisburg, Germany
e-mail: thanh.tran@uni-due.de

M. Staat (✉)
Institute for Bioengineering, Biomechanics Laboratory, Aachen University of Applied Sciences, Jülich Campus, Heinrich-Mußmann-Str. 1, 52428 Jülich, Germany
e-mail: m.staat@fh-aachen.de

P. Fuschi et al. (eds.), *Direct Methods for Limit and Shakedown Analysis of Structures*,
DOI 10.1007/978-3-319-12928-0_14

certain evaluation of structural performance can be conducted only if the uncertainty of the actual load-carrying capacity of the structure is taken into consideration since all resistance and loading variables are random in nature. To ensure the safety of the structures to be designed, two approaches are normally used [1]. The classical approach fixes the values of the safety factors and chooses the values of the design variables to satisfy the safety conditions. All the variables involved are then assumed to be deterministic and fixed to particular quantiles, i.e. mean value or characteristic values [2]. The probability-based approach deals directly with realistic random variables to find the global probability of failure as the basic design criterion. Obviously, the later problem is more difficult since the evaluation of the probability of failure is not an easy task, especially when the structure has more than one failure mode (multimode failure or multiple design points). In this case, analysis of the structural system is required to evaluate the safety of the structure as a whole [1]. To handle problems of this kind, the real structure is sometimes modelled by an equivalent system in such a way that all relevant failure modes can be treated [2].

The reliability analysis of plates and shells with respect to plastic collapse or to inadaptation was formulated on the basis of limit and shakedown theorems [3]. The technique was based upon an upper bound approach using a re-parameterized exact Ilyushin yield surface and nonlinear optimization procedures. Based on a direct definition of the limit state function, the non-linear problems may be efficiently solved by using the First and Second Order Reliability Methods (FORM and SORM). In order to get the design point, a non-linear optimization was implemented. FORM and SORM match particularly well with direct plasticity methods because they render the problem time invariant.

The non-linear optimization algorithm developed in [3] is guaranteed to converge to a minimum-distance point on the limit state surface, provided that the limit state function is continuous and differentiable. However, as with any non-convex optimization problem, it is not guaranteed that the solution point is the global minimum-distance point when the system has more than one failure mode. This paper aims at extending the method developed in [3] for the probabilistic shakedown analysis of multimode-failure plate and shell structures. A method to successively find the multiple design points of a component reliability problem, when they exist on the limit state surface, is presented. Each design point corresponds with an individual failure mode or mechanism. FORM and SORM approximations are applied at each design point followed by a series system reliability analysis to lead to improved estimates of the system failure probability.

2 Deterministic Shakedown Analysis of Shells

2.1 Governing Equations

In this study, we restrict ourselves to the case of homogeneous materials and shells of constant thickness in which the yield limit σ_y and thickness h are the same at every

Gaussian point of the structure. So we can write

$$\sigma_y = Y\sigma_0, \quad h = Zh_0, \tag{1}$$

where σ_0, h_0 are constant reference values and Y, Z are random variables. Let E denote the Young's modulus, ν is the Poisson's ratio, $N_{\alpha\beta}, M_{\alpha\beta}, \bar{\varepsilon}_{\alpha\beta}, \kappa_{\alpha\beta} (\alpha, \beta \in \{1, 2\})$ are the physical membrane forces, bending moments, mid-plane strains and curvatures, respectively (Fig. 1). We define the non-dimensional stress and strain resultant vectors as follows

$$\boldsymbol{\sigma} = \left[\mathbf{n} \ \mathbf{m} \right]^T, \quad \boldsymbol{\varepsilon} = \left[\bar{\mathbf{e}} \ \mathbf{k} \right]^T, \tag{2}$$

with

$$\begin{aligned} \mathbf{n} &= \frac{1}{N_0} \left[N_{11} \ N_{22} \ N_{12} \right]^T, \quad \mathbf{m} = \frac{1}{M_0} \left[M_{11} \ M_{22} \ M_{12} \right]^T, \\ \bar{\mathbf{e}} &= \frac{1}{\varepsilon_0} \left[\bar{\varepsilon}_{11} \ \bar{\varepsilon}_{22} \ 2\bar{\varepsilon}_{12} \right]^T, \quad \mathbf{k} = \frac{1}{\kappa_0} \left[\kappa_{11} \ \kappa_{22} \ 2\kappa_{12} \right]^T, \end{aligned} \tag{3}$$

and $N_0 = \sigma_0 h_0$, $M_0 = \sigma_0 h_0^2/4$, $\varepsilon_0 = \sigma_0(1 - \nu^2)/E$ and $\kappa_0 = 4\varepsilon_0/h_0$ are the normalized quantities.

The quadratic strain intensities are defined by

$$\begin{aligned} P_\varepsilon &= \frac{3}{4} \left(d\boldsymbol{\varepsilon}^p \right)^T \mathbf{P}_1 d\boldsymbol{\varepsilon}^p, \quad (\geq 0) \\ P_{\varepsilon\kappa} &= 3 \left(d\boldsymbol{\varepsilon}^p \right)^T \mathbf{P}_2 d\boldsymbol{\varepsilon}^p, \\ P_\kappa &= 12 \left(d\boldsymbol{\varepsilon}^p \right)^T \mathbf{P}_3 d\boldsymbol{\varepsilon}^p, \quad (\geq 0) \end{aligned} \tag{4}$$

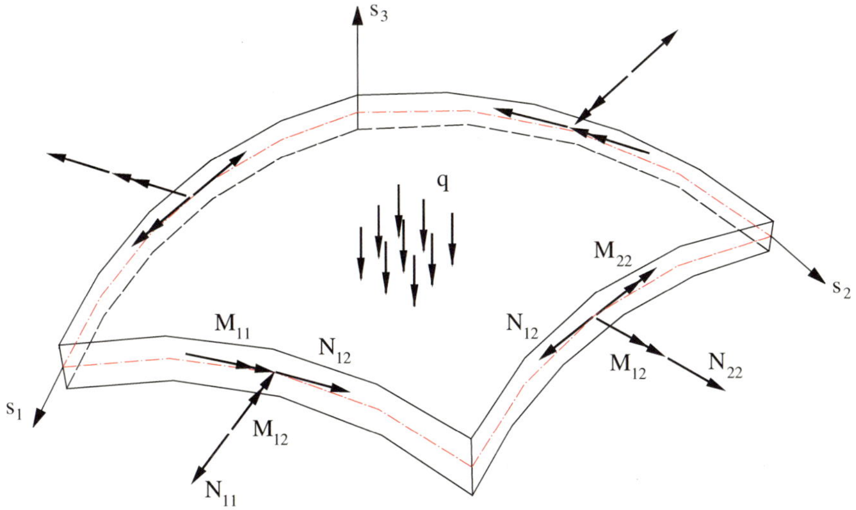

Fig. 1 Static shell quantities

where $d\boldsymbol{\varepsilon}^p$ denotes the increment of the plastic strain resultant vector, and $\mathbf{P}_i\,(i = 1, 2, 3)$ are

$$\mathbf{P}_1 = \begin{pmatrix} \mathbf{P} & \mathbf{0} \\ \mathbf{0} & \mathbf{0} \end{pmatrix}, \quad \mathbf{P}_2 = \begin{pmatrix} \mathbf{0} & \mathbf{P}/2 \\ \mathbf{P}/2 & \mathbf{0} \end{pmatrix}, \quad \mathbf{P}_3 = \begin{pmatrix} \mathbf{0} & \mathbf{0} \\ \mathbf{0} & \mathbf{P} \end{pmatrix}, \quad \mathbf{P} = \begin{pmatrix} 4/3 & 2/3 & 0 \\ 2/3 & 4/3 & 0 \\ 0 & 0 & 1/3 \end{pmatrix}. \tag{5}$$

By introducing some parameters as

$$\upsilon = \frac{P_\varepsilon}{P_\kappa}, \quad \beta = -\frac{P_{\varepsilon\kappa}}{P_\kappa} \quad \text{and} \quad \gamma = \upsilon - \beta^2, \tag{6}$$

then, the plastic dissipation function for a shell structure may be written in the form [4]

$$D^p(\dot{\boldsymbol{\varepsilon}}^p) = YN_0\varepsilon_0\sqrt{\frac{P_\kappa}{3}}\left(\beta_1\sqrt{\beta_1^2+\gamma} + \beta_2\sqrt{\beta_2^2+\gamma} + \gamma K_0\right), \tag{7}$$

with

$$\beta_1 = \frac{Z}{2} - \beta, \quad \beta_2 = \frac{Z}{2} + \beta, \quad K_0 = \ln\left(\frac{\sqrt{\beta_1^2+\gamma}+\beta_1}{\sqrt{\beta_2^2+\gamma}-\beta_2}\right). \tag{8}$$

2.2 Upper Bound Shakedown Analysis

Consider a convex polyhedral load domain $\mathcal{L}$ and a special loading path consisting of all load vertices $\hat{P}_k$ $(k = 1, \ldots, m)$ of $\mathcal{L}$. Let $\boldsymbol{\sigma}^E$ denote the fictitious elastic stress resultant vector, $\dot{\mathbf{u}}$ denote the velocity fields, $\mathbf{B}_i$ denote the deformation matrix, and w_i be the weighting factor of the ith Gauss point $(i = 1, \ldots, NG)$. According to the kinematic shakedown theorem, the upper bound shakedown load multiplier may be formulated as [4]

$$\alpha_{\lim} = \min \sum_{i=1}^{NG}\sum_{k=1}^{m} w_i YN_0\varepsilon_0\sqrt{\frac{P_\kappa}{3}}\left(\beta_1\sqrt{\beta_1^2+\gamma} + \beta_2\sqrt{\beta_2^2+\gamma} + \gamma K_0\right),$$

$$\text{s.t.:} \begin{cases} \sum\limits_{k=1}^{m} \dot{\boldsymbol{\varepsilon}}_{ik} = \mathbf{B}_i\dot{\mathbf{u}} & \forall i = 1, \ldots, NG, \\ \sum\limits_{i=1}^{NG}\sum\limits_{k=1}^{m} w_i N_0\varepsilon_0 \dot{\boldsymbol{\varepsilon}}_{ik}^T \sigma_{ik}^E = 1. \end{cases} \tag{9}$$

By introducing some new notations

$$\dot{\mathbf{e}}_{ik} = w_i\dot{\boldsymbol{\varepsilon}}_{ik}, \quad \mathbf{t}_{ik} = N_0\varepsilon_0\sigma_{ik}^E, \quad \hat{\mathbf{B}}_i = w_i\mathbf{B}_i, \tag{10}$$

and replacing $d\boldsymbol{\varepsilon}^p$ in Eq. (4) by $\dot{\mathbf{e}}$ for calculating the quadratic strain intensities P_ε, $P_{\varepsilon\kappa}$ and P_κ, the formulation in (9) is simplified as follows

$$\alpha_{\lim} = \min \sum_{i=1}^{NG} \sum_{k=1}^{m} YN_0\varepsilon_0\sqrt{\frac{P_\kappa}{3}}\left(\beta_1\sqrt{\beta_1^2+\gamma}+\beta_2\sqrt{\beta_2^2+\gamma}+\gamma K_0\right),$$

$$\text{s.t.:}\quad \begin{cases} \sum\limits_{k=1}^{m} \dot{\mathbf{e}}_{ik} = \hat{\mathbf{B}}_i\dot{\mathbf{u}} & \forall i = 1,\ldots,NG, \\ \sum\limits_{i=1}^{NG}\sum\limits_{k=1}^{m} \dot{\mathbf{e}}_{ik}^T\mathbf{t}_{ik} = 1. \end{cases} \tag{11}$$

Note that β_i and γ in (11) are the same as those in (9) after replacing $\dot{\boldsymbol{\varepsilon}}$ or $d\boldsymbol{\varepsilon}^p$ by $\dot{\mathbf{e}}$. Dealing with the nonlinear constrained optimization problem (11), an efficient iterative algorithm for large-scale optimization problems which is successfully applied in [4, 5] is employed. Details of the iterative algorithm can be found in [5]. It is noted that when $m = 1$, the formulation (11) reduces to limit analysis.

3 Probabilistic Formulation

Let $\mathbf{X} = (X_1, X_2, \ldots, X_n)$ be an n-dimensional random variable vector characterizing uncertainties in loads, material strength and shell thickness, and $\mathbf{x} = (x_1, x_2, \ldots, x_n)$ be *realizations* of $\mathbf{X}$. The deterministic safety margin is based on the comparison of a structural resistance (threshold) R and loading S. With R, S are functions of $\mathbf{X}$, the structure fails for any realization with non-positive *failure function or limit state function*, i.e.

$$g(\mathbf{X}) = R(\mathbf{X}) - S(\mathbf{X}) \begin{cases} <0 & \text{for failure,} \\ =0 & \text{for limit state,} \\ >0 & \text{for safe structure.} \end{cases} \tag{12}$$

The limit state function $g(\mathbf{x}) = 0$, defines the limit state hyper-surface ∂F which separates the failure region $F = \{\mathbf{x} |\, g(\mathbf{x}) < 0\}$ from the safe region. The failure probability P_f is the probability that $g(\mathbf{X})$ is non-positive, i.e.

$$P_f = P\left(g(\mathbf{X}) \le 0\right) = \int_F f_{\mathbf{X}}(\mathbf{x})d\mathbf{x}, \tag{13}$$

where $f_{\mathbf{X}}(\mathbf{x})$ is the n-dimensional joint probability-density function. In general, evaluation of the integral in (13) for an arbitrary failure region may not be possible. Therefore, approximation methods are obviously needed. In this study, the First- and Second-Order Reliability Methods (FORM and SORM) are used to evaluate the failure probability. FORM and SORM are the most effective methods if gradient information is available [6]. The details of probabilistic limit and shakedown analysis for one design point can be found in [7, 8].

3.1 Definition of the Limit State Function

As mentioned above, the limit state function contains the parameters of structural resistance and loading. By normalization with the actual load P, the limit state function can be expressed by the form [3, 4]

$$g = \alpha_{\lim} - 1, \tag{14}$$

in which $\alpha_{\lim}$ is the shakedown load factor calculated by (11). The limit state function is the function of the yield stress variable $Y = X_{n-1}$, thickness variable $Z = X_n$ and load variables X_j $(j = 1, \ldots, n-2)$. The actual load Q, in general, can be defined by its components as follows

$$Q = \sum_{j=1}^{n-2} x_j Q_j^0, \tag{15}$$

where x_j and Q_j^0 are the realization and constant reference load of the jth basic load variable X_j, respectively. By that way, the actual fictitious elastic stress resultants $\boldsymbol{\sigma}^E$ and $\mathbf{t}$ can also be decomposed as

$$\boldsymbol{\sigma}^E = \sum_{j=1}^{n-2} x_j \boldsymbol{\sigma}_j^0, \quad \mathbf{t} = \sum_{j=1}^{n-2} x_j \mathbf{t}_j^0. \tag{16}$$

The Jacobian and the Hessian of the limit state function, $\partial g/\partial \mathbf{x}$ and $\partial^2 g/\partial \mathbf{x}^2$, are needed for the FORM and SORM as well as for finding the most likely failure points. They must be first calculated at each probabilistic iteration in the physical $\mathbf{x}$-space. Then it is mapped into the standardized Gaussian $\mathbf{u}$-space by using the chain rule

$$\begin{aligned} \nabla_u g(\mathbf{u}) &= \nabla_u g(\mathbf{x}) = \nabla_x g(\mathbf{x}) \nabla_u \mathbf{x}, \\ \nabla_u^2 g(\mathbf{u}) &= \nabla_u \left(\nabla_x g(\mathbf{x}) \nabla_u \mathbf{x}\right) = \left(\nabla_u \mathbf{x}\right)^T \nabla_x^2 g(\mathbf{x}) \nabla_u \mathbf{x} + \nabla_x g(\mathbf{x}) \nabla_u^2 \mathbf{x}. \end{aligned} \tag{17}$$

The calculation of the Jacobian and the Hessian in the physical $\mathbf{x}$-space is based on a sensitivity analysis. Details of the calculation can be found in [4].

3.2 First- and Second-Order Reliability Methods

It is well-known that the most essential contributions to the failure probability come from the vicinity of the most likely failure point if the distance from the origin in the standardized Gaussian space to this point is suitably large [9]. The most likely failure point or the design point is the point on the limit state surface that has the shortest distance to the origin in the **u**-space. FORM and SORM are analytical probability integration methods in which the limit state function is approximated by a linear or second-order surface at the design point in the **u**-space. If the limit state function is not strictly non-linear, the probability of failure P_f can be determined with good accuracy by FORM, where $g(\mathbf{U})$ with $g(\mathbf{0}) > 0$ is approximated by its Taylor expansion $g_L(\mathbf{u}) = g(\mathbf{u}^*) + \nabla_u g(\mathbf{u}^*)^T(\mathbf{u} - \mathbf{u}^*)$ at the so-called design point $\mathbf{u}^*$ (so that $g(\mathbf{u}^*) = 0$, Fig. 2)

$$g_L(\mathbf{u}) = \beta_{HL} + \boldsymbol{\alpha}^T\mathbf{u},\ \boldsymbol{\alpha} = \frac{\nabla_u g(\mathbf{u}^*)}{|\nabla_u g(\mathbf{u}^*)|},\ \beta_{HL} = -\boldsymbol{\alpha}^T\mathbf{u}^*. \tag{18}$$

The vector $\boldsymbol{\alpha}$ is proportional to the sensitivities $\nabla_u g(\mathbf{u}^*)$. The failure event is equivalent to the event $\{\boldsymbol{\alpha}^T\mathbf{u} \le -\beta_{HL}\}$, such that an approximation of the failure probability

$$P_{f,I} \approx P\left(\boldsymbol{\alpha}^T\mathbf{U} \le -\beta_{HL}\right) = \Phi(-\beta_{HL}) = \frac{1}{\sqrt{2\pi}}\int_{-\infty}^{-\beta_{HL}} e^{-0.5z^2}dz, \tag{19}$$

where β_{HL} is the distance from the origin in the **u**-space to the design point on the limit state surface and $\Phi(.)$ is the standard Gaussian distribution function. β_{HL} is defined as the shortest distance from origin to the limit state surface ∂F, i.e.

$$\beta_{HL} = \min_{g(\mathbf{u})=0} \sqrt{\mathbf{u}^T\mathbf{u}} \tag{20}$$

A non-linear optimization algorithm which is based on the sequential quadratic programming (SQP) is adopted to solve the optimization problem in (20). Details of the algorithm can be found in [3].

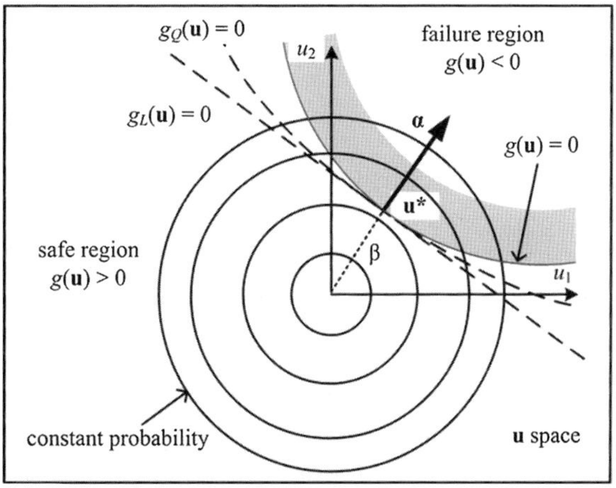

Fig. 2 Safe and failure regions with their linear and quadratic approximations in U-space

As an attempt to improve the accuracy of FORM, the limit state surface in SORM is approximated by a quadratic hypersurface (Fig. 2)

$$g_Q(\mathbf{u}) = \beta_{HL} + \boldsymbol{\alpha}^T \mathbf{u} + \frac{1}{2}\left(\mathbf{u} - \mathbf{u}^*\right)^T \mathbf{H}\left(\mathbf{u} - \mathbf{u}^*\right), \tag{21a}$$

$$\boldsymbol{\alpha} = \frac{\nabla_u g(\mathbf{u}^*)}{|\nabla_u g(\mathbf{u}^*)|}, \quad \mathbf{H} = \frac{\nabla_{uu}^2 g(\mathbf{u}^*)}{|\nabla_u g(\mathbf{u}^*)|}, \quad \beta_{HL} = -\boldsymbol{\alpha}^T \mathbf{u}^*, \tag{21b}$$

where $\mathbf{H}$ is the scaled Hessian matrix of the limit state function. The main curvatures κ_j $(j = 1, \ldots, n-1)$ at the design point are equal to those of the limit state surface in SORM. The failure probability is then calculated as a three term approximation [7, 8, 10]

$$P_{f,II} = S_1 + S_2 + S_3, \tag{22a}$$

with

$$S_1 = \Phi(-\beta_{HL}) \prod_{j=1}^{n-1} \left(1 - \beta_{HL}\kappa_j\right)^{-1/2}, \tag{22b}$$

$$S_2 = [\beta_{HL}\Phi(-\beta_{HL}) - \phi(\beta_{HL})] \times \left\{ \prod_{j=1}^{n-1} \left(1 - \beta_{HL}\kappa_j\right)^{-1/2} - \prod_{j=1}^{n-1} \left(1 - (\beta_{HL} + 1)\kappa_j\right)^{-1/2} \right\} \tag{22c}$$

$$S_3 = (\beta_{HL} + 1)\,[\beta_{HL}\Phi(-\beta_{HL}) - \phi(\beta_{HL})] \times \left\{ \prod_{j=1}^{n-1} \left(1 - \beta_{HL}\kappa_j\right)^{-1/2} - \mathrm{Re}\left[\prod_{j=1}^{n-1} \left(1 - (\beta_{HL} + i)\kappa_j\right)^{-1/2} \right] \right\} \tag{22d}$$

where $i = \sqrt{-1}$, Re [.] represents the real part of the complex argument and $\phi(.)$ is the standard Gaussian probability density function.

4 Multimode Failure

Structural systems can generally be characterized as series or parallel systems or some combination of the two [11]. In series system, the formation of any individual failure mode or mechanism is defined as system failure. For example, in statically determinate or rigid-plastic structures, formation of a collapse mechanism results in failure of the total system and therefore they can be modelled as series system with each element of the series being a failure mechanism. In parallel system, failure in a single element does not result in failure of the system, because the remaining elements may be able to sustain the external loads by redistributing of the loads. A typical example of a parallel system is a statically indeterminate structure. Failure of such structures always requires that more than one element fails before the structure loses integrity and fails.

System-reliability analysis concerns the calculation of the failure probability when the structure has more than one failure mode (multimode failure). Mathematically we encounter system reliability analysis if the limit state surface is composed of more pieces that generally intersect pairwise in sets of singular points. Each of these pieces corresponds with an individual failure mode or mechanism. This section aims at presenting a method to successively find all the design points of a system-reliability problem, if they exist on the limit state surface. FORM and SORM approximations are applied at each the joint density function by a series system reliability analysis to obtain improved estimates of the system failure probability.

4.1 Bounds for the System Probability of Failure

If there are q failure mechanisms and the limit state surface is respectively described by q equations

$$g_i(\mathbf{X}) = g_i(X_1, \ldots, X_n) = 0, \quad i = 1, \ldots, q \tag{23}$$

and if we denote the failure due to the ith mode as the random event $E_i = \{\mathbf{x}|\, g_i(\mathbf{X}) \leq 0\}$, then the probability that the system fails is the probability that any failure mechanism occurs. It means that

$$P_f = P\left(E_1 \cup E_2 \cup \cdots \cup E_q\right) = P\left(\bigcup_{i=1}^{q} E_i\right). \tag{24}$$

If the joint probability density function of the failure events $f_E(\mathbf{e})$ is known, then the system probability of failure can be calculated by the q-dimensional integral

$$P_f = P\left(\bigcup_{i=1}^{q} E_i\right) = \int_{-\infty}^{0} \cdots \int_{-\infty}^{0} f_E(e_1, \ldots, e_q) de_1 \ldots de_q. \tag{25}$$

Generally, evaluation of the system probability of failure through direct integration of (25) may not be feasible, even if an expression exists for the joint density function of failure modes, and all failure modes have been identified. In this case, bounds relieve the necessity of evaluating the q-dimensional integral either analytically, numerically or through Monte Carlo simulation [12, 13]. Several first-order bounds exist (e.g. [14]) which only require knowledge of the individual probabilities of failure resulting directly from the axioms of the probability theory. Unfortunately, these bounds may be quite wide for structural reliability application [15]. More precise second-order bounds can be given in terms of the individual failure probabilities and the joint failure probabilities between any two modes. If we denote the individual failure probabilities as

$$P_i = P\left[g_i(\mathbf{X}) \le 0\right], \quad i = 1, \ldots, q, \tag{26}$$

then the bounds of the system probability of failure for a series system are [16]

$$P_1 + \sum_{i=2}^{q} \max\left\{P_i - \sum_{j=1}^{i-1} P_{ij}, 0\right\} \le P_f \le \sum_{i=1}^{q} P_i - \sum_{i=2}^{q} \max_{j<i}\left(P_{ij}\right), \tag{27}$$

where the notation P_{ij} has been used for the joint failure probability

$$P_{ij} = P\left[g_i(\mathbf{X}) \le 0, g_j(\mathbf{X}) \le 0\right]. \tag{28}$$

Since not all couples of the random events E_i are taken into account in Eq. (27) the ordering of the modes has an effect on the bounds. Practical experience suggested that ordering the failure modes according to decreasing values P_i may correspond to the better bounds. In structural reliability, these bounds are frequently used and are considered sufficiently accurate for most structural systems [15].

4.2 First-Order System Reliability Analysis

In a first-order system reliability analysis, the failure set is approximated by the polyhedral set bounded by the tangent hyper-planes at the design points. Each design point corresponds to a failure mode and they are the points on the limit state surface that have smallest distances to the origin in the **u**-space. We denote the design points in the **u**-space as $\mathbf{u}_i^*$, $i = 1, 2, \ldots, q$ and their distances to the origin as $\beta_{HLi} = \left\|\mathbf{u}_i^*\right\|$, which are the corresponding reliability indices. The individual failure probabilities P_i are determined as

$$P_i = \Phi(-\beta_{HLi}). \tag{29}$$

The first-order approximation to P_{ij} is obtained by approximating the joint failure set by the set bounded by the tangent hyper-planes at the design points for the two failure modes. Figure 3 presents the projection of the limit state surface for the two failure modes on the plane containing the origin and the two design points $\mathbf{u}_i^*$ and $\mathbf{u}_j^*$. The joint failure probability P_{ij} is thus calculated as

$$P_{ij} = \Phi\left(-\beta_{HLi}, -\beta_{HLj}; \rho_{ij}\right) = \int_{-\infty}^{-\beta_{HLi}} \int_{-\infty}^{-\beta_{HLj}} \varphi\left(x, y; \rho_{ij}\right) dxdy, \tag{30}$$

where the correlation coefficients between two failure modes ρ_{ij} are

$$\rho_{ij} = \cos \nu_{ij} = \frac{\left(\mathbf{u}_i^*\right)^T \mathbf{u}_j^*}{\beta_{HLi}\beta_{HLj}}, \tag{31}$$

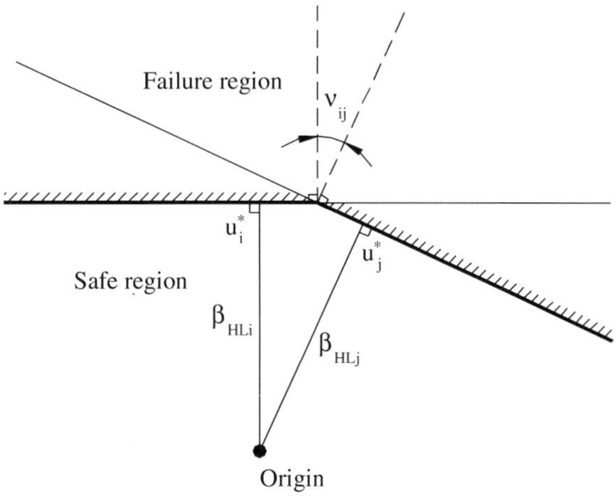

Fig. 3 Geometrical illustration of first-order system failure set

and $\varphi(x, y; \rho)$,

$$\varphi(x, y; \rho) = \frac{1}{2\pi\sqrt{1-\rho^2}} \exp\left[-\frac{1}{2}\frac{x^2+y^2-2\rho xy}{1-\rho^2}\right], \tag{32}$$

is the probability density function for a bivariate normal vector with zero mean values, unit variances and correlation coefficient ρ.

Substituting the density function in (30) by the corresponding cumulative distribution function $\Phi(x, y; \rho)$, which gives

$$\begin{aligned} P_{ij} &= \Phi\left(-\beta_{HLi}, -\beta_{HLj}; 0\right) + \int_0^{\rho_{ij}} \frac{\partial \Phi\left(-\beta_{HLi}, -\beta_{HLj}; z\right)}{\partial \rho} dz \\ &= \Phi\left(-\beta_{HLi}\right) \Phi\left(-\beta_{HLj}\right) + \int_0^{\rho_{ij}} \phi\left(-\beta_{HLi}, -\beta_{HLj}; z\right) dz. \end{aligned} \tag{33}$$

Numerical techniques are available for evaluating the joint failure probability in Eq. (33), e.g. Newton-Codes method. Simple bounds on the joint failure probability, which is based on geometrical illustration of a multimode failure system, can also be given, thus avoiding any numerical integration [11]. It should be noted that the bounds (27) still estimate a solution of the generally unknown region with respect to the exact value of the probability of failure. If we do not know where the values of the probabilities are placed with respect to the exact values, we cannot confirm that the bounds given above estimate the probability of failure. They bound some approximation and we can only more-or-less reasonably expect that the approximation is close to the exact result and the bounds remain meaningful.

4.3 Calculation of the Multiple Design Points

A real structure may have several failure modes or failure mechanisms and the existence of multiple failure modes (or multiple design points) may cause the following problems in FORM and SORM. That is, the optimization algorithm which was developed in [3] may converge to a local design point. In that case, the FORM/SORM solution misses the region of dominant contribution to the failure probability integral and, thus, has large error. Even if the global design point is found, the neighborhoods of the local design points may also have a significant contribution to the failure probability integral. Approximating the limit state surface only at the global design point loses these contributions.

In this section, a simple method is presented for finding the multiple design points of a system reliability analysis problem, when they exist on the limit state surface. The method was developed by Kiureghian and Dakessian [17]. The basic idea of the method is to construct "barriers" around previously found solutions, thus forcing the algorithm to seek a new solution. Once all the design points are known, the failure probability of series system is calculated by using first-order system reliability method and second-order bounds as presented above.

Suppose that the first design point $\mathbf{u}_1^*$ is already found by the non-linear optimization algorithm developed in [3]. A 'barrier' for this point is then constructed by adding a 'bulge' to the limit state surface. Thus, the limit state function for the deformed surface is

$$g_1(\mathbf{u}) = g(\mathbf{u}) + B_1(\mathbf{u}), \tag{34}$$

where $B_1(\mathbf{u})$ defines the bulge fitted at $\mathbf{u}_1^*$. Solving the optimization problem with the new limit state function $g_1(\mathbf{u})$ leads to a second design point $\mathbf{u}_2^*$. In order to seek the third solution point $\mathbf{u}_3^*$, a bulge $B_2(\mathbf{u})$ is now added at $\mathbf{u}_2^*$ resulting in the new limit state function $g_2(\mathbf{u}) = g_1(\mathbf{u}) + B_2(\mathbf{u})$. The process is repeated until all design points are found. The limit state function for finding the qth design point thus, is

$$g_{q-1}(\mathbf{u}) = g(\mathbf{u}) + \sum_{i=1}^{q-1} B_i(\mathbf{u}) \tag{35}$$

Details of the definition of the bulges $B_i(\mathbf{u})$ can be found in [17]. As is shown in Fig. 4, it is possible for the optimization algorithm to converge to the points located at the feet of the bulge, which are actually the spurious minimum-distance points. However, practical experience showed that this occurs only when there is no other genuine design point. Thus, convergence to a spurious point usually means that no other genuine design point exists [17]. This nature can be used as the stopping criterion of the algorithm.

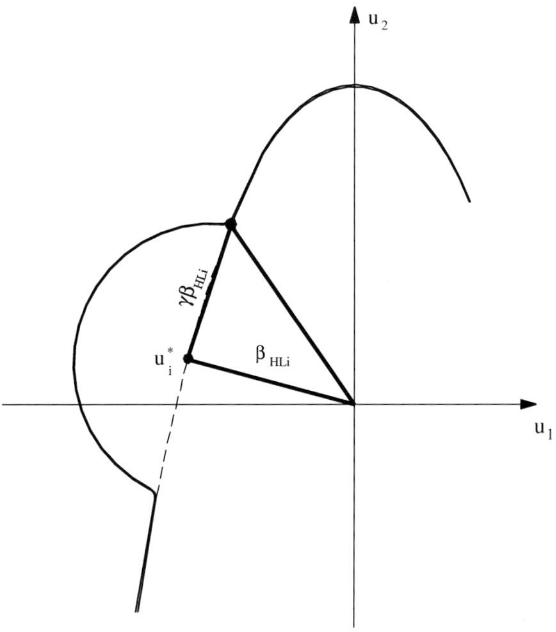

Fig. 4 Definition of a bulge at design point $\mathbf{u}_i^*$

5 Numerical Applications

Numerical examples are presented in this section to assess the performance of the proposed method. In all examples, the structures were made of elastic-perfectly plastic material and flat 4-noded shell element was used to discretize the structural domains. For each test case, analytical solutions are briefly introduced and compared. The probabilistic limit and shakedown analysis of a pipe junction with two failure modes is discussed in detail. Further tests of the method have been made in [18] for the limit analysis of the well-known problem of a portal frame which exhibits three failure modes.

5.1 Plastic Collapse of a Pipe-Junction Subjected to Internal Pressure and Bending Moment

In this example, a pipe-junction subjected to an internal pressure p and a constant bending moment M as shown in Fig. 5 (left) is considered. The essential dimensions were chosen as $D = 285$ mm, $d = 20$ mm, $s = 15$ mm and $t = 7.5$ mm. The limit

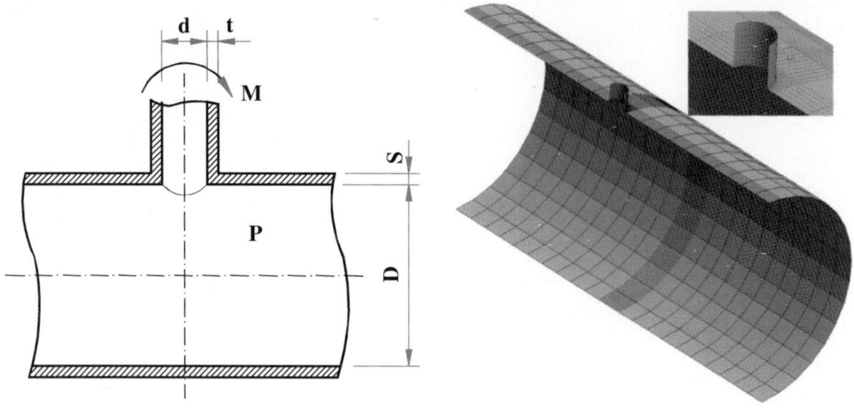

Fig. 5 Geometrical dimensions and FE-mesh of the pipe-junction

and shakedown loads of this pipe-junction were investigated in Vu et al. [19] using the solid 20-node hexahedron element HEXA20 in *Code_Aster* and a primal-dual shakedown algorithm. A deterministic assessment in terms of safety factors was made in Vu et al. [19]. They also showed that the solid 8-node hexahedron element HEXA8 makes the structure much stiffer and large differences in the limit and shakedown loads are observed in Fig. 6. Such differences in the deterministic analyses are critical for the highly sensitive stochastic analyses.

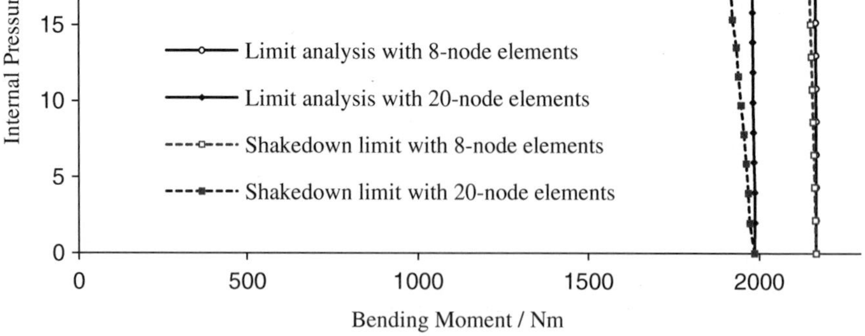

Fig. 6 Interaction diagram of the pipe junction showing the two failure modes of plastic collapse and of non-shakedown and the numerical difference between the solutions with 8-nodes and 20-nodes *solid* elements, Vu et al. [19]

Table 1 Mean values and standard deviations of random variables—pipe-junction

	Internal pressure p (MPa)	Bending moment M (Nm)	Yield stress σ_y (MPa)
Mean value μ	14	720	234
Standard deviation σ	1.4	72	23.4

Loads and the yield stress of the material were supposed to be normally distributed random variables with mean values and standard deviations given in Table 1. It was expected that the pipe-junction has two failure modes: the failure of the large pipe under pressure and the failure of the nozzle under pure bending, see Fig. 6.

Failure of the large pipe:
The exact fully plastic pressure of the large pipe can be obtained by

$$p_{\lim} = \frac{2}{\sqrt{3}}\sigma_y \ln(1 + \frac{2s}{D}) \tag{36}$$

Let X_1, X_2 denote the two load random variables corresponding to pressure and bending moment. The limit state function in the physical $\mathbf{x}$-space is then defined by

$$g(\mathbf{X}) = \frac{2}{\sqrt{3}} \ln(1 + \frac{2s}{D})Y - X_1, \tag{37}$$

is a linear function. Since both yield stress and pressure random variables are normally distributed, the limit state function $g(\mathbf{U})$ in the standardized Gaussian $\mathbf{u}$-space is also a linear function. If $\mu_{p,M,r}$ and $\sigma_{p,M,r}$ are mean values and standard deviations of three random variables: pressure p, bending moment M and yield stress σ_y. The mean and standard deviation of the limit state function can be calculated as follows

$$\mu_{gS} = \frac{2}{\sqrt{3}} \ln(1 + \frac{2s}{D})\mu_r - \mu_p, \quad \sigma_{gS} = \sqrt{\left(\frac{2}{\sqrt{3}} \ln(1 + \frac{2s}{D})\right)^2 \sigma_r^2 + \sigma_p^2}, \tag{38}$$

from which, the reliability index becomes

$$\beta_{HL} = \frac{\mu_{gS}}{\sigma_{gS}} = \frac{\frac{2}{\sqrt{3}} \ln(1 + \frac{2s}{D})\mu_r - \mu_p}{\sqrt{\left(\frac{2}{\sqrt{3}} \ln(1 + \frac{2s}{D})\right)^2 \sigma_r^2 + \sigma_p^2}} = 4.283 \tag{39}$$

Failure of the nozzle:
The exact fully plastic collapse limit bending moment of the nozzle is given by

$$M_{\lim} = W_p \sigma_y, \quad W_p = \frac{1}{6}\left[(d + 2t)^3 - d^3\right], \tag{40}$$

where W_p is the plastic section modulus. The limit state function in the physical **x**-space is then defined by

$$g(\mathbf{X}) = W_p Y - X_2. \tag{41}$$

The limit state function $g(\mathbf{U})$ in the standardized Gaussian **u**-space is also a linear function. Thus, the reliability index is obtained by

$$\beta_{HL} = \frac{W_p \mu_r - \mu_M}{\sqrt{(W_p)^2 \sigma_r^2 + \sigma_M^2}} = 4.160 \tag{42}$$

In our numerical analysis one half of the structure was discretized by 890 4-noded flat shell elements as shown in Fig. 5 (right). The 'barriers' technique developed by Der Kiureghian and Dakessian [17] was performed with the parameters $\gamma = 0.4$, $\delta = 0.3$ to define the bulges in order to find both design points as expected, and numerical results of design points are presented in Table 2.

The global design point $\mathbf{u}_1^* = 4.154\,[0.321 \quad 0.857 \quad -0.404]^T$ with $\beta_{HL1} = 4.154$ was found firstly in the **u**-space (analytically $\beta_{HL1} = 4.160$). Three components of **u** correspond to three basic variables p, M and yield stress σ_y, respectively. The sensitivities a_1, a_2 of the loading variables p, M were positive whereas the sensitivity a_3 of the resistance variable σ_y was negative [13]. This design point corresponds to the failure of the nozzle at its fully plastic bending moment since the effect of M was dominant. After adding a bulge $B_1(\mathbf{u})$ at $\mathbf{u}_1^*$, the algorithm converged to the second design point $\mathbf{u}_2^* = 4.296\,[0.914 \quad 0.137 \quad -0.383]^T$ with $\beta_{HL2} = 4.296$ (analytically $\beta_{HL2} = 4.283$). This local design point corresponds clearly to the fully plastic pressure of the large pipe because p had the greater contribution to the failure of the structure. Then we supposed to proceed further and placed a bulge $B_2(\mathbf{u})$ at $\mathbf{u}_2^*$. Our search algorithm converged to $\mathbf{u}_3^* = 4.432\,[0.633 \quad 0.740 \quad -0.228]^T$ with $\beta_{HL3} = 4.432$. The distance $\left\|\mathbf{u}_3^* - \mathbf{u}_1^*\right\| = 1.640$ between the two design points was less than but close to the radius $r_1 = \gamma\beta_{HL1} = 0.4 \times 4.154 = 1.662$ of the bulge, thus confirming that $\mathbf{u}_3^*$ is a spurious design point. If we further placed a bulge $B_3(\mathbf{u})$ at $\mathbf{u}_3^*$ and continue the algorithm, the point $\mathbf{u}_4^* = 4.745\,[0.831 \quad 0.478 \quad -0.285]^T$ with $\beta_{HL4} = 4.745$ was found. Obviously it was also a spurious design point since the distances $\left\|\mathbf{u}_4^* - \mathbf{u}_2^*\right\| = 1.705$ between the two design points $\mathbf{u}_2^*, \mathbf{u}_4^*$ was also less than but close to the radius $r_2 = \gamma\beta_{HL2} = 0.4 \times 4.296 = 1.718$ of the bulge. Thus, at this stage, we stopped to search and assumed that there are only two design points

Table 2 Multiple design points and search steps—limit analysis ($a_i = u_i / \|\mathbf{u}\|$)

Step	a_1	a_2	a_3	β_{HL}	Nature	Failure
1	0.321	0.857	−0.404	4.154	Global design point	Nozzle
2	0.914	0.137	−0.383	4.296	Local design point	Large pipe
3	0.633	0.740	−0.228	4.432	Spurious design point	–
4	0.831	0.478	−0.285	4.745	Spurious design point	–

for this problem. It should be noted here that, the SQP algorithm worked well to find all the optimal points in this case.

The linear approximation of the failure set was constructed by the tangent hyperplanes at the two design points. The corresponding approximations of failure mode correlations and joint failure mode probabilities are listed in Table 3. The numerical single and system failure probabilities are presented in Table 4. It is observed that the numerical reliability indices compare extremely well with analytical ones obtained from Eqs. (39) and (42). The reliability indices have an error in the order of the deterministic model i.e. of the direct plastic analysis. This demonstrates the importance of good convergence of the deterministic algorithm and of the precision of the shell approximation. It is critically observed that some direct methods, which were compared in Vu et al. [19] failed to converge to an acceptable solution. For highly reliable structures a greater error of the deterministic model makes the probabilistic analysis meaningless.

5.2 *Shakedown of a Pipe-Junction Subjected to Internal Pressure and Bending Moment*

In this continuous example, the pipe-junction investigated in Sect. 5.1 was considered in which the pressure and bending moment vary within the range $[0, p]$ and $[0, M]$, respectively. Consider the case where the maximal load magnitudes p and M are random variables and the minimal magnitudes are fixed to be zero. Numerical deterministic analyses lead to shakedown limits $p_{SD} = 0.079\sigma_y$ (if $M = 0$ Nm) and $M_{SD} = 3.969\sigma_y$ Nm (if $p = 0$ MPa). If all the random variables are supposed to be normally distributed with means and standard deviations given in Table 1, then the semi-analytical reliability indices of the pipe-junction may be calculated as

Table 3 Failure mode correlations and joint failure mode probabilities—limit analysis (first order)

	Failure mode correlations ρ_{ij}		Joint failure probabilities $P_{ij} \times 10^5$	
	1	2	1	2
1	1.0	0.566	1.634	0.0236
2	0.566	1.0	0.0236	0.870

Table 4 Failure probability of the pipe-junction—limit analysis ($P_f \times 10^5$)

Method	$\mathbf{u}_1^*$ alone (Nozzle)		$\mathbf{u}_2^*$ alone (Large pipe)		$\mathbf{u}_1^*$ and $\mathbf{u}_2^*$
	FORM	SORM	FORM	SORM	FORM
Present	1.634	1.622	0.870	0.893	2.480
Analytical	1.591		0.922		–

Table 5 Multiple design points and search steps—shakedown analysis ($a_i = u_i / \|\mathbf{u}\|$)

Step	a_1	a_2	a_3	β_{HL}	Nature	Failure
1	0.235	0.892	−0.386	1.772	Global design point	Nozzle
2	0.902	0.174	−0.395	1.933	Local design point	Large pipe
3	0.576	0.783	−0.234	2.134	Spurious design point	–
4	0.816	0.486	−0.312	2.407	Spurious design point	–

$$\beta_{HL} = \frac{0.079\mu_r - \mu_p}{\sqrt{0.079^2\sigma_r^2 + \sigma_p^2}} = 1.935 \quad \text{for the large pipe} \tag{43}$$

$$\beta_{HL} = \frac{3.969\mu_r - \mu_M}{\sqrt{3.969^2\sigma_r^2 + \sigma_M^2}} = 1.776 \quad \text{for the nozzle} \tag{44}$$

Numerical results for finding both design points are presented in Table 5. The global design point $\mathbf{u}_1^* = 1.772\,[0.235 \quad 0.892 \quad -0.386]^T$ with $\beta_{HL1} = 1.772$ was found firstly in the $\mathbf{u}$-space (semi-analytically $\beta_{HL1} = 1.776$). This design point corresponds to the failure of the nozzle since the effect of M is dominant. After adding a bulge $B_1(\mathbf{u})$ at $\mathbf{u}_1^*$, the algorithm converged to the second design point $\mathbf{u}_2^* = 1.933\,[0.902 \quad 0.174 \quad -0.395]^T$ with $\beta_{HL2} = 1.933$ (semi-analytically $\beta_{HL2} = 1.935$). This local design point corresponds clearly to the failure of the large pipe because p had a greater contribution. Then we supposed to proceed further and placed a bulge $B_2(\mathbf{u})$ at $\mathbf{u}_2^*$. Our search algorithm converged to $\mathbf{u}_3^* = 2.134\,[0.576 \quad 0.783 \quad -0.234]^T$ with $\beta_{HL3} = 2.134$. The distance $\|\mathbf{u}_3^* - \mathbf{u}_1^*\| = 0.703$ between the two design points $\mathbf{u}_1^*$, $\mathbf{u}_3^*$ was less than but close to the radius $r_1 = \gamma\beta_{HL1} = 0.4 \times 1.772 = 0.709$ of the bulge, thus confirming that $\mathbf{u}_3^*$ is a spurious design point. If we further placed a bulge $B_3(\mathbf{u})$ at $\mathbf{u}_3^*$ and continued the algorithm, the point $\mathbf{u}_4^* = 2.407\,[0.816 \quad 0.486 \quad -0.312]^T$ with $\beta_{HL4} = 2.407$ was found. Obviously it is also a spurious design point since the distances $\|\mathbf{u}_4^* - \mathbf{u}_2^*\| = 0.743$ between the two design points $\mathbf{u}_2^*$, $\mathbf{u}_4^*$ was also less than but close to the radius $r_2 = \gamma\beta_{HL2} = 0.4 \times 1.933 = 0.773$ of the bulge. Thus, at this stage, we stopped to search and assumed that there were only two design points for this problem.

The linear approximation of the failure set was then constructed by the tangent hyper-planes at the two found design points. The corresponding approximations of failure mode correlations and joint failure mode probabilities are listed in Table 6. The numerical single and system failure probabilities are presented in Table 7.

Table 6 Failure mode correlations and joint failure mode probabilities—shakedown analysis (first order)

	Failure mode correlations ρ_{ij}		Joint failure probabilities $P_{ij} \times 10^2$	
	1	2	1	2
1	1.0	0.520	3.820	0.684
2	0.520	1.0	0.684	2.662

Table 7 Failure probability of the pipe-junction—shakedown analysis ($P_f \times 10^2$)

Method	$\mathbf{u}_1^*$ alone (Nozzle)		$\mathbf{u}_2^*$ alone (Large pipe)		$\mathbf{u}_1^*$ and $\mathbf{u}_2^*$
	FORM	SORM	FORM	SORM	FORM
Present	3.820	3.812	2.662	2.660	5.798
Semi-analytical	3.787		2.650		–

6 Conclusions

The present work provides an effective plastic analysis method for the integrity assessment of general shells with multimode failure. Practical experience showed that the existence of multimode failure (multiple design points) in component reliability analysis could give rise to great errors in FORM and SORM approximations of the failure probability. In this work, a technique has been performed with an SQP algorithm to successively find the multiple design points of a system reliability problem, if they exist on the limit state surface. This technique was based upon a 'barrier' method by constructing a bulge around previously found design points, thus forcing the algorithm to seek a new design point. Second-order bounds of the reliability of series system were then calculated based on the first-order system reliability analysis. Numerical examples showed that it improved considerably the estimates of the system failure probability.

References

1. Ditlevsen O, Bjerager P (1986) Methods of structural system reliability. Struct Saf 3:195–229
2. Thoft-Christensen P, Baker MJ (1982) Structural reliability theory and its applications. Springer, Berlin
3. Trần TN, Kreißig R, Staat M (2009) Probabilistic limit and shakedown analysis of thin shells. Struct Saf 31:1–18
4. Trần TN, Tinh PP, Khoi VD, Staat M (2009) Reliability analysis of inelastic shell structures under variable loads. In: Weichert D, Ponter ARS (eds) Limit states of materials and structures: direct methods. Springer, Dordrecht, pp 135–156
5. Trần TN, Kreißig R, Vu DK, Staat M (2008) Upper bound limit and shakedown analysis of shells using the exact Ilyushin yield surface. Comput Struct 86:1683–1695
6. Gollwitzer S, Abdo T, Rackwitz R (1988) FORM manual. RCP UmbH, München

7. Trần TN (2008) Limit and shakedown analysis of plates and shells including uncertainties. Dr.-Ing. Dissertation, Technische Universität Chemnitz
8. Staat M (2014) Limit and shakedown analysis under uncertainty. Int J Comput Methods 11(2). Online http://dx.doi.org/10.1142/S0219876213430081
9. Ditlevsen O, Madsen HO (2005) Structural reliability methods. Technical University of Denmark
10. Tvedt L (1983) Two second-order approximations to the failure probability. Veritas Report RDIV/20-004-83, Det Norske Veritas, Oslo
11. Madsen HO, Krenk S, Lind NC (1986) Methods of structural safety. Prentice-Hall, Englewood Cliffs
12. Ramachandran K (1984) Systems bounds: a critical study. Civ Eng Syst 1:123–128
13. Melchers RE (1999) Structural reliability—analysis and prediction. Ellis Horwood, Chichester
14. Cornell CA (1967) Bounds on the reliability of structural systems. J Struct Div, ASCE 93:171–200
15. Grimmelt MJ, Schuëller GI (1982) Benchmark study on methods to determine collapse failure probabilities of redundant structures. Struct Saf 1:93–106
16. Ditlevsen O (1979) Narrow reliability bounds for structural systems. J Struct Mech 7:453–472
17. Kiureghian AD, Dakessian T (1998) Multiple design points in first and second-order reliability. Struct Saf 20:37–49
18. Trần TN, Staat M (2014) Uncertain multimode failure and limit analysis of shells. In: Proceedings 11th world congress on computational mechanics (WCCM XI), 5th European conference on computational mechanics (ECCM V), 6th European conference on computational fluid dynamics (ECFD VI), 20–25 July 2014, Barcelona, Spain
19. Vu DK, Staat M, Tran IT (2007) Analysis of pressure equipment by application of the primal-dual theory of shakedown. Commun Numer Methods Eng 23:213–225

Three-Dimensional Shakedown Solutions for Cross-Anisotropic Cohesive-Frictional Materials Under Moving Loads

Hai-Sui Yu, Juan Wang and Shu Liu

Abstract Based on Melan's lower-bound shakedown theorem, shakedown analysis of cross-anisotropic cohesive-frictional materials under three-dimensional moving Hertz loads is presented. The material behaviour is characterised by a generalised Mohr-Coulomb criterion. By means of a critical self-equilibrated residual stress field, rigorous lower-bound shakedown solutions can be obtained through a simple optimisation procedure. Influences of both elastic anisotropy and plastic anisotropy are investigated. And shakedown solutions for a two-layered pavement system with cross-anisotropic materials are also presented. Results imply that pavement design based on the isotropic assumption may result in unsafe design against rutting.

Keywords Shakedown · Layered pavements · Cross-anisotropic · Mohr-Coulomb criterion

1 Introduction

When an elastic-plastic structure is subjected to cyclic loads (larger than yield limit), it may gradually adapt itself to the applied loads during initial load cycles, thereafter deform purely elastically without any further plastic deformation. This behaviour is termed as 'shakedown'. The maximum load below which shakedown can occur is regarded as 'shakedown limit'. Shakedown analysis can be applied in the field of pavement engineering. If the applied load is lower than the pavement shakedown limit, pavement rutting depth will cease to increase after a number of load passes. Otherwise, plastic deformation accumulates after each load application and excessive rutting will eventually occur [3, 24].

H.-S. Yu · S. Liu
Faculty of Engineering, Nottingham Centre for Geomechanics,
The University of Nottingham, NG7 2RD, Nottingham, UK

J. Wang (✉) · S. Liu
Department of Civil Engineering, The University of Nottingham,
Ningbo 315100, China
e-mail: juan.wang@nottingham.edu.cn

P. Fuschi et al. (eds.), *Direct Methods for Limit and Shakedown Analysis of Structures*,
DOI 10.1007/978-3-319-12928-0_15

The determination of the shakedown limit, as an essential part in the application of the shakedown concept, can be achieved either by means of numerical elastic–plastic analysis (e.g. [29]) or by shakedown analysis. Compared with the numerical elastic–plastic method, where full history of stress–strain curve is calculated, shakedown analysis can directly obtain the shakedown limit by using the static/lowerbound shakedown theorem [16] or the kinematic/upper-bound shakedown theorem [10]. Therefore, it has drawn a lot of attention from engineering researchers.

For the problem of pavements under repeated traffic loads, upper-bound shakedown solutions have been presented for both two-dimensional (2D) pavement model [4, 13] and three-dimensional (3D) pavement model [2]. Meanwhile, 2D lower-bound pavement shakedown solutions [11, 17, 19–21, 27, 33] were proposed. Recently, Yu [23] and Yu and Wang [28] performed lower-bound shakedown analysis for an isotropic, homogenous cohesive-frictional half-space under a 3D moving surface load. It should be noted that static and kinematic shakedown solutions provide lower and upper bounds to the true shakedown limit respectively. This is because the static shakedown theorem satisfies the internal equilibrium equations and the stress boundary conditions, while the kinematic shakedown theorem satisfies the compatibility condition for plastic strain rate and boundary conditions for velocity. Nevertheless, some converged upper and lower bound solutions have been obtained. For instance, in the case of 3D Tresca half-space subjected to a moving load, the lower-bound shakedown solution (as obtained by [23, 28]) is 4.68 while the upper-bound shakedown solution (as obtained by [18]) is 4.7. For conservative pavement design, this paper will however focus on the development of lower-bound shakedown solutions for more realistic cases in practice.

For many ground deformation problems, soils are assumed to have a single vertical axis of symmetry with the same properties in any horizontal direction but different properties in vertical direction. This kind of anisotropy is known as cross-anisotropy or transverse isotropy. In the elastic range, the behaviour of a cross-anisotropic material can be described as follows:

$$\begin{bmatrix} \delta\varepsilon_{xx} \\ \delta\varepsilon_{yy} \\ \delta\varepsilon_{zz} \\ \delta\varepsilon_{xy} \\ \delta\varepsilon_{xz} \\ \delta\varepsilon_{yz} \end{bmatrix} = \begin{bmatrix} 1/E_h & -\upsilon_h/E_h & -\upsilon_{vh}/E_v & & & \\ -\upsilon_h/E_h & 1/E_h & -\upsilon_{vh}/E_v & & & \\ -\upsilon_{hv}/E_h & -\upsilon_{hv}/E_h & 1/E_v & & & \\ & & & 1/2G_h & & \\ & & & & 1/2G_{vh} & \\ & & & & & 1/2G_{vh} \end{bmatrix} \begin{bmatrix} \delta\sigma_{xx} \\ \delta\sigma_{yy} \\ \delta\sigma_{zz} \\ \delta\sigma_{xy} \\ \delta\sigma_{xz} \\ \delta\sigma_{yz} \end{bmatrix}, \quad (1)$$

where the stress increments $\delta\sigma_{ij}$ and strain increments $\delta\varepsilon_{ij}$ are referred to Cartesian axes (i.e. i and j denote x axis, y axis or z axis), with the z axis being vertical; E_h is Young's modulus in horizontal (H) direction; E_v is Young's modulus in vertical (V) direction; G_h is shear modulus in horizontal plane; G_{vh} is shear modulus in VH plane; υ_h is Poisson's ratio (effect of horizontal strain on complementary horizontal strain); υ_{vh} is Poisson's ratio (effect of vertical strain on horizontal strain); υ_{hv} is Poisson's ratio (effect of horizontal strain on vertical strain). There are another two correlations between these parameters (Eqs. 2–3), so that a cross anisotropic material can be fully defined by five independent parameters.

$$G_h = \frac{E_h}{2(1 + \upsilon_h)}, \tag{2}$$

$$\frac{\upsilon_{vh}}{\upsilon_{hv}} = \frac{E_v}{E_h}. \tag{3}$$

Elastic properties of anisotropic soils have been widely explored. For example, typical values of E_v/E_h for clays may range from 0.25 to 1.11 (e.g. [14, 22, 25]). Experimental results for sands and gravel also show some degree of inherent anisotropy with E_v/E_h from 1.06 to 2 (e.g. [8, 9, 12]). Graham and Houlsby [5] proposed that the elastic anisotropy of natural clays can be described by three parameters: E^* and υ^* and α by giving the following definitions: $E_v = E^*$, $E_h = \alpha^2 E^*$, $\upsilon_h = \upsilon^*$, $\upsilon_{vh} = \upsilon^*/\alpha$, $G_{vh} = \alpha E^*/(2 + 2\upsilon^*)$, $G_h = \alpha^2 E^*/(2 + 2\upsilon^*)$.

The property of anisotropy is also an inherent characteristic of asphalt concrete caused by the non-uniform distribution of aggregates. Experimental data [32] show that anisotropy of asphalt concrete under field compaction might also be approximated as cross-anisotropy with $E_v/E_h \approx 3.33$. The significant anisotropy has important implications for pavement design and analysis. Pavement design based on the isotropic assumption may result in unsafe design against rutting and fatigue cracking.

Laboratory tests performed on soil specimens cut at different orientations have also demonstrated the directional dependence of soil shear strength (e.g. [1, 6, 15]). It has been suggested that the variation of soil cohesion with direction due to inherent anisotropy is much more significant than the anisotropy of material friction angle [1].

In this paper, previous 3D lower-bound shakedown analysis of Yu and Wang [28] will be further developed to consider more realistic cases in pavement engineering, including variation of elastic and plastic material properties with direction, and layered pavement system.

2 Problem Definition

This paper considers a cohesive-frictional layered system that is subject to a surface contact loading limited to a circle of raidus a, as shown in Fig. 1. If tensile stresses are treated as positive, then pressure p and traction q on the contact surface are formulated as:

$$p = \frac{3P}{2\pi a^3}\left(a^2 - x^2 - y^2\right), \tag{4}$$

$$q = \frac{3Q}{2\pi a^3}\left(a^2 - x^2 - y^2\right), \tag{5}$$

where P is the total normal load applied in the z-direction (i.e. the vertical direction) and Q is the total shear load applied in the x-direction (i.e. the moving load

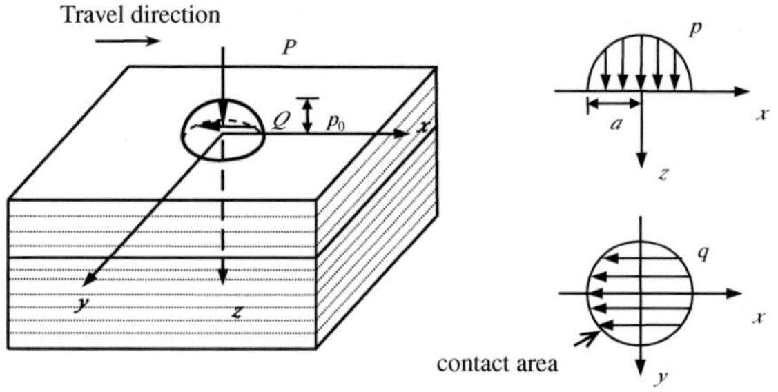

Fig. 1 Problem definition

direction). This load distribution is also known as the 3D Hertz load distribution. It has a maximum pressure $p_0 = 3P/2\pi a^2$ at the centre of contact area ($x = y = z = 0$). The normal and shear loads are assumed to be correlated by a frictional coefficient $\mu = Q/P$.

Materials for this layered system are assumed to be cross-anisotropic with a vertical axis of symmetry. The elastic properties can be described by five parameters. In Hanson [7], five elastic constants $A_{11}, A_{13}, A_{33}, A_{44}, A_{66}$ were used to define the material behaviour as below:

$$\begin{bmatrix} \delta\sigma_{xx} \\ \delta\sigma_{yy} \\ \delta\sigma_{zz} \\ \delta\sigma_{xy} \\ \delta\sigma_{xz} \\ \delta\sigma_{yz} \end{bmatrix} = \begin{bmatrix} A_{11} & A_{11}-2A_{66} & A_{13} & & & \\ A_{11}-2A_{66} & A_{11} & A_{13} & & & \\ A_{13} & A_{13} & A_{33} & & & \\ & & & 2A_{66} & & \\ & & & & 2A_{44} & \\ & & & & & 2A_{44} \end{bmatrix} \begin{bmatrix} \delta\varepsilon_{xx} \\ \delta\varepsilon_{yy} \\ \delta\varepsilon_{zz} \\ \delta\varepsilon_{xy} \\ \delta\varepsilon_{xz} \\ \delta\varepsilon_{yz} \end{bmatrix}. \tag{6}$$

In the present study, the following parameters are considered: E_h (or E_v), E_v/E_h, G_{vh}/G_h, υ_h and υ_{vh}. These five constants correlate with Young's modulus, shear modulus and Poisson's ratio as follows:

$$A_{11} = \frac{1-\upsilon_{vh}\upsilon_{hv}}{E_h E_v \Delta}, \quad A_{13} = \frac{\upsilon_{hv}+\upsilon_h\upsilon_{hv}}{E_h^2\Delta}, \quad A_{33} = \frac{1-\upsilon_h^2}{E_h^2\Delta}, \quad A_{44} = G_{vh},$$

$$A_{66} = G_h = \frac{E_h}{2(1+\nu_h)}, \text{ where } \Delta = \frac{(1+\upsilon_h)(1-\upsilon_h-2\upsilon_{hv}\upsilon_{vh})}{E_h^2 E_v}.$$

In terms of plastic anisotropy, it is assumed that the directional strength, in particular the variation of cohesion with direction, can be described by the following mathematical expression [15]:

$$c_\theta = c_h + (c_v - c_h)\sin^2\alpha, \tag{7}$$

where c_h and c_v are the values of cohesion on the horizontal and vertical planes respectively and c_θ represents the cohesion measured on a plane inclined at an angle α to the horizontal plane. If Mohr-Coulomb material is considered, the failure criterion should be generalised to consider directional strength variation. This gives [26]:

$$f = (\sigma_{zz} - \sigma_{xx} - 2\sigma_{xz}\tan\phi)^2 + (c_v - c_h + 2\sigma_{xz} - \sigma_{xx}\tan\phi + \sigma_{zz}\tan\phi)^2 \\ - (c_v + c_h - \sigma_{xx}\tan\phi - \sigma_{zz}\tan\phi)^2 \le 0, \tag{8}$$

where ϕ is material friction angle.

3 Melan's Lower-Bound Theorem

Melan's lower-bound shakedown theorem states that an elastic-perfectly plastic structure will shakedown under repeated or cyclic loads if the yield condition at any point is not violated by the combination of a self-equilibrated residual stress field and a load-induced elastic stress field. If the applied load is denoted by λp_0 and λ is a dimensionless scale parameter, then all the induced elastic stress components are proportional to λ. Melan's shakedown theorem hence demands:

$$f(\lambda\sigma_{ij}^e + \sigma_{ij}^r) \le 0, \tag{9}$$

where ${\sigma^e}_{ij}$ is the elastic stress field due to the applied pressure p_0, ${\sigma^r}_{ij}$ is the residual stress field and $f(\sigma_{ij}) = 0$ is the yield condition for the material.

4 Shakedown Analysis

Based on Melan's shakedown theorem, the essence of shakedown analysis is to find the maximum admissible load under which a self-equilibrated residual stress field can be found so that the total stress state will not violate the yield criterion.

For the problem considered here, the material is cross-anisotropic with a vertical axis of symmetry, the resulting permanent deformation and therefore the residual stress field will be independent of the travel (x) direction. According to Yu [23], under a moving 3D Hertz load distribution, plane $y = 0$ can be considered as the most critical plane in the half-space and ${\sigma^r}_{xx}$ is the only possible residual stress in this plane. Therefore, the total stresses for a point in the half space under the Hertz load distribution can be defined as the sum of elastic stresses and residual stresses:

$$\begin{aligned} \sigma_{xx} &= \lambda\sigma_{xx}^e + \sigma_{xx}^r, \\ \sigma_{zz} &= \lambda\sigma_{zz}^e, \\ \sigma_{xz} &= \lambda\sigma_{xz}^e. \end{aligned} \tag{10}$$

If the generalised, anisotropic Mohr-Coulomb yield criterion Eq. 8 is utilised to describe the material yield condition and σ_{yy} is the intermediate stress, Eq. 10

can be substituted to the yield criterion Eq. 8 to satisfy the requirement of Melan's shakedown theorem and this gives:

$$f = \left(\sigma^r_{xx} + \mathrm{M}\right)^2 + \mathrm{N} + \mathrm{P} \le 0, \tag{11}$$

where σ^r_{xx} is self-equilibrated,

$$\begin{aligned}
\mathrm{M} &= \lambda\sigma^e_{xx} - \lambda\sigma^e_{zz} + 2\left(c_h - \lambda\sigma^e_{zz}\tan\phi\right)\tan\phi, \\
\mathrm{N} &= 4\left(\tan^2\phi + 1\right)\left[\left(\lambda\sigma^e_{xz}\right)^2 - \left(c_h - \lambda\sigma^e_{zz}\tan\phi\right)^2\right], \\
\mathrm{P} &= 4\left(c_v - c_h\right)\left[\lambda\sigma^e_{xz} - \left(c_h - \lambda\sigma^e_{zz}\tan\phi\right)\right].
\end{aligned}$$

It should be noted that if c_v equals to c_h, then P is zero and Eq. 11 becomes equivalent to the shakedown condition of Yu and Wang [28] for the special case of isotropic materials.

A critical residual stress field that always fulfills the equilibrium condition is then conceived by calculating $\min(-\mathrm{M}_i + \sqrt{-\mathrm{N}_i - \mathrm{P}_i})$ (referred to as 'minimum larger root') or $\max(-\mathrm{M}_i - \sqrt{-\mathrm{N}_i - \mathrm{P}_i})$ (referred to as 'maximum smaller root') at each depth $z = j$ (i is node number). The present shakedown problem can be written as a mathematical formulation:

$$\begin{aligned}
&\text{maximise } \lambda, \\
&\text{subject to} \\
&\begin{cases}
f\left(\sigma^r_{xx}\left(\lambda\sigma^e\right), \lambda\sigma^e\right) \le 0, \\
\sigma^r_{xx}\left(\lambda\sigma^e\right) = \min\limits_{z=j}(-\mathrm{M}_i + \sqrt{-\mathrm{N}_i - \mathrm{P}_i}) \text{ or } \sigma^r_{xx}\left(\lambda\sigma^e\right) = \max\limits_{z=j}(-\mathrm{M}_i - \sqrt{-\mathrm{N}_i - \mathrm{P}_i}).
\end{cases}
\end{aligned} \tag{12}$$

If a single layer of material is considered, Eq. 12 can be rewritten as:

$$\begin{aligned}
&\text{maximise } \lambda, \\
&\text{subject to } f\left(\lambda, \mu, E_h, \frac{E_v}{E_h}, \frac{G_{vh}}{G_h}, \upsilon_h, \upsilon_{vh}, \phi, c_h, c_v\right) \le 0.
\end{aligned} \tag{13}$$

As can be seen, the shakedown limit $\lambda_{\mathrm{sd}}p$ will be dependent on frictional coefficient μ, elastic parameters of materials (i.e. E_h, E_v/E_h, G_{vh}/G_h, υ_h and υ_{vh}), plastic parameters of materials (i.e. c_h, c_v and ϕ). If a layered pavement system is considered, the shakedown limit will also be affected by layer thicknesses and material properties of each layer.

When the load form, materials and layer thicknesses are determined, Eq. 13 can be reduced to the following expression with only one variable λ:

$$\begin{aligned}
&\text{maximise } \lambda, \\
&\text{subject to } f(\lambda) \le 0.
\end{aligned} \tag{14}$$

5 Numerical Technique and Results

In the present study, numerical technique and results are firstly presented and discussed for a special case—homogeneous half-space, and then for cases with two-layered materials.

5.1 Homogenous Materials

A numerical procedure has been developed in Wang and Yu [31] to solve the present shakedown problem Eq. 12, so it will not be repeated here. It as well as the closed-form expressions for elastic stress fields in a homogenous half-space [7] has been programmed into FORTRAN. In this section, lower-bound shakedown limit is defined as $k = \lambda_{sd} p_0 / c_h$.

Analyses were first conducted in consideration of elastic anisotropy only. As mentioned before, the elastic stress fields for a cross-anisotropic material are dependent on five elastic parameters, and the following parameters were used in this study: E_h, E_v/E_h, G_{vh}/G_h, υ_h and υ_{vh}. It has been found through sensitive study that variation of Poisson's ratio υ_h and υ_{vh} only has slightly influences on the shakedown limit. Further investigations show that for the problem of a homogeneous half-space, the shakedown limit is not influenced by Young's modulus in a particular direction (i.e. E_h or E_v), but affected by Young's modulus ratio E_v/E_h. This finding is similar to that for an isotropic homogeneous half-space, in which shakedown solutions do not vary with Young's modulus E.

Figure 2 demonstrates the influence of Young's modulus ratio on lower-bound shakedown limits when the material internal friction angle is 0 and 30°. There exists a peak shakedown limit at a given shear modulus ratio. However, it should be noted that the elastic parameters of pavement materials are somewhat related. A series of analysis then was performed to investigate the interactive effect of E_v/E_h and G_{vh}/G_h on the shakedown limit. Results are presented in a logarithmic scale as shown in Fig. 3. When the frictional coefficient μ is zero, failure always initiates below the surface (i.e. subsurface failure), and the shakedown limit changes more quickly with E_v/E_h than that with G_{vh}/G_h until a peak value is reached. In these figures, the peak value is around 5 for $\phi = 0°$ and around 22 for $\phi = 30°$. When the frictional coefficient $\mu = 0.5$, the critical points mostly lie on the surface (i.e. surface failure) and the change of shakedown limit is dominated by the variation of G_{vh}/G_h, not E_v/E_h, except those at the lower-right corner (their shakedown limits are controlled by subsurface critical points). A maximum value $k = 2$ is also observed for the case $\phi = 0°$. These results imply that when the elastic anisotropy is taken into account, shakedown solutions for cohesive-frictional materials under a 3D Hertz load tend to be controlled by E_v/E_h for subsurface failure cases but by G_{vh}/G_h for surface failure cases.

Figure 4 presents the shakedown limit against the frictional coefficient μ for different values of cohesion ratio c_v/c_h, while the soil friction angle $\phi = 30°$. When c_v/c_h is smaller than 1, the shakedown limit reduces as the frictional coefficient increases.

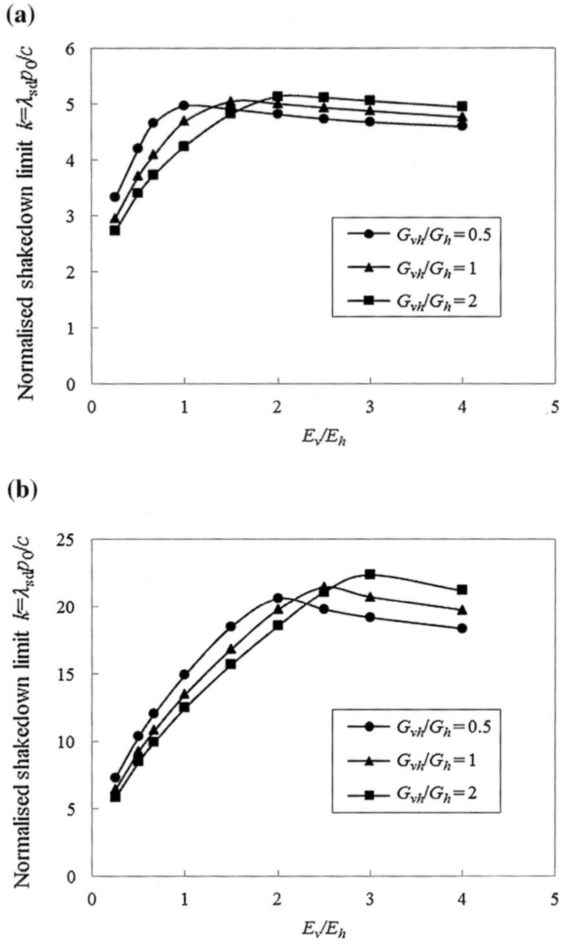

Fig. 2 Influences of Young's modulus ratio on lower-bound shakedown limits when $\mu = 0$. **a** $\phi = 0°$. **b** $\phi = 30°$

This trend follows that for isotropic solutions (i.e. $c_v/c_h = 1$) in [28]. It is interesting to find out that when c_v/c_h is larger than 1, there exists a peak value of shakedown solutions giving an optimum frictional coefficient μ (>0). This is different from the isotropic solutions of which the value is always maximum when μ equals 0.

Results in Fig. 4 are also divided into surface failure cases and subsurface failure cases by a dash line and a dot line. The dot line represents subsurface failure cases of which value is controlled by $1/a$ (Eq. 15). On the left hand side of the dash line, the critical point initiates below the surface and the lower-bound shakedown limit is controlled by $1/b$ (Eq. 16). On the right hand side, the shakedown limit is controlled

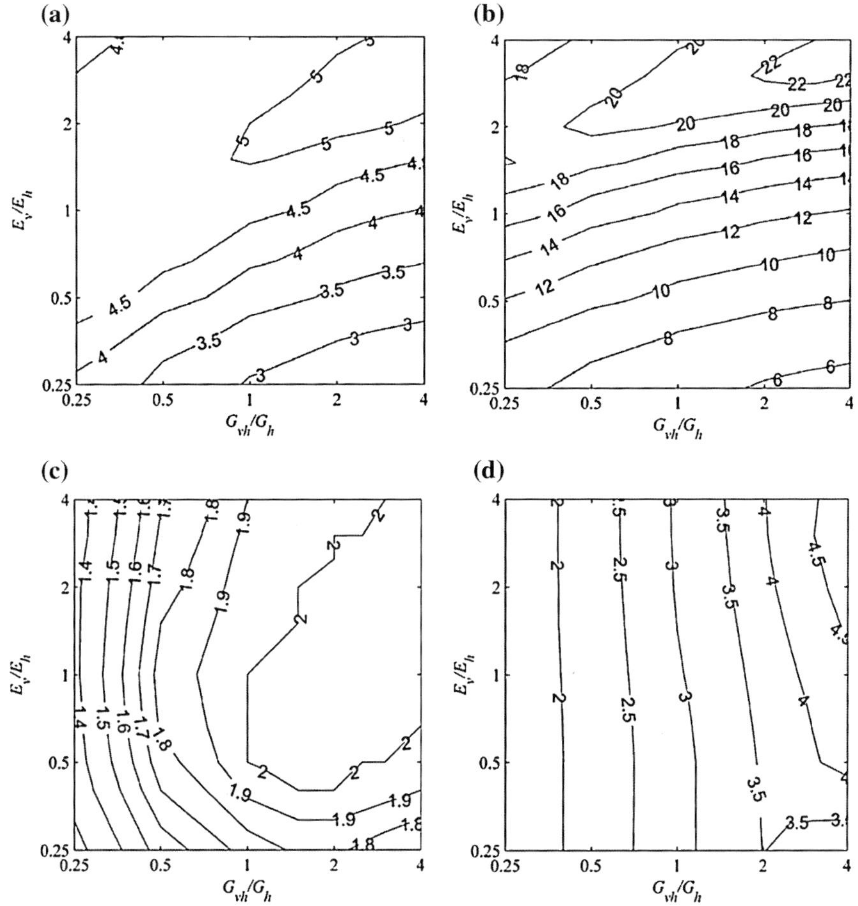

Fig. 3 Contour plot of shakedown limits. **a** $\mu = 0, \phi = 0°$. **b** $\mu = 0, \phi = 30°$. **c** $\mu = 0.5, \phi = 0°$. **d** $\mu = 0.5, \phi = 30°$

by critical point on the surface. More results for different values of material internal frictional angle can be found in Wang and Yu [31].

$$a = \sigma_{zz}^e \tan\phi + \sigma_{xz}^e. \tag{15}$$

$$b = (\sigma_{zz}^e \tan\phi - \sigma_{xz}^e)\frac{1 + \tan^2\phi}{c_v/c_h + \tan^2\phi}. \tag{16}$$

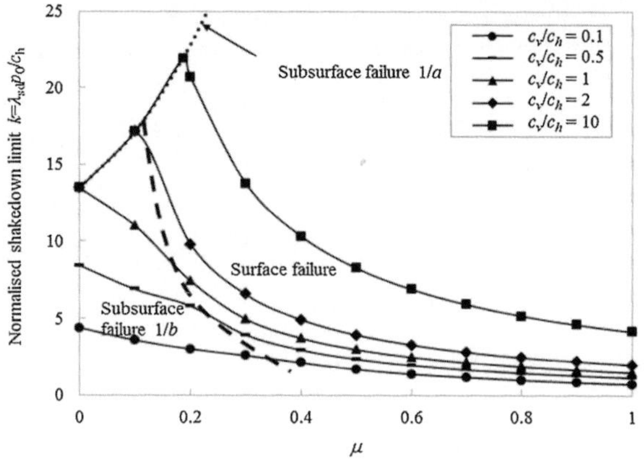

Fig. 4 Influence of plastic anisotropy on shakedown limit when $\phi = 30°$

5.2 Layered Materials

Shakedown limits of a layered system can be solved by means of a MATLAB program of Wang and Yu [30] using elastic stress fields obtained from finite element method. Finite element analyses have been carried out by means of commercial software ABAQUS integrated with a user subroutine UMAT which defines the mechanical behaviour of cross-anisotropic materials.

In this study, a two-layered system is considered as shown in Fig. 5 in which the normalised first layer thickness h_1/D equals 1 (D is diameter of the contact area). Typical mesh of finite element model is shown in Fig. 6. The simulated region has a radius $5D$ and depth $6D$. The front face represents the plane of symmetry $y = 0$. There are restraints on two horizontal directions at the back face and another restraint on vertical movement at the bottom face. The load distributions formulated as Eqs. 4 and 5 are applied on the top surface within a half circle due to the symmetric condition. The simulated region is discretised by twenty-noded, reduced-integrated,

Fig. 5 A two-layered pavement system

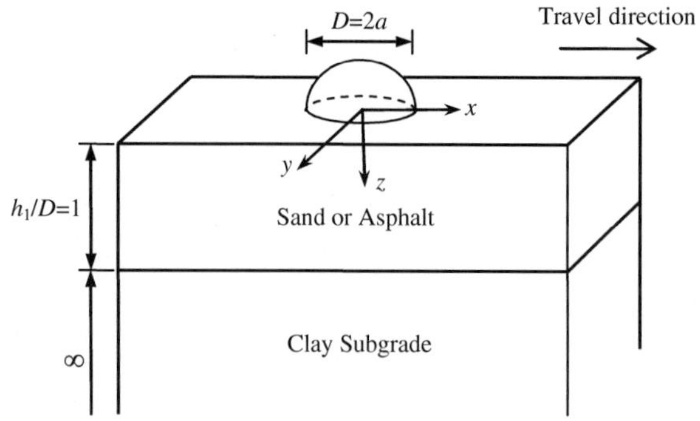

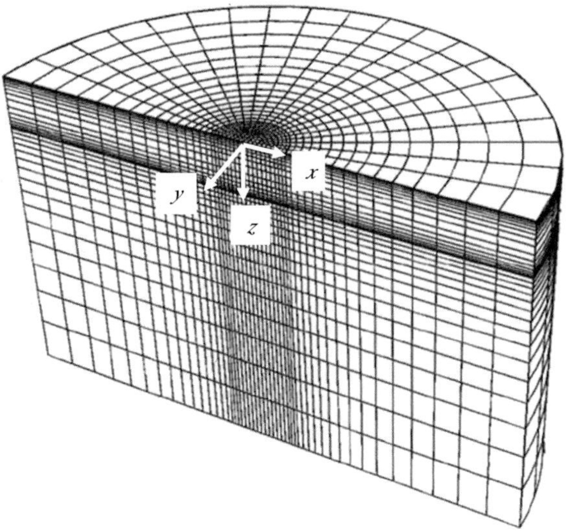

Fig. 6 Typical mesh of a two-layered pavement model

brick elements (C3D20R), and mesh density is relatively high in the vicinity of the loading area and near the interface between two layers. Results presented in this section are for cases with normal loading only. Lower-bound shakedown limit of mth layer is defined as $k_m = \lambda_{sd}^m p_0/c_2$ $(m = 1, 2)$. Therefore, lower-bound shakedown limit of the whole layered system k is the minimum value among all k_m.

In order to verify this numerical technique, comparison has been made between shakedown limits of Wang and Yu [31] for Winnipeg Clay and those from present analysis. It can be seen from Table 1 that good agreements have been achieved with around 3 % error. In addition, elastic stress fields obtained through finite element analyses also agree with analytical solutions of Hanson [7].

Two series of shakedown analysis have been carried out for this layered system: the first one is a sand layer over day subgrade; and the second one is an asphalt layer over clay subgrade. Material plastic anisotropy is not considered in this study. In addition, it is assumed that the five elastic parameters (E_h, E_v/E_h, G_{vh}/G_h, υ_h and

Table 1 Comparison of shakedown limits for cross-anisotropic Winnipeg clay

Case	E_h (MPa)	E_v/E_h	G_{vh}/G_h	υ_{vh}	υ_h	k (Ref)	k (this study)	Difference (%)
1	9.35	0.53	0.73	0.17	0.23	*4.00*	3.87	3.2
2	6.96	0.41	0.64	0.08	0.12	*3.74*	3.63	2.9
3	7.67	0.52	0.72	0.17	0.23	*3.97*	3.85	3.0
4	5.76	0.76	0.87	0.23	0.27	*4.37*	4.24	3.0
Isotropy						4.68	4.56	2.6

Table 2 Summary of material properties and shakedown limits for a sand-clay system

No.	1st layer—sand					2nd layer—clay					k
	$E_{Iv} = 75\,\text{MPa}$, $c_1 = 10\,\text{kPa}$, $\phi_1 = 30°$					$E_{Iv} = 15\,\text{MPa}$, $c_2 = 20\,\text{kPa}$, $\phi_2 = 0°$					
	E_{Iv}/E_{Ih}	G_{Ivh}/G_{Ih}	υ_{Ih}	υ_{Ivh}	k_I	E_{2v}/E_{2h}	G_{2vh}/G_{2h}	υ_{2h}	υ_{2vh}	k_2	
1	1	1	0.2	0.2	10.20	1	1	0.4	0.4	34.87	10.20
2	1	1	0.2	0.2	9.80	2	1.4	0.4	0.49	44.63	9.80
3	1	1	0.2	0.2	10.60	0.5	0.7	0.4	0.28	25.65	10.60
4	1	1	0.2	0.2	11.07	0.3	0.55	0.4	0.24	21.15	11.07
5	1	1	0.2	0.2	11.11	0.25	0.5	0.4	0.2	19.79	11.11
6	1	1	0.2	0.2	11.30	0.2	0.45	0.4	0.18	18.40	11.30
7	1	1	0.2	0.2	11.50	0.15	0.39	0.4	0.16	16.90	11.50
8	3	1.73	0.2	0.35	15.80	0.5	0.7	0.4	0.28	15.00	15.80
9	2	1.4	0.2	0.28	13.57	0.5	0.7	0.4	0.28	17.84	13.57
10	1.5	1.2	0.2	0.24	12.23	0.5	0.7	0.4	0.28	20.54	12.23
11	0.75	0.87	0.2	0.174	9.68	0.5	0.7	0.4	0.28	30.45	9.68
12	0.5	0.71	0.2	0.14	8.46	0.5	0.7	0.4	0.28	39.45	8.46
13	0.3	0.55	0.2	0.1	6.89	0.5	0.7	0.4	0.28	56.56	6.89

υ_{vh}) are approximately related by a cross-anisotropy factor α defined in Graham and Houlsby [5]: $E_v/E_h = 1/\alpha^2$, $G_{vh}/G_h = 1/\alpha$, $\upsilon_{vh} = \upsilon_h/\alpha$.

Table 2 summarises material properties, normalised shakedown limits of each layer k_{m} and normalised shakedown limits of the layered system k for the first series of analysis. It should be noted that case 1 represents isotropy assumption for both sand and clay layer; cases 2–7 assume cross-anisotropic clay layer; and cases 8–13 consider the property of cross-anisotropy for both layers. Results show that lower-bound shakedown limits k are all controlled by critical points in the first layer. Further studies show that all critical points are located beneath the surface. While sand is considered as an isotropic material, the rise of anisotropic factor of clay leads to an obvious decrease of second layer shakedown limit and a slightly increase of first layer shakedown limit (Fig. 8a). However, a reversed trend is observed if sand anisotropy is also considered (Fig. 8b). The rate of change is relatively small in clay anisotropy cases.

Materials and shakedown limits of the second series of analysis for an asphalt-clay system is summarised in Table 3. Case 1 represents isotropy case for both layers; cases 2–6 assume cross-anisotropic clay layer; and cases 7–12 also consider asphalt cross-anisotropy. Compared with Figs. 7, 8 exhibits similar trend of change for shakedown limits of each layer. However, as two lines intersect in Fig. 8, shakedown limit of the current system first rises then drops with increasing factor of clay anisotropy. In reality, because the factor of anisotropy α tends to be smaller than 1 for asphalt (i.e. $E_{\text{v}}/E_{\text{h}} = 1/\alpha^2 > 1$), but larger than 1 for clay (i.e. $E_{\text{v}}/E_{\text{h}} < 1$), the shakedown limit is more likely to be smaller than that under isotropic assumption. That is to say, when the critical point lies on the surface of the second layer, pavement design tends to be unsafe without consideration of cross-anisotropy.

Table 3 Summary of material properties and shakedown limits for an asphalt-clay system

No.	1st layer—asphalt					2nd layer—clay					
	$E_{1v} = 690\,\text{MPa}$, $c_1 = 400\,\text{kPa}$, $\phi_1 = 30°$					$E_{1v} = 15\,\text{MPa}$, $c_2 = 20\,\text{kPa}$, $\phi_2 = 0°$					
	E_{1v}/E_{1h}	G_{1vh}/G_{1h}	υ_{1h}	υ_{1vh}	k_1	E_{2v}/E_{2h}	G_{2vh}/G_{2h}	υ_{2h}	υ_{2vh}	k_2	k
1	1	1	0.3	0.3	148.4	1	1	0.4	0.4	178.6	148.4
2	1	1	0.3	0.3	145.6	2	1.4	0.4	0.49	263.6	145.6
3	1	1	0.3	0.3	151.6	0.5	0.7	0.4	0.28	115.4	115.4
4	1	1	0.3	0.3	155.4	0.3	0.55	0.4	0.24	83.35	83.35
5	1	1	0.3	0.3	156.0	0.25	0.5	0.4	0.2	75.30	75.30
6	1	1	0.3	0.3	157.6	0.2	0.45	0.4	0.18	65.97	65.97
7	3	1.73	0.3	0.49	157.2	0.5	0.7	0.4	0.28	47.90	47.90
8	2	1.4	0.3	0.42	153.6	0.5	0.7	0.4	0.28	65.50	65.50
9	1.5	1.2	0.3	0.36	152.2	0.5	0.7	0.4	0.28	82.43	82.43
10	0.75	0.87	0.3	0.261	151.0	0.5	0.7	0.4	0.28	147.6	147.6
11	0.5	0.71	0.3	0.213	144.8	0.5	0.7	0.4	0.28	211.5	144.8
12	0.3	0.55	0.3	0.165	130.8	0.5	0.7	0.4	0.28	337.0	130.8

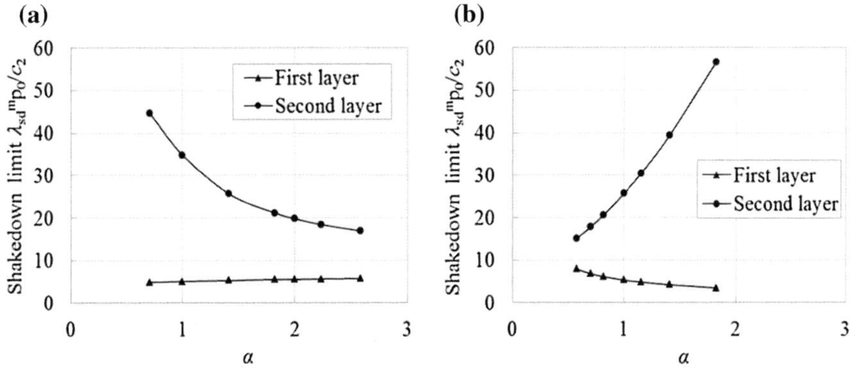

Fig. 7 Influence of factor of anisotropy in a sand-clay system

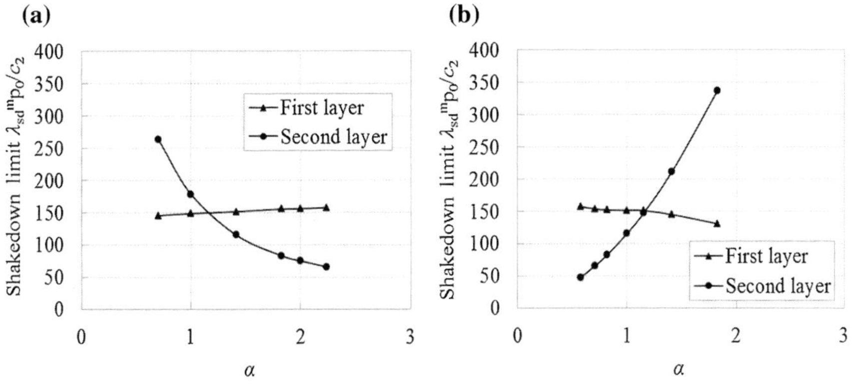

Fig. 8 Influence of factor of anisotropy in an asphalt-clay system

6 Conclusions

Shakedown solutions for cross-anisotropic cohesive-frictional materials under moving surface loads have been developed based on lower-bound shakedown theorem. Elastic anisotropy was taken into account by means of elastic stress fields for cross-anisotropic materials. Directional plastic anisotropy was also considered by introducing a generalised Mohr-Coulomb criterion.

Results show that shakedown limits of the homogenous half-space are dominated by Young's modulus ratio E_v/E_h for the cases of subsurface failure and by shear modulus ratio G_{vh}/G_h for the cases of surface failure. Plastic anisotropy is mainly controlled by material cohesion ratio c_v/c_h, the rise of which increases the shakedown limit until a maximum value is reached. The anisotropic shakedown limit varies with frictional coefficient μ and the peak value may not occur at $\mu = 0$ (i.e. normal load only).

For a two-layered pavement system, the shakedown limit of the second layer decreases with increasing clay anisotropic factor and decreasing first layer anisotropic factor. This is compensated by a reversed trend of the shakedown limit of the first layer.

Pavement design using shakedown theory can be carried out by making sure that the pavement shakedown limit is no less than the pavement design load. The influences of layer thicknesses and material properties on the pavement shakedown limit can be found in Wang and Yu [30]. An optimum pavement design can be chosen by considering other factors such as overall cost.

References

1. Arthur JRF, Menzies BK (1972) Inherent anisotropy in a sand. Géotechnique 22:115–128
2. Boulbibane M, Ponter ARS (2005) Extension of the linear matching method to geotechnical problems. Comput Methods Appl Mech Eng 194:4633–4650
3. Brown SF, Yu HS, Juspi S, Wang J (2012) Validation experiments for lower bound shakedown theory applied to layered pavement systems. Géotechnique 62:923–932
4. Collins IF, Cliffe PF (1987) Shakedown in frictional materials under moving surface loads. Int J Numer Anal Methods Geomech 11:409–420
5. Graham J, Houlsby GT (1983) Anisotropic elasticity of a natural clay. Géotechnique 33:165–180
6. Guo P (2008) Modified direct shear test for anisotropic strength of sand. J Geotech Geoenviron Eng 134:1311–1318
7. Hanson MT (1992) The elastic field for spherical Hertzian contact including sliding friction for transverse isotropy. J Tribol 114:606–611
8. Hoque E, Tatsuoka F, Sato T (1996) Measuring anisotropic elastic properties of sand using a large triaxial specimen. Geotech Test J 19:411–420
9. Jiang GL, Tatsuoka F, Flora A, Koseki J (1997) Inherent and stress-state-induced anisotropy in very small strain stiffness of a sandy gravel. Géotechnique 47:509–521
10. Koiter WT (1960) General theorems for elastic-plastic solids. In: Sneddon IN, Hill R (eds) Progress in solid mechanics. North Holland, Amsterdam, pp 165–221

11. Krabbenhøft K, Lyamin AV, Sloan SW (2007) Shakedown of a cohesive-frictional half-space subjected to rolling and sliding contact. Int J Solids Struct 44:3998–4008
12. Kuwano R, Jardine RJ (2002) On the applicability of cross-anisotropic elasticity to granular materials at very small strains. Géotechnique 52:727–749
13. Li HX, Yu HS (2006) A nonlinear programming approach to kinematic shakedown analysis of frictional materials. Int J Solids Struct 43:6594–6614
14. Lings ML, Penningtonf DS, Nash DFT (2000) Anisotropic stiffness parameters and their measurement in a stiff natural clay. Géotechnique 50:109–125
15. Lo KY (1965) Stability of slopes in anisotropic soils. J Soil Mech Found Div, ASCE 91:85–106
16. Melan E (1938) Der Der Spannungszustand eines Hencky-Mises'schen Kontinuums bei veränderlicher Belastung. Sitzungberichte der Ak Wissenschaften Wien (Ser 2A) 147:73
17. Nguyen AD, Hachemi A, Weichert D (2008) Application of the interior-point method to shakedown analysis of pavements. Int J Numer Methods Eng 75:414–439
18. Ponter ARS, Hearle AD, Johnson KL (1985) Application of the kinematical shakedown theorem to rolling and sliding point contacts. J Mech Phys Solids 33:339–362
19. Raad L, Weichert D, Najm W (1988) Stability of multilayer systems under repeated loads. Transp Res Rec 1207:181–186
20. Sharp RW, Booker JR (1984) Shakedown of pavements under moving surface loads. J Trans Eng, ASCE 110:1–14
21. Shiau SH, Yu HS (2000) Load and displacement prediction for shakedown analysis of layered pavements. Transp Res Rec 1730:117–124
22. Yimsiri S, Soga K (2011) Cross-anisotropic elastic parameters of two natural stiff clays. Géotechnique 61:809–814
23. Yu HS (2005) Three-dimensional analytical solutions for shakedown of cohesive-frictional materials under moving surface loads. Proc R Soc A: Math, Phys Eng Sci 461:1951–1964
24. Yu HS (2006) Plasticity and geotechnics. Springer, Berlin
25. Yu S, Dakoulas P (1993) General stress-dependent elastic moduli for cross-anisotropic soils. J Geotech Eng 119:1568–1586
26. Yu HS, Sloan SW (1994) Limit analysis of anisotropic soils using finite elements and linear programming. Mech Res Commun 21:545–554
27. Yu HS, Hossain MZ (1998) Lower bound shakedown analysis of layered pavements using discontinuous stress fields. Comput Methods Appl Mech Eng 167:209–222
28. Yu HS, Wang J (2012) Three-dimensional shakedown solutions for cohesive-frictional materials under moving surface loads. Int J Solids Struct 49:3797–3807
29. Wang J, Yu HS (2013) Shakedown and residual stresses in cohesive-frictional half-space under moving surface loads. Geomech Geoeng: Int J 8:1–14
30. Wang J, Yu HS (2013) Shakedown analysis for design of flexible pavements under moving loads. Road Mater Pavement Des 14:703–722
31. Wang J, Yu HS (2014) Three-dimensional shakedown solutions for anisotropic cohesive-frictional materials under moving surface loads. Int J Numer Anal Methods Geomech 38:331–348
32. Wang L, Hoyos LR, Wang J, Voyiadjis G, Abadie C (2005) Anisotropic properties of asphalt concrete: characterization and implications for pavement design and analysis. J Mater Civ Eng 17:535–543
33. Zhao J, Sloan SW, Lyamin AV, Krabbenhøft K (2008) Bounds for shakedown of cohesive-frictional materials under moving surface loads. Int J Solids Struct 45:3290–3312

Printed by Printforce, the Netherlands